Proceedings of the

Sixth Meeting on CPT and Lorentz Symmetry

Proceedings of the

Sixth Meeting on CPT and Lorentz Symmetry

Bloomington, USA   17 – 21 June 2013

Editor

V. Alan Kostelecký

Indiana University, USA

World Scientific

NEW JERSEY • LONDON • SINGAPORE • BEIJING • SHANGHAI • HONG KONG • TAIPEI • CHENNAI

Published by

World Scientific Publishing Co. Pte. Ltd.
5 Toh Tuck Link, Singapore 596224
USA office: 27 Warren Street, Suite 401-402, Hackensack, NJ 07601
UK office: 57 Shelton Street, Covent Garden, London WC2H 9HE

British Library Cataloguing-in-Publication Data
A catalogue record for this book is available from the British Library.

CPT AND LORENTZ SYMMETRY
Proceedings of the Sixth Meeting

ISBN 978-981-4566-42-1

Printed in Singapore by World Scientific Printers.

PREFACE

This volume comprises the proceedings of the Sixth Meeting on CPT and Lorentz Symmetry, CPT'13, which was hosted by the Physics Department at Indiana University, Bloomington from Monday June 17 to Friday June 21, 2013. The scope of this triennial conference series covers experimental studies and theoretical descriptions of spacetime symmetries and their violations.

The papers collected here provide summaries of the presentations delivered at the meeting. Experimental topics considered in the talks include accelerator and collider tests, particle propagation and decay, studies of birefringence, dispersion, and anisotropy in astrophysical sources, clock-comparison measurements, searches with resonant cavities, tests of the equivalence principle, matter interferometry, oscillations and propagation of mesons and neutrinos, particle-antiparticle comparisons, post-newtonian tests of gravity, space-based missions, and the spectroscopy of hydrogen and antihydrogen. The theory presentations addressed physical effects potentially arising from CPT and Lorentz violation, related classical and quantum topics including origins and mechanisms, and mathematical foundations including Finsler geometry. The papers are ordered according to the schedule of talks, followed by the posters.

I would like to thank the many people who contributed to the success of the event. Particular appreciation is due to Ralf Lehnert, Neil Russell, and Jordan Tillett, who provided essential assistance with the meeting organization. Day-to-day operational challenges were skillfully handled by Kathy Hirons, Yuri Bonder, Jorge Díaz, André Gomes, and Arnaldo Vargas. Robert Bluhm supplied key advice, and Kevin Kostelecký and Matt Mewes provided crucial help with digital imaging and animations. Financial support for the meeting was granted by the Indiana University Office of the Vice Provost for Research and by the Indiana University Center for Spacetime Symmetries (IUCSS).

Alan Kostelecký
September 2013

CONTENTS

Preface v

Bounds on LLI Violation and Long-Range Spin-Spin Interactions Using Hg, Cs, and the Earth 1
L.R. Hunter et al.

Antihydrogen, CPT, and Naturalness 5
M.C. Fujiwara

Testing Lorentz Symmetry with the Double Chooz Experiment 9
T. Katori and J. Spitz

Galileo Galilei (GG) Space Test of the Weak Equivalence Principle to 10^{-17}: the Key Features 13
A.M. Nobili

Probing Physics Beyond the Standard Model with He/Xe Clock Comparison Experiments 17
F. Allmendinger et al.

Using Binary Pulsars to Test Lorentz Symmetry in the Gravitational Sector 21
J.M. Weisberg

A New Approach to Search for Lorentz Invariance Violation in β Decay 25
H.W. Wilschut

Relevance and Prospects of Magic Telescope Gamma-Ray Observations for Lorentz Symmetry Tests 29
R.M. Wagner

Comments on Lorentz and CPT Violation 33
V.A. Kostelecký

Testing Periodic Local Position Invariance Using Long-Term Comparison of the SYRTE Atomic Fountains and H-Masers ... 37
M.E. Tobar et al.

CPT Test with (Anti)Proton Magnetic Moments Based on Quantum Logic Cooling and Readout 41
M. Niemann et al.

Tests of the Lorentz and CPT Symmetries at the Planck Energy Scale from X-Ray and Gamma-Ray Observations 45
H. Krawczynski et al.

An Odd (Parity) Test of Lorentz Symmetry with Atomic Dysprosium 49
N. Leefer and M. Hohensee

Lorentz and CPT Violation in the Neutrino Sector 53
J.S. Díaz

The Search for Neutrino-Antineutrino Mixing from Lorentz-Invariance Violation Using Neutrino Interactions in MINOS ... 57
S. Mufson and B. Rebel

Terrestrial vs. Spaceborne, Quantum vs. Classical Tests of the Equivalence Principle 61
M. Hohensee and H. Müller

Higher-Order Lorentz Violation 65
M. Mewes

Towards a Measurement of the Free Fall of Antihydrogen with the AEGIS Experiment 69
A. Kellerbauer

Tests of Lorentz Invariance Using High-Energy Astrophysics Observations 73
F.W. Stecker

The KATRIN Experiment: Status and Outlook 77
D.S. Parno

Status of Matter-Gravity Couplings in the SME 81
J.D. Tasson

Lorentz and CPT Violation in the Hydrogen Spectrum 85
G.S. Adkins and T.J. Yoder

Testing the Universality of Free Fall with a Dual-Species Atom Interferometer on STE-QUEST 89
M. Krutzik and A. Peters

Test of Lorentz Invariance with a Rb-^{21}Ne Comagnetometer at the South Pole 95
M.A. Smiciklas and M.V. Romalis

Sensitivity of Atmospheric Neutrinos in Super-Kamiokande to Lorentz Violation 99
T. Akiri

Search for CPT and Lorentz Symmetry Violation in Neutral Kaons at KLOE/KLOE-2 103
A. De Santis

How to Test the SME with Space Missions? 107
A. Hees et al.

Neutrino Velocity Measurements with MINOS 111
I. Anghel

Prospects for a Global Network of Optical Magnetometers for Exotic Physics (GNOME) 115
D.F.J. Kimball et al.

Lorentz-Violating Photons with a Mass Term 119
M. Cambiaso, R. Lehnert, and R. Potting

Testing the Equivalence Principle 10,000 Times Better on a Sounding Rocket 123
J.D. Phillips and R.D. Reasenberg

Covariant Photon Quantization in the SME 127
D. Colladay

Using Microwave Cavities to Test Lorentz Invariance 131
S.R. Parker et al.

Using Quantum Entanglement to Study CP and CPT Violations 135
Y. Shi

A CPT-Even Lorentz-Violating Nonminimal Coupling Between Fermions and Photons 139
R. Casana et al.

Ultra-Stable Cryogenic Optical Resonators for Tests of Fundamental Physics 143
M. Nagel et al.

What We Really Know about Neutrino Speeds 149
B. Altschul

Lorentz Violation and Gravity 153
R. Bluhm

Lorentz Violation in Top-Quark Physics 156
M.S. Berger

Testing for CPT Violation in B_s^0 Semileptonic Decays 160
R. Van Kooten

Nuclear β Decay with Lorentz Violation 164
J.P. Noordmans, H.W. Wilschut, and R.G.E. Timmermans

Remarks on Finsler Geometry and Lorentz Violation 168
N. Russell

Lorentz Violation in a Uniform Gravitational Field 172
Y. Bonder

Obtaining Bounds from Ultra-High Energy Cosmic Rays in Isotropic Modified Maxwell Theory 176
M. Schreck

Modern Ives-Stilwell Experiments at Storage Rings: Large Boosts Meet High Precision 180
G. Gwinner et al.

Renormalization in QED and QFT with a Lorentz- and CPT-Violating Background 184
I.L. Shapiro

Constraints on Violations of Lorentz Symmetry from Gravity Probe B 188
J.M. Overduin, R.D. Everett, and Q.G. Bailey

The Influence of Lorentz Violation on UHE Photon Detection ... 192
G.I. Rubtsov, P.S. Satunin, and S.M. Sibiryakov

Limits on Lorentz Violation in Neutral-Kaon Decay 196
K.K. Vos, H.W. Wilschut, and R.G.E. Timmermans

Search for Lorentz Violation in Top Quark Pair Production and Decay 200
D.W. Whittington

The Källén-Lehmann Representation in Lorentz-Violating Field Theory 204
R. Potting

Time-Varying Nuclear Decay Parameters and Dark Matter 208
J. Nistor et al.

Lorentz Violation and Topological Defects 212
M.D. Seifert

Testing Lorentz Invariance with a Double-Pass Optical Ring Cavity 216
Y. Michimura et al.

Laser Time-Transfer and Space-Time Reference in Orbit 220
P. Berceau and L. Hollberg

Equivalence Principle in Cosmology 224
S. Kopeikin

Unitarity, Ghosts, and Nonminimal Terms in Lorentz-Violating QED 228
C.M. Reyes

Laboratory Search for Spin-Dependent Short-Range Force from Axion-Like Particles Using Optically Polarized ^{3}He Gas 232
P.-H. Chu

Local Lorentz-Symmetry Breaking and Gravity 236
Q.G. Bailey

Generation of Axion Couplings via Lorentz Violation 240
A.F. Ferrari

Finite One-Loop Radiative Corrections in the Lorentz- and CPT-Violating QED Extension 243
D.H.T. Franco and A.H. Gomes

Macroscopic Objects, Intrinsic Spin, and Lorentz Violation 246
D.W. Atkinson, M. Becker, and J.D. Tasson

Application of Laser Frequency Combs and Nitrogen Vacancy Diamond Magnetometers to Searches for New Physics 249
D.F. Phillips and R.L. Walsworth

Using a Rotating Magnetic Guiding Field for the ^{3}He-^{129}Xe Comagnetometer 252
F. Allmendinger et al.

Classification of Relativity Theories by Dispersion Relation 255
M.J. Goodman

Bipartite Finsler Spaces and the Bumblebee Model 258
J.E.G. Silva and C.A.S. Almeida

Microlocal Methods in Quantum Field Theory and Their Use in Proving Renormalizability in SME and Tachyonic Models .. 261
M.J. Radzikowski

Renormalization and Asymptotic States in Lorentz-Violating Quantum Field Theory 264
M. Cambiaso, R. Lehnert, and R. Potting

Constraints on Spin Independent Forces at the ~100 nm Range by Means of a Micromechanical Oscillator 267
R.S. Decca

Cardinal Gravity 270
R. Potting

BPS Maxwell-Chern-Simons-Like Vortices in a Lorentz-Violating Framework 273
R. Casana et al.

Observer and Particle Transformations and Newton's Laws 276
T.H. Bertschinger, N.A. Flowers, and J.D. Tasson

BOUNDS ON LLI VIOLATION AND LONG-RANGE SPIN-SPIN INTERACTIONS USING HG, CS, AND THE EARTH

L.R. HUNTER*, S.K. PECK, D. ANG, D.K. KIM , D. STEIN,
D. ORBAKER, A. FOSS, M.T. HUMMON, AND J.E. GORDON
Physics Department, Amherst College
Amherst, MA 01002, USA
**E-mail: lrhunter@amherst.edu*

J.F. LIN
Department of Geological Sciences, Jackson School of Geoscience
Austin, TX 78712, USA

We report bounds from a second generation LLI violation experiment using Hg and Cs magnetometers. The experiment is mounted on a rotating table that has made it possible to reduce experimental noise and to extract the first bounds on $\tilde{b}_z$ for the proton and neutron. The same experiment can be used in conjunction with a polarized electron spin map of the Earth to place bounds on long range spin-spin interactions. For some of the proposed long-range spin-spin interactions, these new bounds are a million times more restrictive than the bounds derived by other methods. Our upper bound on the magnitude of the spin-spin force between an electron and a neutron is nearly a million times smaller than their gravitational attraction.

1. Introduction

In the first CPT conference held here fifteen years ago, we reported bounds from our first test of local Lorentz invariance (LLI).[1] In that work we used optically pumped light-absorption oscillators to make precise measurements of the spin precession frequencies of ^{199}Hg and Cs atoms and searched for changes in their relative precession frequencies as a function of the orientation of our applied magnetic field with respect to the fixed stars.[2] At the time, this experiment established the best bounds on the Standard-Model Extension parameters $\tilde{b}_\perp$ for the electron, neutron and proton. The suggestion that these bounds might actually be probing physics at the Planck scale created a great deal of excitement within the field.

2. The second generation Hg-Cs LLI experiment

We recently completed a new test of LLI using a Hg-Cs comagnetometer.[3] In this experiment, we have upgraded our ultraviolet light source from a Hg discharge lamp to a 271 nm laser. We have also mounted the entire apparatus on a rotation table so that we can invert the horizontal component of our magnetic field every eight minutes, rather than having to wait for the rotation of the Earth. We collected 12 data runs with this second-generation apparatus over a period of a year, where each run consisted of four or five days of data collection. With the existing bounds on possible electron LLI couplings,[4] we use our data to place the first bounds on $\tilde{b}_z$ for the neutron and the proton: $\tilde{b}_z^n < 7\times10^{-30}$ GeV and $\tilde{b}_z^p < 7\times10^{-29}$ GeV. These bounds are significantly less restrictive than the limits that have been established on $\tilde{b}_\perp$.[5] As such, any signal associated with $\tilde{b}_z$ is likely to be associated with an Earth-bound source. One interesting possibility is that the Earth may provide a source of spin-polarized electrons that might interact with our ^{199}Hg spins via long-range spin-spin interactions.

3. Using our results to limit spin-spin interactions

Long-range spin-spin interactions can arise from a variety of possible extensions beyond the standard model of particle physics. Dobrescu and Mocioiu have enumerated nine possible long range spin-spin interactions that could arise via the virtual exchange of a low mass vector boson such as the Z'.[6] Similarly, the virtual exchange of 'unparticles'[7] has been shown to result in long-range spin-spin interactions.[8] Torsion gravity also quite naturally leads to long-range spin-spin interactions.[9]

If such long-range spin-spin interactions do exist, they would create an interaction between the spin-polarized geoelectrons and the ^{199}Hg nuclear spins in our apparatus. The unpaired electron spins within the Earth acquire a modest polarization in the presence of the geomagnetic field. We have created a model of the polarized geoelectron density throughout the Earth using recent deep-Earth geophysics and geochemistry results. We have combined this model with our experimental results and the results from two other experiments[4,10] to constrain the possible strength of these long-range spin-spin interactions between electrons and electrons, neutrons and protons.[11] For potentials that fall off as 1/r at long range, these new bounds are about a million times more restrictive than those achieved from earlier studies that used spin-polarized laboratory sources.[12–14]

Because of the relatively large velocity of the geoelectrons with respect to our apparatus, we have been able to constrain the velocity-dependent spin-spin interactions suggested in Ref. 6.[15] Of the six velocity-dependent potentials only V_8 was previously bounded.[16] In the long-range limit, our constraint on V_8 improves upon this earlier bound by about 30 orders of magnitude.

4. Future prospects

We anticipate further improvement in all of these bounds. The work underway using a Rb-^{21}Ne comagnetometer at the South Pole by the Princeton group (presented at this conference) and a new experimental comparison of ^{85}Rb and ^{87}Rb at Cal State East Bay[17] are likely to result in interesting new experimental bounds.

The sensitivity of our recent ^{199}Hg-Cs comparison was not limited by the intrinsic sensitivity of our magnetometers, but rather by small changes in the AC light shift that occurred upon the rotation of our apparatus. While the 'Mx' optical-pumping scheme we have used in our magnetometers is convenient, it is unfortunately first-order sensitive to AC light shifts. We intend to change to either a 'Bell-Bloom'[18]or a 'pump-then-probe'[17,19] geometry where the AC light shifts should be greatly suppressed. With this change we hope to achieve at least an order of magnitude improvement.

Further improvement in the long-range spin-spin interactions will require that we address two additional possible sources of systematic error. Our ability to subtract off the Earth's gyroscopic effect is limited by how well we know the direction of our magnetic field. A straightforward solution to this issue is to move the rotation apparatus to the equator. At the equator, the misalignment of an applied magnetic field (pointing north or south) with the Earth's rotation axis z will only produce uncertainties in the gyroscopic shifts that are second order in the magnetic field misalignment with z. The second obstacle to further improvement arises from the fact that we currently must use separate cells for the Hg and Cs magnetometers. With a single stack of three cells (say, a single Hg cell sandwiched between two Cs cells), we are unable to subtract off a differential frequency shift that would be associated with a quadrupole moment of the magnetic field at our cell stack that does not rotate with the apparatus. This issue could be addressed by the addition of two more Cs cells to monitor the the magnetic quadrupole moment of the field.

A more radical change that would address all of these issues would be to create a rotating comagnetometer with three nuclear species in the same cell (e.g. ^{3}He, ^{129}Xe, and ^{21}Ne or ^{199}Hg). One species could be used to regulate the magnetic field. Forming the difference between the frequencies of the other two species would remove the Earth's gyroscopic effect while retaining sensitivity to both neutron and proton couplings to the spin-polarized geoelectrons. Such an apparatus should be limited only by the intrinsic magnetometer sensitivities and could operate at any geographic location.

Acknowledgments

This work was supported by Amherst College and the NSF under grants PHY 0855465 and PHY 1205824. J.F.L. acknowledges support from the NSF EAR-1056670 and EAR-1053446.

References

1. L.R. Hunter *et al.*, in V.A. Kostelecký, ed., *Proceedings of the Meeting on CPT and Lorentz Symmetry*, World Scientific, Singapore, 1999.
2. C.J. Berglund *et al.*, Phys. Rev. Lett. **75**, 1879 (1995).
3. S.K. Peck *et al.*, Phys. Rev. A **86**, 012109 (2012).
4. B.R. Heckel, E.G. Adelberger, C.E. Cramer, T.S. Cook, S. Schlamminger, and U. Schmidt, Phys. Rev. D **78**, 092006 (2008).
5. J.M. Brown, S.J. Smullin, T.W. Kornack, and M.V. Romalis, Phys. Rev. Lett. **105**, 151604 (2010).
6. B.A. Dobrescu and I. Mocioiu, JHEP **11**, 005 (2006).
7. H. Georgi, Phys. Rev. Lett. **98**, 221601 (2007).
8. Y. Liao and J.-Y. Liu, Phys. Rev. Lett. **99**, 191804 (2007).
9. D.E. Neville, Phys. Rev. D **25**, 573 (1982).
10. B.J. Venema, P.K. Majumder, S.K. Lamoreaux, B.R. Heckel, and E.N. Fortson, Phys. Rev. Lett. **68**, 135 (1992).
11. L.R. Hunter, J.E. Gordon, S.K. Peck, D. Ang, and J.F. Lin, Science **339**, 928 (2013).
12. A.G. Glenday, C.E. Cramer, D.F. Phillips, and R.L. Walsworth, Phys. Rev. Lett. **101**, 261801 (2008).
13. G. Vasilakis, J.M. Brown, T.W. Kornack, and M.V. Romalis, Phys. Rev. Lett. **103**, 261801 (2009).
14. C.E. Cramer, *A torsion balance search for spin coupled forces*, Ph.D. thesis, Univ. of Washington (2007).
15. L.R. Hunter and D. Ang, arXiv:1306.1118.
16. M.P. Ledbetter, M.V. Romalis, and D.F.J. Kimball, Phys. Rev. Lett. **110**, 040402 (2013).
17. D.F.J. Kimball *et al.*, Ann. Physik **525**, 514 (2013).
18. W.E. Bell and A.L. Bloom, Phys. Rev. Lett. **6**, 280 (1961).
19. W.C. Griffith *et al.*, Phys. Rev. Lett. **102**, 101601 (2009).

ANTIHYDROGEN, CPT, AND NATURALNESS

MAKOTO C. FUJIWARA
TRIUMF National Laboratory for Particle and Nuclear Physics
Vancouver, British Columbia, V6T 2A3, Canada
and
Department of Physics and Astronomy, University of Calgary
Calgary, Alberta, T2N 1N4, Canada
E-mail: Makoto.Fujiwara@triumf.ca

Studying fundamental symmetries of Nature has proven fruitful in particle physics. I argue that recent results at the LHC, and the naturalness problem highlighted by them, provide a renewed motivation for tests of CPT symmetry as a probe for physics beyond quantum field theory. I also discuss prospects for antihydrogen CPT tests with sensitivities to Planck scale suppressed effects.

1. Introduction

It is clear that testing CPT invariance at highest possible precision is a worthwhile effort, given its fundamental importance in modern physics.[1] Comparisons of antihydrogen atoms ($\overline{\mathrm{H}}$) with their well-studied matter counterpart, atomic hydrogen (H), could provide competitive tests of CPT. (See, e.g., Ref. 2). In this short paper, I will try to put $\overline{\mathrm{H}}$ studies in the larger context of current particle physics, and argue that the recent LHC results provide enhanced motivations for symmetry tests with $\overline{\mathrm{H}}$.

2. Naturalness and CPT

Before going into antihydrogen, let us digress and ask a rather basic question: what is particle physics? According to Grossman,[3] particle physics is about asking a simple question:

$$\mathcal{L} = ? \tag{1}$$

To this simple question, we seem to have a simple answer: the Standard Model (SM), including the recently discovered final piece, a Higgs boson. However, as is well known, the SM has a number of open issues, not least of which is the (technical) naturalness problem. Loosely speaking, the naturalness (also called the hierarchy, or the fine-tuning) problem in this context refers to the lightness of the observed Higgs mass compared to the Planck (or the grand unification) scale, which implies fine tuning of the Higgs parameters by some $O(30)$ within the SM.

In the past decades, the issue of naturalness has been a central guiding principle in particle physics. A number of solutions to the problem have been proposed and studied extensively, the most popular scenario being supersymmetry. These solutions usually require that new phenomena appear at the energy scale near the electroweak scale, leading to the expectation that we would observe new physics beyond the SM at the LHC. The lack of such observations thus far has ruled out many of the most attractive new physics scenarios, and it seems to be putting many particle physicists in the state of 'soul searching.' Of course, there is still room for discoveries as the LHC energy is increased, which may solve the naturalness problem, but most of the surviving scenarios appear rather contrived. In addition to the light Higgs mass, the cosmological constant presents an even greater challenge to our belief in naturalness, naively requiring a fine-tuned cancelation at the $O(120)$ level.

This apparently disparate situation has lead to the increasing popularity of the anthropic principle, which states certain parameters in physics are fine tuned (possibly in a landscape of the 'multiverse') to allow the existence of the observer. Before accepting this controversial (but logically possible) option, however, I wish to step back and ask: are we asking the right question? When the answer we get (i.e., incredible degrees of fine tuning) does not make sense, it is possible that we are asking a wrong question.

In fact, implicit in the question in Eq. (1) is the validity of quantum field theory (QFT), at least as the low energy effective description of Nature. However, it is the effective QFT framework itself that gives rise to the fine-tuning problems in the first place. The question we should be asking may be about the validity of the framework itself, rather than the ingredients in it. The possibility of such an option, although admittedly speculative, motivates putting the QFT framework to stringent experimental tests.

How does one test the validity of QFT? One possibility is to improve the precision of measurements of physical quantities that can be predicted precisely. The electron $g-2$ factor is a leading candidate. However, as a test of QED, it is currently limited by independent knowledge of the fine structure constant. Another approach is to search for a violation of symmetries guaranteed in QFT, such as CPT. $\overline{\text{H}}$-H comparisons fall into the latter category. By directly comparing matter and antimatter, we would be free from the uncertainties due to theory or the constants.

Search for violation of CPT and/or Lorentz invariance has been the subject of considerable recent activities. A general effective field theory framework by Kostelecký *et al.*, known as the Standard-Model Extension (SME), has provided an extremely powerful framework in which to test these symmetries.[1] It has been well known that the CPT theorem is guaranteed in QFT under general assumptions including Lorentz invariance, locality, and unitarity. A notable finding in the past decade is that within the SME (and presumably in any local QFT), CPT violation is always accompanied by Lorentz violation. This implies that limits on Lorentz violation in matter-only experiments also provide constraints on CPT violation. Given that extremely sensitive limits of the SME parameters are being placed via matter-only experiments,[1] is there need for antimatter experiments?

Matter-antimatter comparison experiments confront the entire framework of (effective) QFT by testing CPT-odd, but Lorentz-even interactions, which are forbidden in local QFT. In other words, any violation of CPT, e.g., in comparison of $\overline{\mathrm{H}}$ and H atomic spectra, would indicate physics beyond QFT, forcing a fundamental change in our understanding of Nature.

Of course, it is by no means guaranteed that any potential modifications of QFT would lead to an observable CPT violating effect in $\overline{\mathrm{H}}$ experiments. However, antihydrogen should serve at least as a 'lamp post' case, because of the potentially very high sensitivities it could offer.[2]

Since the CPT'13 meeting, I learned that there are in fact attempts to explain the fine-tuning problem using (mildly) nonlocal theory beyond the conventional QFT, e.g., by considering effects of wormholes in the multiverse.[4] Whether this class of models would have observable implications at low energies is an open question.

3. Antihydrogen experiments at the CERN AD

Several experiments are ongoing or under construction at the CERN Antiproton Decelerator, with the goal of testing fundamental symmetries with $\overline{\mathrm{H}}$. Here, I briefly discuss one of the experiments, ALPHA (Antihydrogen Laser PHysics Apparatus). Since the CPT'10 meeting, ALPHA has made significant progress. Indications of trapped $\overline{\mathrm{H}}$, which I reported at CPT'10,[5] has now been unambiguously confirmed.[6] The confinement times have been extended to as long as 1000 seconds.[7] First proof-of-principle demonstration of spectroscopic measurement on $\overline{\mathrm{H}}$ has been performed by driving hyperfine transitions via microwaves.[8] An entirely new trap (ALPHA-2) has been constructed and is being commissioned to allow laser and improved microwave spectroscopy.

Laser spectroscopy of the 1s-2s level is the golden mode for a CPT test with $\overline{\mathrm{H}}$, since the same transition in H is measured to parts in 10^{-15}. Hyperfine spectroscopy offers a complementary test.[2,9] In this respect, recent developments in measurements of the bare antiproton g-factor are encouraging. While the latter probes the long-distance magnetic property of the antiproton, $\overline{\mathrm{H}}$ hyperfine splitting can probe the antiproton's internal structure via the contact interaction of the positron and the antiproton.[9] Recall that while the proton electric charge is very well known, there is a puzzle in its charge distribution. The gravitational interaction of antimatter is another increasingly active subject. See, e.g., Ref. 10.

A benchmark for CPT tests may be the sensitivity to Planck scale suppressed effects, $\Delta E \sim m^2_{\mathrm{proton}}/M_{\mathrm{Planck}} \sim 10^{-18}$ GeV.[2] In frequency units, this corresponds to the precision of $\Delta f \sim 100$ kHz. This is within the reach of current antihydrogen experiments.

Acknowledgments

I wish to thank Alan Kostelecký and the organizing committee for a stimulating meeting. I thank the members of ALPHA for fruitful collaboration, Art Olin and Dave Gill for a critical reading of the manuscript. This work is supported in part by Canada's NSERC, and TRIUMF.

References

1. *Data Tables for Lorentz and CPT Violation,* V.A. Kostelecký and N. Russell, 2013 edition, arXiv:0801.0287v6.
2. M.C. Fujiwara *et al.*, AIP Conf. Proc. **1037**, 208 (2008).
3. Y. Grossman, arXiv:1006.3534.
4. D.L. Bennet, C.D. Froggatt, and H.B. Nielsen, arXiv:hep-ph/9504294; C.D. Froggatt and H.B. Nielsen, Phys. Lett. B **368**, 96 (1996); H. Kawai and T. Okada, Prog. Theor. Phys. **127**, 689 (2012).
5. M.C. Fujiwara *et al.*, in V.A. Kostelecký, ed., *Proceedings of the Fifth Meeting on CPT and Lorentz Symmetry,* World Scientific, Singapore, 2010.
6. G.B. Andresen *et al.*, Nature **468**, 673 (2010).
7. G.B. Andresen *et al.*, Nature Physics **7**, 558 (2011).
8. C. Amole *et al.*, Nature **483**, 439 (2012).
9. E. Widmann *et al.*, nucl-ex/0102002.
10. C. Amole *et al.*, Nature Comm. **4**, 1785 (2013).

TESTING LORENTZ SYMMETRY WITH THE DOUBLE CHOOZ EXPERIMENT

TEPPEI KATORI* AND JOSHUA SPITZ

Laboratory for Nuclear Science
Massachusetts Institute of Technology
Cambridge, MA, 02139, USA
**E-mail: katori@fnal.gov*

The Double Chooz reactor-based oscillation experiment searches for an electron antineutrino disappearance signal to investigate the neutrino mass matrix mixing angle θ_{13}. Double Chooz's reported evidence for this disappearance is generally interpreted as mass-driven mixing through this parameter. However, the electron antineutrino candidates collected by the experiment can also be used to search for a signature of the violation of Lorentz invariance. We study the sidereal time dependence of the antineutrino signal rate and probe Lorentz violation within the Standard-Model Extension (SME) framework. We find that the data prefer the sidereal time independent solution, and a number of limits are applied to the relevant SME coefficients, including the first constraints on those associated with Lorentz violation in the e-τ mixing sector.

1. Double Chooz

The neutrino Standard Model can successfully describe all precision neutrino oscillation measurements to date. Knowledge of the mixing angle θ_{13} represents the last requirement before a measurement of CP violation in the lepton sector can proceed. A measurement of θ_{13} is therefore critical, and a worldwide effort in the form of short-baseline reactor-based[1] and long-baseline accelerator-based[2] experiments has been undertaken to accomplish this goal.

The Double Chooz reactor antineutrino experiment employs two 4.25 GW reactors in the north of France near the border with Belgium as an antineutrino source. A liquid-scintillator-based far detector, located about 1050 m southeast of the cores, is used to detect the antineutrinos.

We briefly describe the main features of the Double Chooz far detector below. Details of the Double Chooz experiment can be found elsewhere.[3] Double Chooz is designed to detect the interaction of reactor electron antineutrinos with free protons via inverse beta decay ($\bar{\nu}_e + p \to e^+ + n$) resulting in the coincidence of a fast positron annihilation and delayed neutron capture on a gadolinium or hydrogen nucleus.[4]

The detector is made up of four layers of concentric cylinders. The innermost region is the antineutrino target volume, where a 10.3 m^3 acrylic tank is filled with gadolinium-doped liquid scintillator. The next layer is the 'gamma-catcher,' where a 55 cm thick volume of liquid scintillator is used to fully reconstruct gamma rays originating in the antineutrino target region. The third layer is a 105 cm thick mineral oil buffer, where 390 10-inch PMTs are located. Then, after a stainless steel wall, there is a 50 cm thick inner veto region, where 78 8-inch PMTs are installed in the liquid scintillator to detect particles originating from the outside.

The dominant backgrounds are spallation products (^{9}Li and ^{8}He), stopping muons, and cosmic- and radioactivity-induced fast neutrons. However, these backgrounds are directly constrained from Double Chooz reactor-off data.[5]

2. Sidereal variation analysis

Double Chooz has excluded the no-oscillation hypothesis at the 2.9σ level.[3] The analysis reported here is based on the same data set and is used to look for a sidereal time dependence among the 8249 antineutrino-induced inverse beta decay candidates. An observed sidereal time dependence of an experimental observable is widely considered a smoking gun of Lorentz violation. The analysis is performed under the SME formalism.[6] The relatively small observed oscillation signal allows the effective hamiltonian to be expanded with each oscillation channel written as one matrix element.[7] The disappearance can be written in terms of two oscillation channels, assuming there are no neutrino-antineutrino oscillations:

$$P(\bar{\nu}_e \to \bar{\nu}_e) = 1 - P(\bar{\nu}_e \to \bar{\nu}_\mu) - P(\bar{\nu}_e \to \bar{\nu}_\tau). \tag{1}$$

This allows access to the e-τ sector of SME coefficients for the first time. The disappearance signal is found to be compatible with the time-independent solution and limits on combinations of SME coefficients in the e-μ and e-τ sectors are extracted.[8]

3. SME coefficient limits

Using the limits reported in the Double Chooz analysis,[8] we extract the upper limits on each individual SME coefficient. Table 1 shows the results.

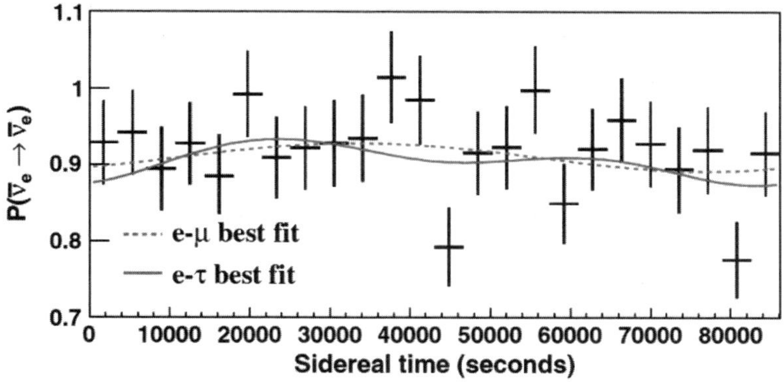

Fig. 1. The electron antineutrino disappearance probability as a function of sidereal time overlaid with the best fit curves.[8] The solid curve (e-τ best fit) assumes all Lorentz violating oscillations occur in $\bar{\nu}_e \to \bar{\nu}_\tau$, and the dashed curve ($e$-$\mu$ best fit) assumes all Lorentz violating oscillations occur in $\bar{\nu}_e \to \bar{\nu}_\mu$. Both fits are dominated by the sidereal time independent terms.

Table 1. The extrapolated SME coefficients from the Double Chooz analysis. We use 4.2 MeV as the average antineutrino energy.

SME coefficients	e-τ fit	e-μ fit
$\mathrm{Re}(a_L)^T$ or $\mathrm{Im}(a_L)^T$	7.8×10^{-20} GeV	—
$\mathrm{Re}(a_L)^X$ or $\mathrm{Im}(a_L)^X$	4.4×10^{-20} GeV	1.6×10^{-21} GeV
$\mathrm{Re}(a_L)^Y$ or $\mathrm{Im}(a_L)^Y$	9.0×10^{-20} GeV	6.1×10^{-20} GeV
$\mathrm{Re}(a_L)^Z$ or $\mathrm{Im}(a_L)^Z$	2.7×10^{-19} GeV	—
$\mathrm{Re}(c_L)^{XY}$ or $\mathrm{Im}(c_L)^{XY}$	3.4×10^{-18}	—
$\mathrm{Re}(c_L)^{XZ}$ or $\mathrm{Im}(c_L)^{XZ}$	1.8×10^{-17}	—
$\mathrm{Re}(c_L)^{YZ}$ or $\mathrm{Im}(c_L)^{YZ}$	3.8×10^{-17}	—
$\mathrm{Re}(c_L)^{XX}$ or $\mathrm{Im}(c_L)^{XX}$	3.9×10^{-17}	—
$\mathrm{Re}(c_L)^{YY}$ or $\mathrm{Im}(c_L)^{YY}$	3.9×10^{-17}	—
$\mathrm{Re}(c_L)^{ZZ}$ or $\mathrm{Im}(c_L)^{ZZ}$	4.9×10^{-17}	—
$\mathrm{Re}(c_L)^{TT}$ or $\mathrm{Im}(c_L)^{TT}$	1.3×10^{-17}	—
$\mathrm{Re}(c_L)^{TX}$ or $\mathrm{Im}(c_L)^{TX}$	5.2×10^{-18}	—
$\mathrm{Re}(c_L)^{TY}$ or $\mathrm{Im}(c_L)^{TY}$	1.1×10^{-17}	—
$\mathrm{Re}(c_L)^{TZ}$ or $\mathrm{Im}(c_L)^{TZ}$	3.2×10^{-17}	—

The fit is done separately by assuming $\bar{\nu}_e \to \bar{\nu}_\tau$ oscillation only or $\bar{\nu}_e \to \bar{\nu}_\mu$ oscillation only. Therefore, there are two sets of relevant SME coefficient limits. With regard to the latter case, since a Lorentz violation analysis with the MINOS near detector[9] reports an order of magnitude higher-sensitivity for $(c_L)^{\mu\nu}_{e\mu}$ due to their higher energy beam, we set these coefficients to zero. Therefore, we do not report limits on $(c_L)^{\mu\nu}_{e\mu}$. Also, there is no time-independent limit in the e-μ sector reported since Double Chooz does not provide a limit on the time-independent amplitude.

To extract each limit, we set all SME coefficients but one to zero. In this way, the real part and imaginary part have the same constraints. With this study, all oscillation channels have associated SME coefficient constraints, and it can now be considered challenging to discover Lorentz violation with a terrestrial-based neutrino experiment. However, there is still room to search for Lorentz violation with neutrinos. That is, there are manifestations of Lorentz violation which do not affect neutrino mixing and cannot be constrained with neutrino oscillation experiments. The most famous example of this is neutrino time-of-flight measurements.[10] Also, if we admit Lorentz violation is only a second-order effect of neutrino oscillations, behind neutrino mass, a perturbative approach in searching for a small Lorentz violation may be more efficient.[11]

In conclusion, a search for Lorentz violation has been performed using the Double Chooz data. No evidence for this process has been observed. The results are used to extract limits on the relevant SME coefficients, and Lorentz violation in the e-τ mixing sector are constrained for the first time. With the addition of this analysis amongst the world's data, Lorentz violation has been tested in all oscillation channels.

References

1. Y. Abe *et al.*, Phys. Rev. Lett. **108**, 131801 (2012); F.P. An *et al.*, Phys. Rev. Lett. **108**, 171803 (2012); J.K. Ahn *et al.*, Phys. Rev. Lett. **108**, 191802 (2012).
2. K. Abe *et al.*, Phys. Rev. Lett. **107**, 041801 (2011); P. Adamson *et al.*, Phys. Rev. Lett. **107**, 181802 (2011); A. Hatzikoutelis, J. Phys. Conf. Ser. **410**, 012146 (2013).
3. Y. Abe *et al.*, Phys. Rev. D **86**, 052008 (2012).
4. Y. Abe *et al.*, Phys. Lett. B **723**, 66 (2013).
5. Y. Abe *et al.*, Phys. Rev. D **87**, 011102(R) (2013).
6. V.A. Kostelecký and M. Mewes, Phys. Rev. D **69**, 016005 (2004).
7. V.A. Kostelecký and M. Mewes, Phys. Rev. D **70**, 076002 (2004).
8. Y. Abe *et al.*, Phys. Rev. D **86**, 112009 (2012).
9. P. Adamson *et al.*, Phys. Rev. Lett. **101**, 151601 (2008); Phys. Rev. D **85**, 031101 (2012).
10. V.A. Kostelecký and M. Mewes, Phys. Rev. D **85**, 096005 (2012).
11. J.S. Díaz, V.A. Kostelecký, and M. Mewes, Phys. Rev. D **80**, 076007 (2009); J.S. Díaz *et al.*, IUHET 578.

GALILEO GALILEI (GG) SPACE TEST OF THE WEAK EQUIVALENCE PRINCIPLE TO 10^{-17}: THE KEY FEATURES

A.M. NOBILI

Department of Physics 'E. Fermi,' University of Pisa and INFN
Largo Bruno Pontecorvo 3, 56127 Pisa, Italy
E-mail: nobili@dm.unipi.it

Galileo Galilei (GG) is a small satellite to fly in a low altitude, near circular, Sun-synchronous orbit around the Earth with the goal of testing in its field the weak equivalence principle to 10^{-17}. It would improve the best tests with slowly rotating torsion balances by four orders of magnitude. The sensor is a differential accelerometer with two concentric coaxial test cylinders of different composition rotating together with the whole satellite around the symmetry axis, and weakly coupled in the plane perpendicular to it as a 2D mechanical oscillator. One axis rotation at 1 Hz, to be provided once and for all at launch, stabilizes the satellite and up-converts the signal from the low orbital frequency of $\simeq 1.7 \cdot 10^{-4}$ Hz to the much higher 1 Hz rotation frequency, needing neither motor nor bearings (passive rotation). For test masses suspended in orbit the major gain is a signal from Earth almost 500 times stronger than for the same masses suspended on ground. In GG both the sensor and the spacecraft are designed to fully exploit the properties of space in order to push the test four orders of magnitude better than slowly rotating torsion balances on ground.

1. The universality of free fall, Einstein and Eötvös

General Relativity (GR) is founded on the equivalence between inertial and gravitational mass reported by Newton in the opening paragraph of the *Principia* on the basis of his own experimental tests. It was later referred to as the weak equivalence principle (WEP) and it amounts to stating the universality of free fall (UFF).

Einstein puts UFF/WEP at the basis of GR.[1] In one of the first pages of his fundamental 1916 paper entitled *The foundation of the general theory of relativity,*[2] he writes that the *"...the existence of a field of force, namely the gravitational field, which possesses the remarkable property of imparting the same acceleration to all bodies"* is a *"teaching of experience."* Thus, it is an experimental fact and as such it needs to be confirmed by experiments. In the same years, by coupling the test masses on a torsion balance, Eötvös improved Bessel's pendulum results to $\simeq 10^{-8}$. Contrary to what is reported,[3] Einstein was aware of Eötvös's experimental work and added a specific footnote on the same page which reads: *"Eötvös has proved experimentally that the gravitational field has this property in great accuracy."*[2]

2. Lessons from the ground

In the horizontal plane a violation from UFF/WEP with the Earth as the source body is a constant differential effect in the North-South direction; if the balance arm is in the East-West direction it gives a constant deflection. However, if the source body is the Sun, the diurnal (passive) rotation of the Earth provides a 24 hr modulation which compensates by far for the slightly weaker driving signal (by almost a factor of 3). This fact was first exploited in the early 1960s by Dicke and his students,[4] who tested UFF/WEP in the field of the Sun to about 10^{-11}, improved almost a decade later to 10^{-12} by Braginsky and Panov.[5] About 30 years after Dicke's efforts, the Eöt-Wash group has developed actively rotated torsion balances which — in a remarkable series of experiments — have reached several parts in 10^{13} in the field of the Sun[6] and 10^{-13} in the field of the Earth.[7]

Galileo-like mass dropping tests have been unable to compete with rotating torsion balances despite the much stronger driving signal — in the field of the Earth — by almost a factor 600. The major limitation is due to errors in releasing the test masses at initial time, on ground as well as in space and independent of the kinds of masses being dropped (macroscopic or atoms).[8–10] In a very innovative 8 m drop test the falling masses were coupled to form a vertical disk with two halves of different composition.[11] At the end of fall the rotation angle was measured very precisely with corner cube reflectors and laser interferometry. It should have been zero if UFF/WEP holds. In almost 130 drops (half with the disk reverted) it reached a sensitivity not better than $7.2 \cdot 10^{-10}$. By dropping a single Al disk this was proved to be due to disk rotation errors at release.

3. Lessons from space

Moving a torsion balance-type sensor in low Earth orbit would combine the advantage of a drop test (a driving signal from Earth only slightly lower than on ground and almost 500 times stronger than for the balance on ground), with those of the torsion balance, mainly, no release errors issue and high sensitivity. However, at $0g$ the torsion balance loses the property it has at $1g$ of rejecting almost 'perfectly' effects which, unlike the violation signal, point to the same direction. A new 'space balance' design is needed.

In space, each test mass orbiting the Earth is a 2-body problem with 2 degrees of freedom (in the orbit plane).[10] The sensor should have the same dimensionality, which is achieved by coupling the test bodies as weakly as possible in a plane close to the orbital one. Classical tidal (differential) effects should be reduced by making their centers of mass to coincide as well as possible. The balance thus formed should rotate around the axis perpendicular to the orbit/sensitive plane so as to up-convert the signal to a frequency higher than the orbital one (the higher the better).

In essence, the sensor should be a 2D mechanical oscillator with very low natural coupling frequency (for high sensitivity) and very high rotation rate (for high frequency modulation). This is by definition a rotor in 'supercritical' regime, which is well known in rotordynamics and for a wide variety of applications. In 2D (unlike in 1D) it is known to provide a stable solution by self-centering. In space a low natural frequency, comparable to the torsion frequency of a torsion balance with a very thin suspension fiber, is achieved thanks to the absence of weight. In space the whole lab (the satellite) is isolated; given a cylindrical symmetry it can be passively stabilized by one-axis rotation. After initial spin-up, rotation is maintained by angular momentum conservation, requiring neither a motor nor bearings (passive rotation). It up-converts the signal to higher frequency. Most importantly, in 2D (unlike in 1D) this occurs with no signal reduction.[12,13]

The GG sensor and spacecraft have been designed along these guidelines.[13,14] The GG satellite is stabilized by rotation around the symmetry axis at 1 Hz. The signal is up-converted at this frequency where thermal noise from losses in the suspensions is much lower than at the orbital frequency.[12] Including all losses, the integration time required to reduce thermal noise amounts to a few hours only.[15] Thus, many measurements can be performed in different dynamical configurations of the orbit plane relative to the Earth (resulting from the Sun-synchronous orbit) while the satellite attitude is fixed in space. This allows many systematics (caused by the monopole moment of the Earth, like the signal) to be discriminated with certainty from a violation signal because of their known, different, signature.[16]

In the rotating frame disturbances due to mass anomalies in the concentric test cylinders are DC, while the signal is close to the 1 Hz spin frequency. The test cylinders form a balance whose arms will be balanced in flight for common mode rejection. Their quadrupole mass moments can be measured and reduced prior to launch. The experiment is at room temperature. Thus, precision manufacture requirements of the test masses are not strict and a very wide choice of materials is possible based on their fundamental atomic properties, which is a crucial asset of the test.[17] GG can even consider testing Pb and H rich composites such as CH_2 polymers.[14]

Sensitivity of the concentric test cylinders in 2D and rotation around the symmetry axis allows a full scale prototype of GG on the ground (GGG). GGG is much less sensitive than in absence of weight and, unlike GG, it is affected by local terrain tilt noise and by motor/bearings noise. As a prototype of the space sensor it has reached a sensitivity slightly better than 10^{-11}.[13] For the reasons discussed above, test cylinders of unusual compositions can be considered for UFF/WEP tests in the field of the Sun should they be found to be relevant in the framework of the Standard-Model Extension[18] along the lines presented in Ref. 17.

References

1. A.M. Nobili *et al.*, Am. J. Phys. **81**, 527 (2013).
2. A. Einstein, Ann. Physik (ser. 4), **49**, 769 (1916).
3. C.C. Speake and C.M. Will, Class. Quantum Grav. **29**, 180301 (2012).
4. P.G. Roll, R. Krotkov, and R.H. Dicke, Ann. Phys. **26**, 442 (1964).
5. V.B. Braginsky and V.I. Panov, Sov. Phys. JETP **34**, 463 (1972).
6. S. Baeßler *et al.*, Phys. Rev. Lett. **83**, 3585 (1999).
7. S. Schlamminger *et al.*, Phys. Rev. Lett. **100**, 041101 (2008).
8. J.P. Blaser, Class. Quantum Grav. **18**, 25094 (2001).
9. G.L. Comandi *et al.*, Phys. Lett. A **318**, 251 (2003).
10. A.M. Nobili *et al.*, Gen. Rel. Gravit. bf 40, 1545 (2008).
11. S. Carusotto *et al.*, Phys. Rev. Lett. **69**, 1722 (1992).
12. R. Pegna *et al.*, Phys. Rev. Lett. **107**, 200801 (2011).
13. A.M. Nobili *et al.*, Class Quantum Grav. **29**, 184011 (2012).
14. A.M. Nobili *et al..*, *GG Report of Mission Study (Phase A-2)*, ASI (2009).
15. R. Pegna *et al.*, submitted for publication.
16. A.M. Nobili *et al.*, in preparation.
17. M.A. Hohensee, H. Müller, and R.B. Wiringa, submitted for publication.
18. V.A. Kostelecký and J.D. Tasson, Phys. Rev. D **83**, 016013 (2011).

PROBING PHYSICS BEYOND THE STANDARD MODEL WITH HE/XE CLOCK COMPARISON EXPERIMENTS

F. ALLMENDINGER and U. SCHMIDT*

Physikalisches Institut, Ruprecht-Karls-Universität Heidelberg
69120 Heidelberg, Germany
** E-mail: ulrich.schmidt@physi.uni-heidelberg.de*

W. HEIL, S. KARPUK, A. SCHARTH, Y. SOBOLEV, K. TULLNEY and S. ZIMMER

Institut für Physik, Johannes Gutenberg-Universität
55099 Mainz, Germany

The comparison of the free precession of co-located ^{3}He-^{129}Xe spins (clock comparison) enables us to search for very tiny nonmagnetic spin interactions. With our setup we could establish new limits for Lorentz invariance violating interactions of spins with a relic background field which permeates the Universe and points in a preferred direction in space.

1. Experimental setup and principle of measurement

In our experiments we employ polarized ^{3}He and ^{129}Xe as nuclear spin clocks. The two polarized gas species at pressure of order mbar are filled in a low-relaxation spherical glass cell with diameters between 6 cm and 10 cm together with nitrogen, which acts as buffer gas. The cell is positioned in the center of a homogeneous static magnetic field of about 400 nT. The field is generated by means of Helmholtz coils, which are mounted inside the 7 layer magnetic shield of BMSR-2 at the PTB in Berlin. Details of the setup are given in Refs. 1 and 2. At that field strength, the Larmor frequency of ^{3}He is about $\omega_{He} \approx 2\pi \cdot 13$ Hz and $\omega_{Xe} \approx 2\pi \cdot 4.7$ Hz respectively. To measure these low precession frequencies, we use low-noise low-temperature DC-SQUID gradiometers as magnetic flux detectors. Due to the very low field gradients of order pT/cm at the location of our cell, we reached T_2^* times of 8h for ^{129}Xe and up to 100h for ^{3}He.[3] Therefore we can measure the free induction decay over a period T of one day, thanks to a measured signal-to-noise ratio of SNR $>$ 3000:1 per $\sqrt{\text{Hz}}$. This long measurement period is important because the final statistical frequency error of our clock comparison experiments is proportional to $T^{-3/2}$ according to the Cramer Rao lower bound (CRLB).[2]

To be sensitive to tiny nonmagnetic interactions, we must calculate the weighted difference of the respective Larmor frequencies of the co-located spin samples or the corresponding time integral, the Larmor phases,

$$\Delta\omega = \omega_{\rm He} - \frac{\gamma_{\rm He}}{\gamma_{\rm Xe}}\omega_{\rm Xe} = 0 \quad , \quad \Delta\Phi = \Phi_{\rm He} - \frac{\gamma_{\rm He}}{\gamma_{\rm Xe}}\Phi_{\rm Xe} = {\rm const.} \tag{1}$$

The weighting coefficient is $\gamma_{\rm He}/\gamma_{\rm Xe} = 2.75408159(20)$. In doing so, magnetic field fluctuations are cancelled. For the purpose of data analysis, we cut each run in subdata sets of 3.2 s time span. For each subdata set we extract both phase and amplitude of each spin species by fitting appropriate functions (for details, see Ref. 4). From these phase values we can extract the accumulated phases and with it the corresponding phase difference $\Delta\Phi(t)$. We reached with our last measurements in 2012 a phase sensitivity of $\Delta\Phi$=0.3 mrad or $\Delta\omega = 2\pi\cdot 40$p Hz. On a closer look, at that accuracy level, $\Delta\Phi$ is not a constant in time, as Eq. (1) may suggest. Instead higher order effects have to be take into account which can be parameterized as

$$\begin{aligned}\Delta\Phi(t) = {\rm const} + a_{\rm lin}t + a_{\rm He}e^{-t/T^*_{2,\rm He}} + a_{\rm Xe}e^{-t/T^*_{2,\rm Xe}} \\ +b_{\rm He}e^{-2t/T^*_{2,\rm He}} + b_{\rm Xe}e^{-2t/T^*_{2,\rm Xe}}.\end{aligned} \tag{2}$$

In Eq. (2), the linear term $a_{\rm lin}$ has two contributions. One is trivial and stems from Earth's rotation. The other is caused by chemical shift, a correction for the finite density of the gases. The terms with a_{He} and a_{Xe} account for the Ramsey-Bloch-Siegert shift.[5] Each spin feels the magnetic moments of the other precessing spins of the same gas species that are slightly detuned in frequency due to magnetic field gradients (self-shift). Its value is proportional to the particular net magnetization of the spin species and decays therefore with the corresponding effective T_2^* time of the free induction decay. Finally the last two terms of Eq. (2) take into account the shift due to the interaction of the spin species among each other (cross-talk). This interaction can be seen as a generalization of the well known Bloch-Siegert shift[6] and therefore its value is proportional to the magnetization squared. We determine the T_2^* times independently by fitting the amplitude data of a run to an exponential. With the knowledge of both, the exact SQUID geometry and the geometry of the cell, we are also able to calculate from these fit results the b coefficients of the Bloch-Siegert shift (see Eq. (2)). In contrast, we have no precise enough model for calculating the corresponding a coefficients. As a consequence, we have to include the amplitudes of the linear term and the self-shift terms as free parameters while fitting the phase data. Now we consider a search for a finite nonmagnetic interaction. We include this interaction by adding an appropriate term $\Delta\Phi_{\text{nonmag}}(t)$ to Eq. (2). Depending on the time dependence of $\Delta\Phi_{\text{nonmag}}(t)$, the amplitude of this term may be strongly correlated with the values of the a coefficients of Eq. (2), which we have also to include as free parameters into the fit, too. Therefore, careful correlation analysis of the fit parameters is mandatory. For high correlations, the sensitivity to nonmagnetic very weak interaction is reduced compared to the CRLB calculation.[2]

2. Constraints on a Lorentz invariance violating interaction

In the context of the Standard-Model Extension (SME), possible interactions between the spin of a bound neutron and a relic background field are discussed. To determine the leading-order effects of a Lorentz violating potential V, it suffices to use a nonrelativistic description for the particles involved given by $V = \tilde{b}^n_{x,y,z} \cdot \vec{\sigma}^n_{x,y,z}$.[7] We search for sidereal variations of the frequency of co-located spin species while the Earth and hence the laboratory reference frame rotates with respect to a preferred inertial frame. The observable to trace possible tiny sidereal frequency modulations is again the combination of measured Larmor frequencies (see Eq. (1)) and the weighted phase differences, respectively. Details of the setup and data analysis of our first run are given in Ref. 4. For the purpose of data analysis we had to add a sidereal modulation to the fitting function of Eq. (2) with

$$\Delta\Phi_{\text{nonmag}}(t) = a_s \sin\left[\Omega_{SD}\left(t - t_0\right) + \varphi_{SD}\right] - a_c \cos\left[\Omega_{SD}\left(t - t_0 + \varphi_{SD}\right)\right]. \tag{3}$$

In this equation, Ω_{SD} is the angular frequency of the sidereal day and φ_{SD} represents the phase offset at the beginning t_0 of the first run. From that, the fit results for the amplitudes a_s and a_c as well as the RMS magnitude of the sidereal phase amplitude $\Phi_{SD} = \sqrt{a_s^2 + a_c^2}$ could be extracted. In terms of the SME[7] we can express the sidereal phase amplitudes according to $a_{s(c)} = 2\pi/\Omega_{SD} \cdot \upsilon_{X(Y)}$ with $2\pi|\delta\upsilon_{X,Y}|\hbar = |2\left(1 - \gamma_{He}/\gamma_{Xe}\right)\sin\chi\, \tilde{b}^n_{X,Y}|$, where χ is the angle between the Earth's rotation axis and the quantization axis of the spins. In Table 1 upper limits of the equatorial component $\tilde{b}^n_{\perp} = \sqrt{(\tilde{b}^n_x)^2 + (\tilde{b}^n_y)^2}$ of the background tensor field interacting with the spin of the bound neutron are given together with recent results from Ref. 8.

Table 1. Results for coefficients for Lorentz violation.

Coefficient	Data 2009[4]	Romalis *et al.*[8]	Preliminary data, March 2012 run
$\tilde{b}_x^n$ [GeV] (1σ)	$(3.4\pm1.7)\cdot10^{-32}$	$(0.1\pm1.6)\cdot10^{-33}$	$(4.1\pm4.7)\cdot10^{-34}$
$\tilde{b}_y^n$ [GeV] (1σ)	$(1.4\pm1.3)\cdot10^{-32}$	$(2.5\pm1.6)\cdot10^{-33}$	$(2.9\pm6.2)\cdot10^{-34}$
$\tilde{b}_\perp^n$ [GeV] (68% C.L.)	$<4.7\cdot10^{-32}$	$<3.7\cdot10^{-33}$	$<6.7\cdot10^{-34}$
$\tilde{b}_\perp^n$ [GeV] (95% C.L.)	$<6.6\cdot10^{-32}$	$<5.5\cdot10^{-33}$	$<1.3\cdot10^{-33}$

3. Conclusion

Nuclear spin clocks, based on the detection of free spin precession of gaseous, nuclear polarized ^{3}He or ^{129}Xe samples with a SQUID as magnetic flux detector can be used as ultra-sensitive probe for nonmagnetic spin interactions, since the magnetic dipole interaction (Zeeman term) drops out in case of co-located spin samples. With the long spin-coherence times, measurements of uninterrupted precession of $T \sim 1$ day can be achieved at the present stage of investigation. With an appropriate setup we also establish new limits for the pseudoscalar spin unpolarized matter interaction mediated by axion-like particles.[9] As the next challenging step, we want to employ this method for the search for an electric dipole moment of ^{129}Xe.

Acknowledgments

We thank our colleagues M. Burghoff, W. Kilian, S. Knappe-Grüneberg, A. Schnabel, F. Seifert and L. Trahms from PTB in Berlin for their support during measurement time at PTB and for helpful discussions.

References

1. C. Gemmel *et al.*, Eur. Phys. J. D **57**, 303 (2010).
2. W. Heil *et al.*, Ann. Phys. (Berlin), 1-11 (2013), in press.
3. F. Allmendinger *et al.*, these proceedings.
4. C. Gemmel *et al.*, Phys. Rev. D **82**, 111901(R) (2010).
5. N.F. Ramsey, Phys. Rev. **100**, 1191 (1955).
6. F. Bloch and A. Siegert, Phys. Rev. **57**, 522 (1940).
7. D. Colladay and V.A. Kostelecký, Phys. Rev. D **58**, 116002 (1998).
8. J.M. Brown *et al.*, Phys. Rev. Lett. **105**, 151604 (2010).
9. K. Tullney *et al.*, arXiv:1303.6612v2.

USING BINARY PULSARS TO TEST LORENTZ SYMMETRY IN THE GRAVITATIONAL SECTOR

J.M. WEISBERG
Department of Physics and Astronomy, Carleton College
Northfield, MN 55057, USA
E-mail: jweisber@carleton.edu

I review some of the major developments in the theoretical background and experimental uses of binary pulsars to explore local Lorentz invariance in the gravitational sector and its possible violation.

1. Binary pulsars

A binary pulsar (PSR) consists of a spinning neutron star (NS) with a radio beam, and a companion, in a mutual orbit. The companion is usually another compact object — either another NS or a white dwarf (WD). A typical NS is a highly relativistic object. With mass $M \sim 1.4 M_{\rm Sun}$ and radius $R \sim 10$ km, the NS has $GM/(c^2R) \sim 0.2$. Furthermore, the PSR's radio pulses serve as a highly precise spin-powered clock, enabling the accurate measurement of the spin and orbital parameters of the binary system and the testing of relativistic gravitational effects in strong-field conditions. For example, the double NS system PSR B1913+16 has provided the first evidence for the existence of gravitational waves,[1] while NS-WD systems tend to be the best binary PSRs for probing violations of Lorentz invariance.

2. PPN extensions and PFE in binary PSRs

The weak-field, parametrized post-newtonian (PPN) framework was extended in Ref. 2 to include the possibility of preferred-frame effects (PFE), via parameters α_1, α_2, and α_3. This analysis was further augmented to encompass the strong fields of binary PSRs,[3] as follows.

A gravitational N-body post-newtonian lagrangian was derived with terms for $\hat{\alpha}_1$ and $\hat{\alpha}_2$, where the 'hat' over a parameter indicates that it is a strong-field modification of the PPN version; commensurate with the

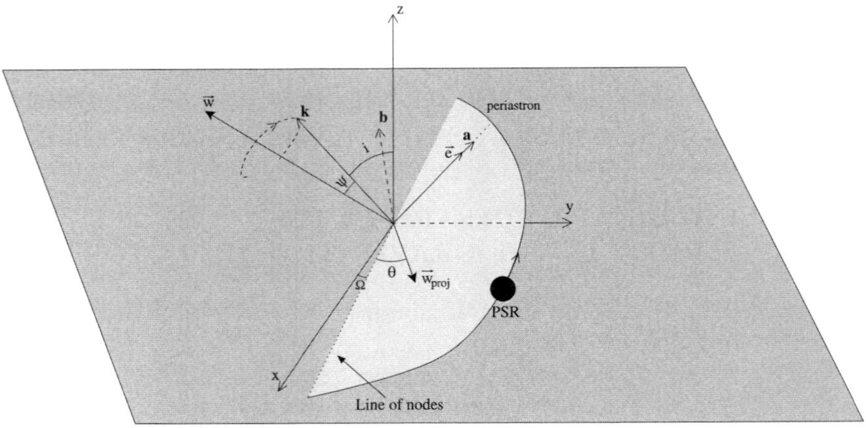

Fig. 1. PSR orbital geometry. Unit vectors $\{\mathbf{a}, \mathbf{b}, \mathbf{k}\}$ define an orthogonal coordinate system centered on the binary's center of mass, with $\mathbf{a}$ and $\mathbf{b}$ in the orbital plane (lightly shaded) and $\mathbf{k}$ pointing in the direction of orbital angular momentum. Coordinate $\mathbf{a}$ points toward PSR periastron, and $\mathbf{a} \times \mathbf{b} = \mathbf{k}$. The eccentricity e is incorporated into the *vector* $\vec{\mathbf{e}} = e\ \mathbf{a}$. The unit vectors $\{\mathbf{x}, \mathbf{y}, \mathbf{z}\}$ define another orthogonal coordinate system, with $\mathbf{x}$ and $\mathbf{y}$ in the plane of the sky (heavily shaded), and $\mathbf{z}$ pointing *from* the observer *to* the binary's center of mass. Unit vector $\mathbf{x}$ points toward celestial north. This right-handed coordinate system is opposite to the observers' convention for the plane of the sky. The velocity of the binary system's center of mass with respect to the preferred frame and its projection onto the orbital plane are $\vec{\mathbf{w}}$ and $\vec{\mathbf{w}}_{\rm proj}$, respectively. Adapted from Ref. 4.

highly relativistic environment of a binary PSR system. The $\hat{\alpha}_1$-part of the lagrangian includes a dependence of the vector velocity $\vec{\mathbf{w}}$ of the binary center of mass with respect to the preferred frame, leading to some potentially measurable effects. For small-eccentricity binary PSRs, the eccentricity vector $\vec{\mathbf{e}} \equiv e\ \mathbf{a}$ (see Fig. 1) undergoes the following time evolution:

$$\vec{\mathbf{e}}(t) = \vec{\mathbf{e}}_{\rm Rel}(t) + \vec{\mathbf{e}}_{\rm Fixed}; \tag{1}$$

where the first term results from the usual general relativistic advance of periastron whose mean rate is $\langle \dot{\omega}_{Rel} \rangle$; but the second is a nonzero constant 'polarizing" term resulting from PFE, and proportional to $G_{\rm orb}$ (a function of orbital parameters and component masses), $\hat{\alpha}_1$ and $\vec{\mathbf{w}}_{\rm proj}$ (the projection of $\vec{\mathbf{w}}$ onto the orbital plane):

$$\vec{\mathbf{e}}_{\rm Fixed} = G_{\rm orb} \hat{\alpha}_1 \mathbf{k} \times \vec{\mathbf{w}}_{\rm proj}, \tag{2}$$

(which points in the orbital plane, perpendicular to $\vec{\mathbf{w}}_{\rm proj}$). Hence the measurement of eccentricity over time affords the possibility of separating the two terms in Eq. (1) and then, for a given choice of preferred frame of

$\vec{\mathbf{w}}$ (e.g., the cosmic microwave background), determining the desired PFE parameter $\hat{\alpha}_1$ from Eq. 2.

However, some observational difficulties remain. First, the velocity of the binary system's center of mass with respect to Earth is difficult to measure, which limits the accuracy of $\vec{\mathbf{w}}$ (both its direction and magnitude). Also the angle Ω is not measurable from PSR timing, thereby further limiting the accuracy of its projection $\vec{\mathbf{w}}_{\rm proj}$. In general, statistical arguments must be employed in order to ameliorate these limitations, although much progress has also been made in overcoming them observationally. (See below.) The measured eccentricity for the NS-WD PSR B1855+09 system was then combined with the above analysis to provide an upper limit on $|\hat{\alpha}_1|$ of 5.0×10^{-4}, which was comparable to contemporaneous solar system limits, and (unlike the solar system case) measured in *strong* gravitational fields.

Building upon the work of Ref. 3, it was shown[4] that *both* $\hat{\alpha}_1$ *and* $\hat{\alpha}_2$ are determinable from timing measurements of low-eccentricity binary PSRs, and the new analysis was applied to two additional NS-WD binary PSR systems: PSR J1012+5307 and PSR J1738+0333, as follows.

A nonzero $\hat{\alpha}_2$ will cause precession of the orbital pole, $\mathbf{k}$, about $\vec{\mathbf{w}}$ (see Fig. 1). This precession leads to an observable consequence in the PSR timing measurable $x \equiv a_{\rm PSR} \sin i$. Here $a_{\rm PSR}$ is the semimajor axis of the PSR orbit, and i is the 'inclination' angle between $\mathbf{k}$ and $\mathbf{z}$:

$$\left(\frac{\dot{x}}{x}\right)_{\hat{\alpha}_2} = -\frac{\hat{\alpha}_2}{4}\left(\frac{2\pi}{P_{\rm b}}\right)\left(\frac{w}{c}\right)^2 \cot i \sin 2\psi \cos\theta, \tag{3}$$

where angles ψ and θ specify the instantaneous direction of $\vec{\mathbf{w}}$ in the (precessing) $\{\mathbf{a}, \mathbf{b}, \mathbf{k}\}$ coordinate system, as shown in Fig. 1. However, there are observational difficulties in determining w and the angles, similar to those discussed above in the determination of $\hat{\alpha}_1$. While significant progress has been made, the nature of PSR timing observations prevents the determination of Ω and hence accurate values of w, θ, and ψ, so it remains necessary to use statistical arguments to extract the PFE parameter $\hat{\alpha}_2$. A joint analysis from both pulsars leads to $|\hat{\alpha}_2| < 1.8 \times 10^{-4}$. This limit is far weaker than the solar system limit of Ref. 5, but does have the benefit of probing strong fields. The analysis of $|\hat{\alpha}_1|$ of Ref. 3 was also revisited. With the benefit of additional kinds of observational data that leave only Ω undetermined, an upper limit on $\hat{\alpha}_1$ of $(-0.4\ [+3.7, -3.1]) \times 10^{-5}$ was found for PSR J1738+0333, making it the best extant constraint for this parameter.

The determination of PPN parameter α_3 from binary PSRs was derived theoretically in Ref. 6; Ref. 7 reports an exquisitely precise null measurement, as expected in semiconservative theories of gravitation.

3. SME coefficicents for Lorentz violation in binary PSRs

The first extensive theoretical SME analysis of the signal of Lorentz violation in binary PSRs was published in Ref. 8. In developing the general theory, twenty independent Lorentz-violating coefficients were found, including nine in the field $\bar{s}^{\mu\nu}$. It was shown that three of the PPN parameters are expressible in terms of the single coefficient $\bar{s}^{00}$; and that $\alpha_1 = 4\alpha_2$, indicating that (at least in the weak-field limit), they are not independent. More importantly, it was shown that binary PSR secular *measurables* $\langle\dot{\omega}_{Rel}\rangle, \langle de/dt\rangle$, and $\langle di/dt\rangle$ can be expressed in terms of combinations of projections of the SME $\bar{s}^{\mu\nu}$ along the orbital axes $\{\mathbf{a}, \mathbf{b}, \mathbf{k}\}$, raising the prospect of determining these SME Lorentz-violating coefficients via binary PSR timing observations. Therefore wholly new tests of Lorentz violation become available which are not accessible in the PPN formulation.

4. Future prospects

Observationally, the most promising direction involves the detection and measurement of additional binary PSR systems. These would increase the parameter-space coverage of current tests, which could lead for example to disentangling the currently covariant SME Lorentz-violation coefficients,[8] and increasing the sky coverage of Lorentz-violation tests.[4] Planned new, highly sensitive radiotelescopes will detect numerous such systems.[9] Theoretically, the most exciting prospect includes the generation of additional SME measurables in binary PSR systems.

Acknowledgments

This research was supported by NSF grant AST-0807556.

References

1. J.M. Weisberg, D.J. Nice, and J.H. Taylor, Ap. J. **722**, 1030 (2010).
2. C.M. Will and K. Nordtvedt Jr., Ap. J. **177**, 757 (1972).
3. T. Damour and G. Esposito-Farèse, Phys. Rev. D **46**, 4128 (1992).
4. L. Shao and N. Wex, Class. Quantum Grav. **29**, 215018 (2012).
5. K. Nordtvedt, Ap. J. **320**, 871 (1987).
6. J.F. Bell and T. Damour, Class. Quantum Grav. **13**, 3121 (1996).
7. M.E. Gonzalez *et al.*, Ap. J. **743**, 102 (2011).
8. Q.G. Bailey and V.A. Kostelecký, Phys. Rev. D **74**, 045001 (2006).
9. J.M. Cordes *et al.*, New Astron. Rev. **48**, 1413 (2004).

A NEW APPROACH TO SEARCH FOR LORENTZ INVARIANCE VIOLATION IN β DECAY

H.W. WILSCHUT

Kernfysisch Versneller Instituut, Rijksuniversiteit Groningen
9747AA Groningen, Netherlands
E-mail: wilschut@kvi.nl

Lorentz invariance violation (LIV) in the weak interaction is studied, both experimentally and theoretically, at the Kernfysisch Versneller Instituut. Here, it is discussed which new observables can be measured, and which experimental methods can be used. An outlook towards high statistics experiments is given.

1. Ways to observe LIV in β decay

The weak interaction has not yet gained much attention in searches of LIV, even though the weak interaction determines the most significant properties of the Standard Model (SM). As discussed elsewhere in these proceedings, aspects of neutrino propagation and K- and B-meson oscillations have gained considerable attention, but not weak decays. Here we focus on rotational invariance in β decay and apply the framework of a modified W propagator introduced in Ref. 1. Consider β decay as depicted in Fig. 1: β decays could be anisotropic with respect to the direction of the β-particle velocity $\vec{\beta}$, the spin of the parent nucleus $\vec{J}$, or the spin of the β particle $\vec{\sigma}$. The preferred directions of these obervables are denoted by $\hat{n}_i$ with $i = 1, 2, 3$, respectively. The relative decay width is then given by

$$\frac{d\Gamma}{\Gamma_0} = 1 + \vec{\beta} \cdot [A_\beta \frac{\langle \vec{J} \rangle}{J} + \xi_1 \hat{n}_1 + G\vec{\sigma}] + \xi_2 \frac{\langle \vec{J} \rangle}{J} \cdot \hat{n}_2 + \xi_3 \vec{\sigma} \cdot \hat{n}_3. \quad (1)$$

The relative strength of the LIV contributions is given by ξ_i. There are two components that result from the SM due to parity violation: they are the β asymmetry, characterized by the parameter A_β, and the longitudinal polarization of the β particles by G.

Only the first LIV anisotropy has been measured[2,3] more than 30 years ago. These results did not receive much attention, possibly because no

Fig. 1. Three basic ways of observing violation of rotational invariance in β decay.

theoretical framework was available at the time. Of the other two anisotropies $\xi_2\hat{n}_2$ is independent of $\vec{\beta}$. This anisotropy is measured in our experimental work. In parallel to the experimental work, theory development started. A rather general extension concerning the W boson was chosen motivated by the fact that the propagation of the leptons is already tackled in other experimental and theoretical work as reported at this conference. Lorentz violation is parameterized by modifying the W propagator with a (complex, possibly momentum dependent) tensor $\chi^{\mu\nu}$ such that

$$\langle W^{\mu +}(p)W^{\nu -}(-p)\rangle = \frac{-i(g^{\mu\nu}+\chi^{\mu\nu})}{M_W^2} \,. \tag{2}$$

The consequences of this modification have been evaluated for (allowed) β decay, revealing a rich structure which allows placing limits on $\chi^{\mu\nu}$ using the specific properties of the various decay modes of β decay.[1] The relevance for forbidden β decay and neutral K-meson decay is also considered as reported in this conference by J.P. Noordmans[4] and by K.K. Vos,[5] respectively. In Table 1 we give the expressions for $\xi_i\hat{n}_i$ in terms of $\chi^{\mu\nu}$.

Table 1. Expressions for ξ_i in terms of $\chi^{\mu\nu}$. In the last row, $\mp$ refers to $\beta^\mp$ emission.

ξ	Fermi	Gamow-Teller	Forbidden
$\xi_1\hat{n}_1^l$	$2\chi_r^{0l}$	$\frac{2}{3}(\chi_r^{l0}+\epsilon^{lmk}\chi_i^{mk})$	Ref. 4
$\xi_2\hat{n}_2^l$	n.a.	$A\epsilon^{lmk}\chi_i^{mk}$	-
$\xi_3\hat{n}_3^l$	$\mp\sqrt{(1-(\alpha Z)^2)(1-\beta^2)}\xi_1\hat{n}_1$	$\mp\sqrt{(1-(\alpha Z)^2)(1-\beta^2)}\xi_1\hat{n}_1$	-

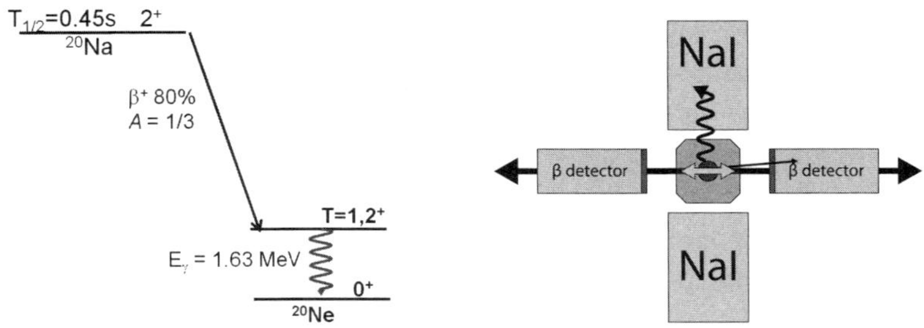

Fig. 2. Left: relevant part of the decay scheme of ^{20}Na. Right: schematic setup.

2. Experimental method

To put a limit on $\xi_2 \hat{n}_2$ we exploit the properties of ^{20}Na shown in Fig. 2. The short-lived isotope ^{20}Na is produced with the KVI cyclotron AGOR in the kinematic inverse reaction ^{1}H(^{20}Ne,^{20}Na)n and removed from the beam by the magnetic separator. The fast Na particles are stopped in a gas cell filled with neon and polarized by optical pumping. The parity violating β emission is used to determine the value of $\langle \vec{J} \rangle / J$. Using either up-down sequences of polarization or east-west sequences, the γ decay is measured to determine the lifetime τ. The γ decay is in principle not sensitive to the sign of the polarization and can thus be used to establish the LIV asymmetry

$$\mathcal{A}_\gamma = -\frac{\tau^\uparrow - \tau^\downarrow}{\tau^\uparrow + \tau^\downarrow} = \frac{|\langle \vec{J} \rangle|}{J} \xi_2 \hat{n}_2 \cdot \hat{p}, \tag{3}$$

where $\pm \hat{p}$ are the up-down or east-west polarization direction, respectively, while $|\langle \vec{J} \rangle / J|$ is constant and is determined from the β asymmetry. More experimental details can be found in Ref. 6.

The experiment evaluated until now has polarization in the up-down directions. The preferred direction in the laboratory is then

$$\hat{n}_2(t) = N_1 \cos\theta_l \cos\omega_\oplus t + N_2 \cos\theta_l \sin\omega_\oplus t + N_3 \sin\theta_l. \tag{4}$$

Here N_k $(k = 1, 2, 3)$ is the time independent preferred direction in the Sun-centered frame, θ_l is the latitude of the location of the measurement, and $\omega_\oplus$ is the sidereal frequency. Therefore, $\mathcal{A}_\gamma$ can have an amplitude varying with sidereal time, while the third term of Eq. (4) implies a constant value. Whereas the sidereal dependence can be measured with an accuracy limited by statistics, the measurement of the constant term has a large systematic error that can not easily be eliminated. Consequently, the up-down measurement limits only $|\xi_2 N_1|$ and $|\xi_2 N_2|$ with a value of 3×10^{-3}.

The corresponding limits on χ_i (cf. Table 1), which in the Sun-centered frame is denoted by X_i, are three times larger, because the β asymmetry $A_\beta = 1/3$. This work is the first measurement to establish a direct limit on a few of the parameters making up the tensor χ. In fact, measurements have been made in the past that can limit χ but require that theory be evaluated for these cases. Examples of this are the calculations in Refs. 4,5.

3. Outlook

The present experiment is limited by statistics. Indeed a new experiment is currently being evaluated that will give an order of magnitude improvement. Major improvements can only be made by considerably increasing the intensity of the source used. For example, the experiment by Newman and Wiesner[2] used a 10 Ci source by which one gains several orders of magnitude. Working which such strong sources has become complicated because of modern regulations concerning radioactive materials. A number of sources could be considered, e.g., a pure Gamow-Teller decay without γ emission (e.g., ^{32}P) could be used in an effective measurement. Other possibilities with high intensity are sources of muons and charged pions which are used for a variety of precision physics. Another road to follow is using beams with a high boost factor that would become available from β-beam facilities that have been proposed. Semileptonic and nonleptonic decays observed at LHC will be of interest provided they are decaying with high rapidity, so the LHCb experiment may be advantageous for this work.

Acknowledgments

This work is a collaborative effort of the KVI experimental group 'Fundamental Interactions and Symmetries' and the theory group. This research was supported by the Stichting voor Fundamenteel Onderzoek der Materie (FOM) under program 114 (TRIμP) and FOM projectruimte 08PR2636-1.

References

1. J.P. Noordmans, H.W. Wilschut, and R.G.E. Timmermans, Phys. Rev. C **87**, 055502 (2013).
2. R. Newman and S. Wiesner, Phys. Rev. D **14**, 1 (1976).
3. J.D. Ullman, Phys. Rev. D **17**, 1750 (1978).
4. J.P. Noormans *et al.*, these proceedings; in preparation.
5. K.K. Vos *et al.*, these proceedings; in preparation.
6. H.W. Wilschut *et al.*, Ann. Phys. (Berlin) DOI 10.1002/andp.201300076; S.E. Müller *et al.*, in preparation.

RELEVANCE AND PROSPECTS OF MAGIC TELESCOPE GAMMA-RAY OBSERVATIONS FOR LORENTZ SYMMETRY TESTS

R.M. WAGNER

Max Planck Institute for Physics, Föhringer Ring 6, D-80805 München, Germany

Oskar Klein Centre for Cosmoparticle Physics, Stockholm University
AlbaNova University Centre, Roslagstullbacken 21, SE-106 91 Stockholm, Sweden
E-mail: wagner@fysik.su.se

On behalf of the MAGIC Collaboration

MAGIC is a gamma-ray telescope system located on the Canary Island of La Palma. With it, the northern very-high energy (VHE) gamma-ray sky can be observed at >50 GeV, bridging the gap in energy between satellite and ground-based measurements. In this overview, we particularly highlight source classes whose VHE gamma-ray observations bear potential of contributing to Lorentz invariance tests: pulsars, active galactic nuclei, and gamma-ray bursts.

1. The MAGIC telescopes

The MAGIC (Major Atmospheric Gamma-Ray Imaging Cherenkov) telescopes comprise a stereoscopic system of two 17-m diameter Imaging Atmospheric Cherenkov Telescopes (IACT). They have been operating in stereoscopic mode since 2009 with a low trigger threshold of 50 GeV and a sensitivity for point-like sources >290 GeV of 0.8 Crab units in 50 hours.[1]

2. Lorentz invariance tests

The simplest Lorentz invariance test accessible to IACT are tests of photon time dispersion, that can be formulated as an obtained delay, including cosmological effects,[2]

$$\Delta t = \xi \frac{\Delta E}{E_{\mathrm{Pl}}} \frac{L}{c} = \xi \frac{\Delta E}{E_{\mathrm{Pl}}} H_0^{-1} \int_0^z (1+z') \frac{\mathrm{d}z'}{\sqrt{\Omega_\Lambda + \Omega_M (1+z')^3}}. \tag{1}$$

To measure these, fast transient phenomena are required, providing well-understood time structures at the cosmic accelerator that produced them.

The relevant figure of merit is given by the sensitivity to a given energy scale, $\propto \xi \frac{L}{c} \frac{E}{\Delta t}$. The time resolution of the instrument and the structure in the γ-ray signal is described by Δt (which, ultimately, is also connected to the amount of photons available for an analysis). L denotes the luminosity distance. While effects like dispersion, vacuum birefringence, anomalous threshold reactions, threshold shifts in standard reactions, or reactions affected by 'speed limits' (e.g., synchrotron radiation) might also be expected, time dispersion currently is the one most robustly tested with IACT.

3. Quantification of time delays

In X-rays and VHE γ-rays, photon data are characterized by arrival directions and energies, which are analyzed on a statistical basis. Besides cross-correlating light curves of different photon-energy bandpasses,[3,4] which does not take into account the full available information of each individual photon,[5] more appropriate methods have been developed, including energy-cost-function methods[5,6] that assume that any dispersive medium will increase the spread of the arrival time distribution;[a] a log-likelihood fit, describing the flare form, the instrumental energy smearing, the energy spectrum of the flare, and the photon arrival times,[7,8] applied to unbinned data; a cross correlation of oversampled light curves[9] allowing for resolving features in the light curves below the size of the flux bins; and a wavelet analysis, suppressing noise in the data and identifying variation points.[10]

4. Present reach using very high-energy gamma-rays

4.1. *Pulsed emission from the Crab nebula*

Pulsars have distinct periodic light-curve shapes, that can be sampled accurately. The Crab pulsar's pulsations are well aligned in time from radio through X-ray wavelengths, so it seems likely that photons of different energies are produced simultaneously. Recently, pulsed emission from the Crab was detected up to $E = 400$ GeV.[11] At a distance of 2.0 ± 0.5 kpc,[12] a quadratic limit of $E_2 > 7 \times 10^9$ GeV was achieved.[13] Observations at even higher energies may yield a competitive limit of $E_2 > 2.1 \times 10^{10}$ GeV; work is in progress. As the Crab pulsar provides a stable, non-transient source of gamma-rays, statistics of the limit will accumulate as more data is taken.

[a]The dispersion does not needs to be source extrinsic, thus even dispersive effects inside γ-ray sources can be modeled.

Also, chances to detect the pulsed signal at higher energies at statistically significant levels will further increase the deduced limit.

4.2. *Gamma-ray bursts*

Gamma-ray bursts[14] (GRB) have rather complex individual time profiles. They have been found at redshifts exceeding $z = 5$ (7% of the presently known GRB; average $z = 2.8$). They are routinely observed by Fermi at energies up to $\mathcal{O}$(GeV), yielding a sensitivity of $E_{\rm Pl} \approx 10^{19}$ GeV and above,[15] while no detection by ground-based instruments could be achieved yet. GRB discovered with IACT will likely be few, if any, during the lifetimes of IACT. This is due to (a) certainly fast, but non-negligible slewing time for IACT towards the GRB location and (b) the rather large distances of GRB, which result in attenuation of photons of a few hundred GeV or more by the extragalactic background light.[16] GRB observation criteria include good weather conditions, an acceptably low moon phase and distance, a zenith angle of 60 degrees or less, and a GRB start time less than 4 hrs ago. IACT, which are sensitive in the >30 GeV region, are still awaiting the first GRB detection.[17,18] Of the 491 alerts received so far from the GRB coordination network, 47 bursts[19] were followed up along the outlined criteria by MAGIC.

4.3. *Flaring active galactic nuclei*

The majority of the currently known VHE γ-ray AGNs are blazars,[20] characterized by high variability at all wavelengths. Leptonic and hadronic particle acceleration models explain their spectral energy distribution and its time evolution. Bright, highly variable γ-ray flares with energies up to $E_\gamma \approx$ 10 TeV with sufficient statistics and a time resolution (given by the flare timescales) of $\approx$ 50-100 s enable a sensitivity up to $E_{\rm Pl} \approx 10^{19}$ GeV. In particular, MAGIC recorded fluxes exceeding four times the Crab-nebula flux from Mkn 501 ($z = 0.034$) in 2005, and revealed rapid flux changes with doubling times of <3 minutes[4] twice. The two flares behaved differently: while the June 30 flare is only seen in 250 GeV-1.2 TeV, the July 9 flare is apparent in all energy bands (120 GeV$< E <$1.2 TeV). A zero-delay probability of $P = 0.026$, a marginal time delay $\Delta t = \tau_l E$ with $\tau_l = (0.030 \pm 0.012)$ s GeV^{-1} towards higher energies was found in the latter.[5,8] For a quadratic effect, $\Delta t = \tau_q E^2$ with $\tau_q = (3.71 \pm 2.57) \times 10^{-6}$ s GeV^{-2} is obtained. Acceleration models may explain a source-intrinsic delay of γ-rays, e.g., by gradually accelerated electrons,[4] an accelerating substructure in the relativistic jet,[21] or by a brief episode of increased particle injection at low energies.[22] When assuming a simultaneous emission of the γ-rays (of different energies) at the source at $z = 0.034$, a time delay can be converted in $E_1 = 1.445 \times 10^{16}\mathrm{s}/\tau_l$ and $E_2 = 1.222 \times 10^8(\mathrm{s}/\tau_q)^{1/2}$, respectively. As the measured time delay is marginal, they translate to a lower limit of $E_1 > 0.21 \times 10^{18}$ GeV (95% C.L.) for the linear case and $E_2 > 0.26 \times 10^{11}$ GeV (95% C.L.) for a quadratic dependence on energy.[5]

5. Future prospects

Future improvements in VHE searches are expected from the future Cherenkov Telescope Array observatory, which aims at ncreasing the instrumental sensitivity by a factor of 10, while extending the energy range with respect to current IACT. The overall increase in the sensitivity for Lorentz invariance tests to be hoped for is about a factor of 100, thus lies well above the Planck energy scale.

References

1. J. Aleksic *et al.*, Astropart. Phys. **35**, 435A (2012).
2. U. Jacob, T. Piran, JCAP **0801**, 031 (2008).
3. D.L. Band, Ap. J. **486**, 928 (1997); S.D. Biller *et al.*, Phys. Rev. Lett. **83**, 2108 (1999).
4. J. Albert *et al.*, Ap. J. **669**, 862 (2007).
5. J. Albert *et al.*, Phys. Lett. B **668**, 253 (2008).
6. J.D. Scargle, J.P. Norris, and J.T. Bonnell, Ap. J. **673**, 972 (2006).
7. R. Lamon, JCAP **0808**, 022 (2008).
8. M. Martínez and M. Errando, Astropart. Phys. **31**, 226 (2008).
9. T.-P. Li *et al.*, Chin. J. Astron. Astrop. **4**, 583 (2004).
10. J. Ellis *et al.*, Astron. Astrophys. **402**, 409 (2003).
11. E. Aliu *et al.*, Science **322**, 1221 (2008); Science **334**, 69 (2011); J. Aleksic *et al.*, Astron. Astrophys. **540**, 69 (2012).
12. D.L. Kaplan *et al.*, Ap. J. **677**, 1201 (2008).
13. A.N. Otte, Proc. of the 32nd Intl. Cosmic Ray Conference **7**, 256 (2011).
14. T. Piran, Rev. Mod. Phys. **76**, 1143 (2005).
15. F.W. Stecker, these proceedings.
16. E. Dwek and F. Krennrich, Astropart. Phys. **43**, 112 (2012).
17. J. Albert *et al.*, Ap. J. **641**, L9 (2006); Ap. J. **667**, 358 (2007).
18. F. Aharonian *et al.*, Astron. Astrophys. **495**, 505 (2009).
19. J. Albert *et al.*, Ap. J. **667**, 358 (2009); J. Aleksic *et al.*, Astron. Astrophys. **517**, A5 (2010).
20. P. Padovani and P. Giommi, Ap. J. **444**, 567 (1995).
21. W. Bednarek and R.M. Wagner, Astron. Astrophys. **486**, 679 (2008).
22. A. Mastichiadis and K. Moraitis, Astron. Astrophys. **491**, L37 (2008).

COMMENTS ON LORENTZ AND CPT VIOLATION

V. ALAN KOSTELECKÝ
Physics Department, Indiana University
Bloomington, IN 47405, USA

This contribution to the CPT'13 meeting briefly introduces Lorentz and CPT violation and outlines two recent developments in the field.

1. Introduction

The idea that small observable violations of Lorentz symmetry could provide experimental access to Planck-scale effects continues to draw attention across several subfields of physics. In the three years since the previous meeting in this series, considerable progress has been made on both experimental and theoretical fronts. This contribution to the CPT'13 proceedings contains a brief introduction, followed by comments on two topics of recent interest: nonminimal fermion couplings and Riemann-Finsler geometry.

2. Basics

A satisfactory theoretical description of Lorentz violation must incorporate coordinate independence, realism, and generality. A powerful approach uses effective field theory, starting with General Relativity coupled to the Standard Model and adding to the Lagrange density all observer-invariant terms that contain Lorentz-violating operators combined with controlling coefficients. This yields the comprehensive realistic effective field theory for Lorentz violation called the Standard-Model Extension (SME).[1,2] The SME also describes general CPT violation, which in the context of realistic effective field theory is accompanied by Lorentz violation.[3] The full SME contains operators of arbitrary mass dimension d, while the minimal SME restricts attention to operators of renormalizable dimension $d \leq 4$.

Observable signals of Lorentz violation are governed by the SME coefficients. Experiments typically search for particle interactions with background coefficient values, which can produce effects dependent on the particle velocity, spin, flavor, and couplings. Many investigations of this type have been performed,[4] achieving impressive sensitivities that in some cases exceed expectations for suppressed Planck-scale effects. If the SME coefficients are produced by spontaneous Lorentz violation,[5] as is necessary when gravity is based on Riemann geometry,[2] then they are dynamical quantities that must incorporate massless Nambu-Goldstone modes.[6] These modes have numerous interpretations, including serving as an alternative origin for the photon in Einstein-Maxwell theory[6] and the graviton in General Relativity,[7] or representing new spin-dependent[8] or spin-independent[9] forces, among other possibilities. Massive modes can also appear.[10]

3. Nonminimal fermion sector

In the nonminimal sector of the SME, the number of Lorentz-violating operators grows rapidly with the mass dimension d. Systematically cataloguing and characterizing the possibilities is therefore indispensible in the search for Lorentz violation. In the CPT'10 proceedings, I outlined some features appearing in the treatment of quadratic operators of arbitrary d in the photon Lagrange density.[11] In the intervening three-year period, investigations of the quadratic fermion sector for arbitrary d have also been performed. The Lagrange density for propagation and mixing of any number of fermions has been developed and applied to describe general Lorentz violation in the neutrino sector.[12] More recently, quadratic operators of arbitrary d have been studied for a massive Dirac fermion.[13]

Many nonminimal operators generate effects that are in principle observable, and each such operator generates a distinct experimental signal. For quadratic operators, which characterize particle propagation and phase-space features of particle decays, the Lorentz-violating effects can include direction dependence (anisotropy), wave-packet deformation (dispersion), and mode splitting (birefringence). In the neutrino sector, for example, some operators control flavor-dependent effects in neutrino and antineutrino mixing, producing novel energy and direction dependences that involve both Dirac- and Majorana-type couplings. Others govern species-independent effects, which can differ for neutrinos and antineutrinos and can produce propagation times varying with energy and direction, in some cases exceeding that of light. A few operators produce 'countershaded' effects[14] that cannot be detected via oscillations or propagation but change interaction properties in processes such as beta decay.[15]

Analogous effects appear in the description of a massive Dirac fermion in the presence of Lorentz violation, for which the exact dispersion relation for arbitrary d is known in closed and compact form.[13] For example, the fermion group velocity is anisotropic and dispersive, while the fermion spin exhibits a Larmor-like precession caused by birefringent operators. Using field redefinitions to investigate observability reveals that many operators of dimension d produce no effects or are physically indistinguishable from others of dimensions d or $d \pm 1$. Nonetheless, the number of observable coefficients grows as the cube of d. To date, almost all the nonminimal coefficient space for fermions is experimentally untouched. This offers an open arena for further exploration with a significant potential for discovery.

4. Riemann-Finsler geometry

The surprising 'no-go' result that the conventional Riemann geometry of General Relativity and its extension to Riemann-Cartan geometry are both incompatible with explicit Lorentz violation raises the questions of whether an alternative geometry is involved and, if so, whether a corresponding gravitational theory exists. The obstruction to explicit Lorentz violation, which disappears for the spontaneous case, can be traced to the generic incompatibility of the Bianchi identities with the external prescription of coefficients for Lorentz violation. It is therefore reasonable to conjecture that a natural geometrical setting would include metric distances depending locally on the coefficients in addition to the Riemann metric.[2]

Support for this conjecture has recently emerged with the discovery that a fermion experiencing explicit Lorentz violation tracks a geodesic in a pseudo-Riemann-Finsler geometry rather than a conventional geodesic in pseudo-Riemann spacetime.[16] Riemann-Finsler geometry is a well-established mathematical field with numerous physical applications (see, e.g., Ref. 17), such as the famous Zermelo navigation problem of obtaining the minimum-time path for a ship in the presence of ocean currents. A large class of Riemann-Finsler geometries is determined by the geodesic motion of particles in the SME.[16,18] Among these are the canonical Randers geometry, which is related to the 1-form SME coefficient a_μ, and numerous novel geometries of simplicity comparable to the Randers case. One example of the latter is b space, a calculable Riemann-Finsler geometry that also is based on a 1-form and has Finsler structure complementary to that of Randers space. Physically, this geometry underlies the geodesic motion of a fermion in curved spacetime in the presence of chiral CPT-odd Lorentz violation. All the SME-inspired geometries exhibit mathematically interesting features connected to physical properties. For instance, when the SME coefficients are covariantly constant, the trajectories are conventional

Riemann geodesics and special Riemann-Finsler geometries known as Berwald spaces result. Many open challenges remain in this area, ranging from more technical questions such as resolving singularities or classifying geometries to physical issues such as generalizing Zermelo navigation or uncovering implications for the SME. The prospects appear promising for further insights to emerge from this geometrical approach to Lorentz violation.

Acknowledgments

This work was supported in part by U.S. D.o.E. grant DE-FG02-13ER42002 and by the Indiana University Center for Spacetime Symmetries.

References

1. D. Colladay and V.A. Kostelecký, Phys. Rev. D **55**, 6760 (1997); Phys. Rev. D **58**, 116002 (1998).
2. V.A. Kostelecký, Phys. Rev. D **69**, 105009 (2004).
3. O.W. Greenberg, Phys. Rev. Lett. **89**, 231602 (2002).
4. *Data Tables for Lorentz and CPT Violation,* V.A. Kostelecký and N. Russell, 2013 edition, arXiv:0801.0287v6.
5. V.A. Kostelecký and S. Samuel, Phys. Rev. D **39**, 683 (1989); Phys. Rev. Lett. **63**, 224 (1989); Phys. Rev. D **40**, 1886 (1989); V.A. Kostelecký and R. Potting, Nucl. Phys. B **359**, 545 (1991); Phys. Rev. D **51**, 3923 (1995); V.A. Kostelecký and R. Lehnert, Phys. Rev. D **63**, 065008 (2001).
6. R. Bluhm and V.A. Kostelecký, Phys. Rev. D **71**, 065008 (2005).
7. S.M. Carroll *et al.*, Phys. Rev. D **80**, 025020 (2009); V.A. Kostelecký and R. Potting, Phys. Rev. D **79**, 065018 (2009); Gen. Rel. Grav. **37**, 1675 (2005).
8. N. Arkani-Hamed *et al.*, JHEP **0507**, 029 (2005).
9. J. Alfaro and L.F. Urrutia, Phys. Rev. D **81**, 025007 (2010); B. Altschul *et al.*, Phys. Rev. D **81**, 065028 (2010); V.A. Kostelecký and J.D. Tasson, Phys. Rev. D **83**, 016013 (2011).
10. R. Bluhm *et al.*, Phys. Rev. D **77**, 065020 (2008).
11. V.A. Kostelecký and M. Mewes, Ap. J. Lett. **689**, L1 (2008); Phys. Rev. D **80**, 015020 (2009); Phys. Rev. Lett. **110**, 201601 (2013).
12. J.S. Díaz *et al.*, arXiv:1308.6344; V.A. Kostelecký and M. Mewes, Phys. Rev. D **85**, 096005 (2012).
13. V.A. Kostelecký and M. Mewes, arXiv:1308.4973.
14. V.A. Kostelecký and J.D. Tasson, Phys. Rev. Lett. **102**, 010402 (2009).
15. J.S. Díaz *et al.*, arXiv:1305.4636.
16. V.A. Kostelecký, Phys. Lett. B **701**, 137 (2011).
17. S.-S. Chern and Z. Shen, *Riemann-Finsler Geometry*, World Scientific, Singapore, 2005.
18. D. Colladay and P. McDonald, Phys. Rev. D **85**, 044042 (2012); V.A. Kostelecky, N. Russell, and R. Tso, Phys. Lett. B **716**, 470 (2012); V.A. Kostelecký and N. Russell, Phys. Lett. B **693**, 2010 (2010).

TESTING PERIODIC LOCAL POSITION INVARIANCE USING LONG-TERM COMPARISON OF THE SYRTE ATOMIC FOUNTAINS AND H-MASERS

M.E. TOBAR,* P.L. STANWIX, and J.J. McFERRAN

School of Physics, University of Western Australia
Crawley, WA 6009, Australia
**E-mail: michael.tobar@uwa.edu.au*

J. GUENA†, M. ABGRALL, S. BIZE, A. CLAIRON, PH. LAURENT, P. ROSENBUSCH, D. ROVERA, and G. SANTARELLI

LNE-SYRTE, Observatoire de Paris
CNRS, UPMC 75014, Paris, France
†E-mail: Jocelyne.Guena@obspm.fr

The frequencies of Cs and Rb fountain clocks have been compared to various H-masers to search for periodic changes correlated with the gravitational potential and boost with respect to the cosmic microwave background. The data span about eight years and the main sources of long-term noise are the offsets and linear drifts associated with the H-masers. To circumvent these effects we apply a numerical derivative to the data, which significantly reduces the standard error. We determine a limit for the Local Position Invariance (LPI) coefficient with respect to gravity to be 4.8×10^{-6} and 10^{-5} for a Cs-H and Rb-H comparison, respectively. From the same data the boost LPI coefficients were measured to parts in 10^{11}. From these results and others, independent limits on all coefficients of the boost violation vector with respect to fundamental constant invariance were determined to parts in 10^{10}.

1. Clock comparison

The LNE-SYRTE clock ensemble at the Paris Observatory combines high precision atomic clocks and oscillators at microwave and optical frequencies, and has been described in detail elsewhere.[1] In this work we concentrate on the long-term comparison between H-masers and Cs and Rb fountains (details of this experiment were recently published in Ref. 2). Data were recorded every 100 seconds, producing a large amount of data after several years of comparisons. Since the periods of the signals of interest are much longer than the sampling interval, the data were averaged over longer periods to reduce the amount of data for subsequent processing. This is important when implementing computationally intensive noise optimizing algorithms like weighted least squares, which whitens the noise and reduces the effects of the noise in the search for small signals. To search for annual variations the data sets are averaged over 95,000 second intervals and to search for sidereal variations the data sets are averaged over 2,500 second intervals.

Figure 1 shows the measured fractional frequency between the H-masers and the fountain clocks. The dominant noise source masking any periodic variation is due to offsets and drifts generated by the H-masers. This is why in general all three Cs comparisons overlap one another as a function of time, as the H-maser used in the comparison is common. It has been show that there is an advantage to analysing the derivative of the beat frequency over the beat frequency directly, because it naturally filters out non-stationary effects such as systematic jumps and drifts, leading to a more sensitive measurement.[3] The technique converts frequency jumps and other systematic offsets to outliers. In general if drifts are small they manifest as relatively small offsets compared to the remaining statistical noise. This technique has been implemented for both sets of data in Fig. 1, and is presented in Figs. 2 and 3 for the annual and sidereal time scales, respectively.

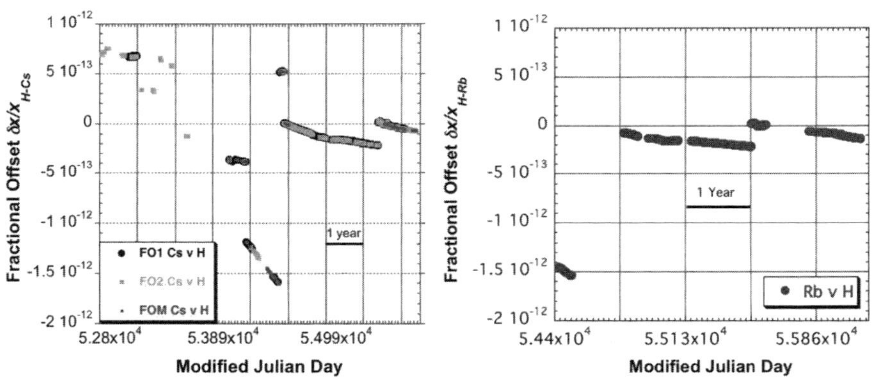

Fig. 1. Left: measured fractional offset variations of the frequency ratio between three Cs fountains and various H-masers. Long-term results span from 2/7/2003 to 2/11/2011, which is 8 years and 4 months. Right: measurement between the FO2 Rb fountain and various H-masers over nearly a 5 year span.

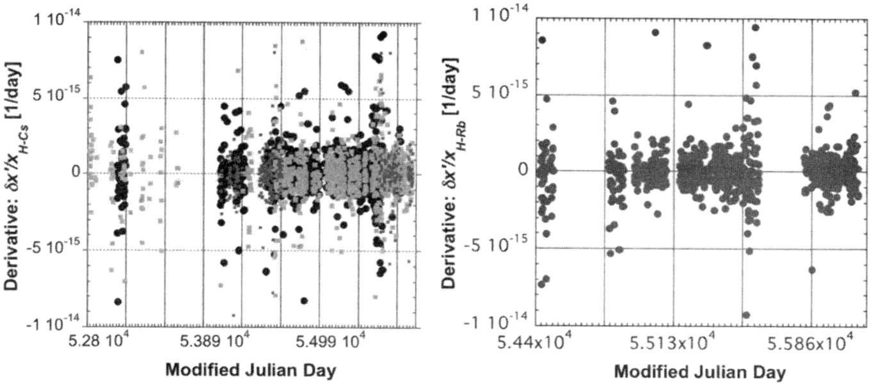

Fig. 2. Left: the derivative with respect to time of the Cs versus H data shown in Fig. 1. Each frequency measurement is averaged over 95,000 seconds before the derivative is taken. Right: Rb versus H data after following the same procedures. These data are used for searching for variations at the annual period.

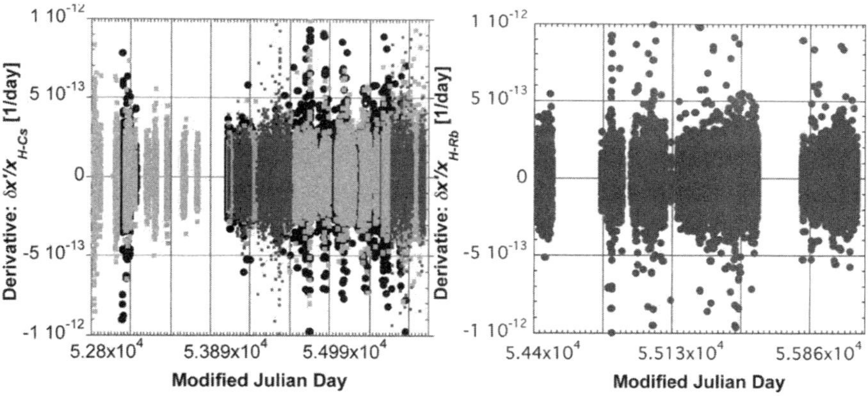

Fig. 3. The derivative with respect to time of the data shown in Fig. 1 (left: Cs versus H; right: Rb versus H). Each frequency measurement is averaged over 2,500 seconds before the derivative is taken. These data are used for searching for variations at the sidereal period.

2. Search for periodic variations and LPI coefficients

The weighted least squares algorithm was used to filter the data in Figs. 2 and 3 at the frequencies of interest where signals due to LPI violations should exist.[2] Results showed no significant departure from zero within the standard error, which indicated a null result and that the effects of unknown systematic errors are most likely minimal, or at least of the order presented.

Table 1. Determination of each component of the boost violation vector.

Component	H-maser/CSO	Cs/H-maser	Rb/H-maser
B_x	$-12.6(9.1) \times 10^{-11}$	$1.6(1.1) \times 10^{-11}$	$-2.2(2.6) \times 10^{-11}$
B_y	$6.5(9.8) \times 10^{-11}$	$-7.6(3.3) \times 10^{-11}$	$-0.8(3.6) \times 10^{-11}$
B_z	$-46.8(48.5) \times 10^{-11}$	$-15(15) \times 10^{-11}$	$-5.6(9.7) \times 10^{-11}$

Table 2. Decomposition of the boost violation vectors in Table 1 to limits on the invariance of fundamental constants with respect to boost.

Fundamental constant boost violation vector	$i = x$	$i = y$	$i = z$
$B_{\alpha i}$	$1.1(0.9) \times 10^{-10}$	$-1.6(1.4) \times 10^{-10}$	$-2.0(4.8) \times 10^{-10}$
B_{ei}	$-5.3(3.4) \times 10^{-10}$	$6.2(5.2) \times 10^{-10}$	$1.4(18) \times 10^{-10}$
B_{qi}	$-7.3(6.7) \times 10^{-10}$	$5.9(9.8) \times 10^{-10}$	$1.4(29) \times 10^{-10}$

Due to the low frequencies of interest (sidereal and annual) it is difficult to determine completely the exact effects of systematic errors. However, the approach presented here only requires a measurement of the relative stability and not an absolute measurement of the frequency. Thus only systematic errors that correlate with the frequencies of interest are important. This has been discussed in detail in Ref. 3. LPI coefficients with respect to gravity were measured to be $|\beta_H - \beta_{Cs}| \leqslant 4.8 \times 10^{-6}$ and $|\beta_H - \beta_{Cs}| \leqslant 10^{-5}$. From the same data boost LPI coefficients were measured to a precision of up to parts in 10^{11} with respect to the cosmic microwave background rest frame and are shown in Table 1. For the first time independent limits on all nine coefficients of the boost violation vector with respect to fundamental constant invariance were determined to a precision of up to parts in 10^{10} and are shown in Table 2.

Acknowledgments

This work was funded by the Australian Research Council.

References

1. J. Guena *et al.*, IEEE Trans. on UFFC, **59**, 391 (2012).
2. M.E. Tobar *et al.*, Phys. Rev. D **87**, 122004 (2013).
3. M.E. Tobar *et al.*, Phys. Rev. D **81**, 022003 (2010).

CPT TEST WITH (ANTI)PROTON MAGNETIC MOMENTS BASED ON QUANTUM LOGIC COOLING AND READOUT

M. NIEMANN,* A.-G. PASCHKE,* T. DUBIELZIG,*
S. ULMER,† and C. OSPELKAUS*,‡,§

*Institute of Quantum Optics, Leibniz Universität Hannover
Welfengarten 1, 30167 Hannover, Germany
†RIKEN Advanced Science Institute
Hirosawa, Wako, Saitama 351-0198, Japan
‡Physikalisch-Technische Bundesanstalt
Bundesallee 100, 38116 Braunschweig, Germany
§E-mail: christian.ospelkaus@iqo.uni-hannover.de

Dehmelt and VanDyck's famous 1987 measurement of the electron and positron g-factor is still the most precise g-factor comparison in the lepton sector, and a sensitive test of possible CPT violation. A complementary g-factor comparison between the proton and the antiproton is highly desirable to test CPT symmetry in the baryon sector. Current experiments, based on Dehmelt's continuous Stern-Gerlach effect and the double Penning-trap technique, are making rapid progress. They are, however, extremely difficult to carry out because ground state cooling using cryogenic techniques is virtually impossible for heavy baryons, and because the continous Stern-Gerlach effect scales as μ/m, where m is the mass of the particle and μ its magnetic moment. Both difficulties will ultimately limit the accuracy. We discuss experimental prospects of realizing an alternative approach to a g-factor comparison with single (anti)protons, based on quantum logic techniques proposed by Heinzen and Wineland and by Wineland *et al.* The basic idea is to cool, control and measure single (anti)protons through interaction with a well-controlled atomic ion.

1. Classical approach

Comparisons of g-factors between particles and their antiparticles are a sensitive test of CPT symmetry.[1,2] With leptons, the most sensitive tests carried out so far are based on the continuous Stern-Gerlach effect developed by Dehmelt and coworkers.[3] In essence, a g-factor measurement is realized by measuring the Larmor frequency $\omega_L = (g/2)(q/m)B$, which is a measure of the energy splitting between the two spin states at a given magnetic field B, and the free cyclotron frequency $\omega_C = (q/m)B$. The

g-factor is then given by $g = 2(\omega_L/\omega_C)$. Measurements of these frequencies are typically carried out in Penning traps, which feature a large, static B field for radial confinement. For the electron, frequencies exceed 100 GHz and allow cryogenic cooling of the cyclotron frequency to the ground state when in equilibrium with a dilution refrigerator. Motional frequencies can be measured using the image charges induced by the particle's motion in the trap electrodes. The well-known invariance theorem[4] relates the measured motional frequencies of the trapped particle to the free cyclotron frequency ω_C. Superimposing a slight axial magnetic field curvature makes the trap frequency spin-dependent; readout of the spin state as required for a Larmor frequency measurement can then be accomplished by a measurement of the axial trap frequency. For single (anti)protons, the large mass of the particles necessitates the application of much stronger field curvatures than for the electron, because the continuous Stern-Gerlach effect is proportional to μ/m. These field inhomogeneities are detrimental to a clean Larmor frequency measurement through a spatially varying resonance condition. Current efforts for the proton and antiproton therefore spatially separate the region of strong magnetic field curvature for spin state detection from a homogeneous magnetic field region, where spin transitions are driven and where the cyclotron frequency is measured ('double Penning-trap technique'[5]). Recently, single-proton single-spin-flip detection has been achieved,[6] a first direct measurement on antiprotons has been realized,[7] and the double Penning-trap technique has been demonstrated experimentally for protons.[8] Still, due to the difficulties outlined above, experimental cycle times for the unambiguous detection of single-proton spin flips exceed an hour, limited by the typical time constants (100 s) of resistive cyclotron cooling used in the state of the art experiments.[6] The coupling of the cyclotron motion's magnetic moment to the magnetic field curvature leads to fluctuations in the measured axial frequency ν_z, where the stability of ν_z scales with the square root of the average cyclotron quantum number n_+. In addition, differences in motional states and observed magnetic fields between the cyclotron and Larmor frequency resonance measurement may ultimately limit the accuracy. It is therefore highly desirable to ground-state-cool the cyclotron mode with highest efficiency, and to find alternative methods for spin state detection.

2. Quantum logic approach

Already in 1990, Heinzen and Wineland[9] proposed to measure g-factors of subatomic particles through their coupling to an atomic ion as a means of cooling the particle and detecting its spin state. The 1990 paper[9] envisaged storing the two particles in separate traps, and coupling their motional degrees of freedom through the image charges induced in a shared trap electrode. In 1998,[10] a direct coupling in a double-well potential was considered. Trapping the particles in spatially separate wells, coupled either directly or through a shared electrode, prevents a positively charged atomic ion and a negatively charged antiproton from annihilating. For the required spin-motional coupling, sideband transitions on the (anti)proton's spin-flip transition were considered, enabled by close proximity to a conductor ('near field').

The basic ingredients of this scheme have recently been demonstrated at NIST with two atomic ions, rather than with an atomic ion and an (anti) proton. In a first experiment,[11] the motion of two $^9Be^+$ ions in separate wells of a double-well potential was coupled at the single quantum level via the Coulomb interaction. In a second experiment,[12] microwave near-fields were used to drive motional sidebands on a hyperfine transition in $^{25}Mg^+$, allowing the implementation of an entangling quantum logic gate. In the context of quantum information processing, compared to the typical laser-based quantum logic gates, this has important advantages in terms of integration, scalability, ease of control and projected fidelities.

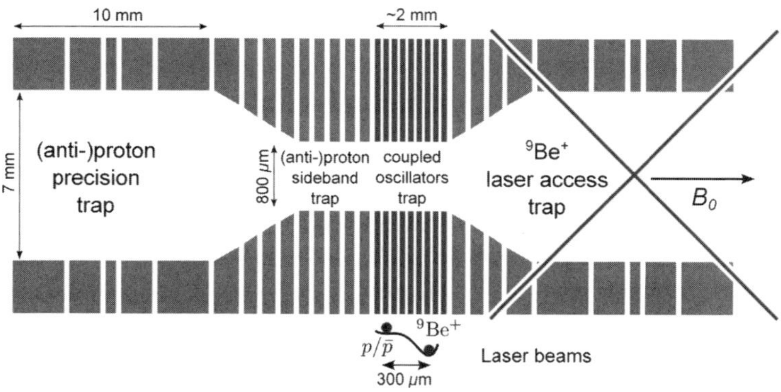

Fig. 1. Conceptual picture of Penning trap array for quantum logic readout and cooling of single (anti)protons (Dimensions not to scale)

We are currently planning and constructing a setup based on the proposal by Heinzen and Wineland[9,10] and on the above NIST experiments to realize a quantum logic based g-factor measurement apparatus for single (anti)protons. Our setup is based on a cylindrically symmetric Penning trap array (see Fig. 1). The different steps in the quantum logic cooling and readout procedure are carried out in spatially separate zones of the trap, a technique which has been used extensively in ion trap quantum computing with Paul traps. Our preliminary design features a narrow (800 μm diameter) section of the trap with a double-well potential for a proton and a $^9Be^+$ ion separated by 300 μm and a zone for spin-motional sideband transfer pulses for the (anti)proton. Our design also features sections with a large diameter for driving spin-flip transitions on single (anti)protons and for laser access to $^9Be^+$. Key advantages of the approach are: the projected short single-proton single spin-flip detection time of significantly less than 1 s (typically exceeding one hour for the classical techniques) with important consequences in the presence of magnetic-field drifts; the ability to ground-state cool, with a corresponding reduction in magnetic-field-inhomogeneity induced uncertainties; finally the ability to use the atomic ion's Larmor frequency as a magnetic-field dependent flywheel in a g-factor comparison independent of the measurement of motional frequncies. The setup is currently under construction at Hannover. An initial demonstration of the approach will be done with protons. If successful, the laser systems and trap can be installed at the beamline of the BASE Collaboration, which is currently setting up an antiproton g-factor experiment based on the classical techniques and the double Penning trap technique at CERN.

Acknowledgments

We acknowledge funding by QUEST, LUH, PTB and RIKEN.

References

1. R. Bluhm, V.A. Kostelecký, and N. Russell, Phys. Rev. Lett. **79**, 1432 (1997).
2. R. Bluhm, V.A. Kostelecký, and N. Russell, Phys. Rev. D **57**, 3932 (1998).
3. H. Dehmelt and P. Ekstrom, Bull. Am. Phys. Soc. **18**, 72 (1973).
4. G. Gabrielse, Phys. Rev. Lett. **102**, 172501 (2009).
5. H. Häffner *et al.*, Eur. Phys. J. D **22**, 163 (2003).
6. A. Mooser *et al.*, Phys. Rev. Lett. **110**, 140405 (2013).
7. J. DiSciacca *et al.*, Phys. Rev. Lett. **110**, 130801 (2013).
8. A. Mooser *et al.*, Phys. Rev. Lett. **723**, 78 (2013).
9. D.J. Heinzen and D.J. Wineland, Phys. Rev. A **42**, 2977 (1990).
10. D.J. Wineland *et al.*, J. Res. Natl. Inst. Stand. Technol. **103**, 259 (1998).
11. K.R. Brown *et al.*, Nature **471**, 196 (2011).
12. C. Ospelkaus *et al.*, Nature **476**, 181 (2011).

TESTS OF THE LORENTZ AND CPT SYMMETRIES AT THE PLANCK ENERGY SCALE FROM X-RAY AND GAMMA-RAY OBSERVATIONS

HENRIC KRAWCZYNSKI, FABIAN KISLAT, MATTHIAS BEILICKE, and ANNA ZAJCZYK

Physics Department and McDonnell Center for the Space Sciences
Washington University, St. Louis, MO 63130, USA

X-ray and gamma-ray observations of astrophysical objects at cosmological distances can be used to probe the energy dependence of the speed of light with high accuracy and to search for violations of Lorentz invariance and CPT symmetry at the Planck energy scale. In this conference contribution, we discuss these searches in the theoretical framework of the Standard-Model Extension. We present new limits on the dispersion relation governed by operators of mass dimension $d = 5$ and $d = 6$, and we discuss avenues for future progress.

1. Search for new physics at the Planck energy scale

The Standard-Model Extension (SME) provides a theoretical framework for studying and constraining Lorentz invariance and CPT symmetry violations.[1] It is an effective field theory that assumes that our theoretical framework — the Standard Model of particle physics and the theory of General Relativity — is the low-energy limit of a more fundamental theory that describes physics at higher energies, possibly the Planck energy scale ($E_{\rm P} = \sqrt{c^5\hbar/G} \approx 1.22 \times 10^{19}$ GeV). The action of the Standard Model is considered to be the zeroth-order term in an expansion approximating the full theory. The SME considers additional terms in the action, and makes it possible to use observational data to constrain the magnitude of the additional terms and thus to quantify the accuracy to which the Standard Model has been tested.[2] The additional terms are ordered according to the mass dimension d of the operators, and operators of dimension $d > 4$ lead to a modified photon dispersion relation. As the deviation of the group velocity from its low-energy value is expected to scale with $(E/E_{\rm P})^{d-4}$ (E being the photon energy), experimental constrains on the leading-order terms ($d = 5, 6$) are the most interesting ones. Even though the observations use photons with $E \ll E_P$, the observations are sensitive to new physics at the Planck energy scale as tiny deviations accumulate over large distances.[3]

Gamma-ray burst (GRB) observations with the Fermi γ-ray telescope have been used to set sensitive limits on the deviations of the speed of light from SME operators with $d = 5$. Writing the energy dependence of the group velocity as $\delta v = \zeta^{(d)0}(E/E_P)^{d-4}$ (with δv in units of c) the Fermi time-of-flight observations imply $|\zeta^{(5)0}| < 0.13$.[4] Here and in what follows, we adopt the notation of Kostelecký & Mewes,[5] but suppress the direction dependence of $\zeta^{(d)0}$. We will do the same below for the coefficients $|\zeta^{(d)a}|$ (with $a = 3$ or $a = +$) describing the birefringent propagation of light. Similar constraints on $|\zeta^{(d)0}|$ follow from AGN observations with the H.E.S.S., MAGIC, and VERITAS experiments.[6] These observations are in tension with the expectation $\delta v \sim 1$ at $E \approx E_P$ (which implies $|\zeta^{(5)0}| \sim 1$).

In the SME (and in many quantum gravity theories[3]) the dispersion relation depends not only on the photon energy and propagation direction but also the photon helicity. For $d = 5$ ($\delta v \propto E$), *time dispersion and birefringence are governed by the same SME coefficients*, and polarization observations constrain the coefficients much more sensitively than time-of-flight measurements. Denoting the birefringent modifications of the group velocity governed by operators of dimension d as $\delta v = \zeta^{(d)a}(E/E_P)^{d-4}$ with $a = 3, +$, the observation of polarized UV/optical emission from the afterglows of two GRBs implies $\zeta^{(5)a} < 2\times10^{-7}$.[7] The detection of polarized GRB emission in the γ-ray band can give even more sensitive constraints, but the significance of the detections so far is still marginal.[5]

2. Constraining $d = 6$ coefficients with γ-ray observations

The strong suppression of the corrections to the dispersion relation from operators with $d = 5$ may find its natural explanation in the fact that these operators break CPT symmetry. The leading-order corrections may be caused by the CPT-conserving operators of mass dimension $d = 6$. Also for the case $d = 6$, polarimetric observations give more sensitive constraints than time-of-flight measurements. However, in the case of $d = 6$ the phenomena of time dispersion and birefringence are not governed by the same SME operators, justifying a dedicated search for energy dependent time delays proportional to E^2. The function $\zeta^{(6)0}$ can be expanded into 25 spherical harmonics.[1] The 25 complex expansion coefficients satisfy a reality condition and are thus given by 25 real coefficients. At the conference, rigid constraints on the 25 real coefficients based on the analysis of 25 sources were presented for the first time. The analysis uses published re-

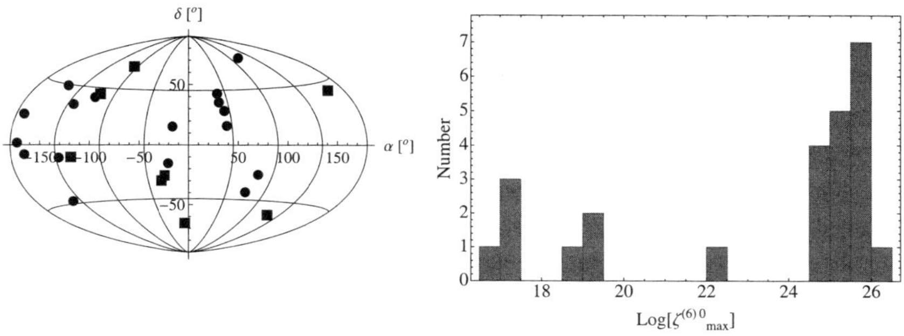

Fig. 1. Left: sky distribution of the sources (circles: Fermi AGNs, squares: other sources). Right: 95% confidence level upper limits on $\zeta^{(6)0}$.

sults of four GRBs detected with Fermi,[4] one GRB detected with RHESSI, and three AGNs detected with the γ-ray telescopes H.E.S.S., MAGIC, and Whipple.[6] We derive additional constraints from the publicly available light curves of 17 Fermi AGNs in two energy bands (300 MeV-1 GeV and 1 GeV-300 GeV). The analysis uses a simple discrete correlation function analysis to constrain the time lag between the two energy bands.

Figure 1 shows the distribution of the 25 sources in the sky and the derived 95% upper limits on the $\zeta^{(6)0}$ (lower limits were derived as well, but are not shown here). The most sensitive constraints come from the Cherenkov telescope observations of AGNs and from the Fermi observation of a bright GRB. Based on these results lower and upper limits on the 25 real coefficients describing the direction dependence of $\zeta^{(6)0}$ were derived. Presently, these latter limits are heavily impacted by the Fermi AGN limits and are of the order of 10^{27}. Work is underway to improve on the Fermi AGN limits with a dispersion cancellation method that uses the full information about the energy and arrival times of the detected photons.[8] Note that the results do not yet exclude order-unity modifications of the dispersion relation at the Planck energy scale. The same holds true for the constraints from the UV/optical polarimetry observations mentioned in the previous section which fall short by more than 18 orders of magnitude.

3. Conclusions and outlook

As mentioned before, polarimetric observations rule out a modification of the photon dispersion relation of order unity at the Planck-scale from operators with $d = 5$ by more than six orders of magnitude. In contrast, neither time-of-flight measurements nor polarimetric observations do so for the case

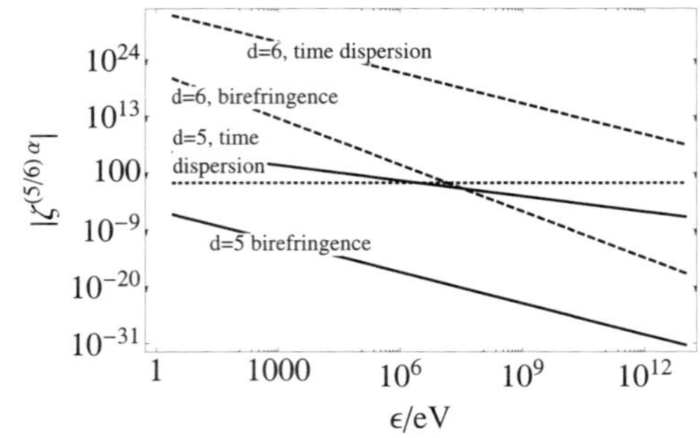

Fig. 2. Estimates of the limits on $\zeta^{(5)\alpha}$ and $\zeta^{(6)\alpha}$ ($\alpha = 0, 3, +$) that can be derived from future time dispersion and birefringence observations at energy ε. Results at and below the dotted line ($\zeta^{(5/6)\alpha} = 1$) constrain effects at the Planck energy scale.

of $d = 6$. It is instructive to evaluate how much better future time-lag and polarization measurements will do in this regard. For this purpose we assume that observations of GRBs at $z = 1$ can constrain the time-of-flight difference of photons of energies $E_1 = E$ and $E_2 = 0.1E$ with an accuracy of 1 msec, and succeed to detect a polarized signal from these GRBs. Figure 2 shows the resulting constraints. Interestingly, the time-of-flight measurements will not have the sensitivity required to constrain new physics at the Planck scale for the case of $d = 6$. Polarization observation do better, but require the detection of polarized signals at >20 MeV energies. Such detections might be possible with a next-generation Compton telescope.

References

1. D. Colladay and V.A. Kostelecký, Phys. Rev. D **55**, 6760 (1997); Phys. Rev. D **58**, 116002 (1998); V.A. Kostelecký, Phys. Rev. D **69**, 105009 (2004); V.A. Kostelecký and M. Mewes, Phys. Rev. D **80**, 015020 (2009).
2. *Data Tables for Lorentz and CPT Violation,* V.A. Kostelecký and N. Russell, 2013 edition, arXiv:0801.0287v6.
3. G. Amelino-Camelia *et al.*, Nature **393**, 319 (1998); L.J. Garay, Phys. Rev. Lett. **80**, 2508 (1998); R. Gambini and G. Pullin, Phys. Rev. D **59**, 124021 (1999); D. Mattingly, Liv. Rev. Rel. **8**, 5 (2005); T. Jacobson, S. Liberati, and D. Mattingly, Ann. Phys. **321**, 150 (2006).
4. V. Vasileiou *et al.*, Phys. Rev. D **87**, 122001 (2013).
5. V.A. Kostelecký and M. Mewes, Phys. Rev. Lett. **110**, 201601 (2013).
6. S.E. Boggs *et al.*, Ap. J. **611**, L77 (2004); A. Abramowski *et al.*, Astrop. Phys. **34**, 738 (2011); J. Albert *et al.*, Phys. Lett. B **668**, 253 (2008); S.D. Biller *et al.*, Phys. Rev. Lett. **83(11)**, 2108 (1999).
7. Y.Z. Fan, D.M. Wei, D. Xu, MNRAS **376**, 1875 (2007).
8. J.D. Scargle, J.P. Norris, J.T. Bonnell, ApJ, **673**, 972 (2008).

AN ODD (PARITY) TEST OF LORENTZ SYMMETRY WITH ATOMIC DYSPROSIUM

NATHAN LEEFER* and MICHAEL HOHENSEE

Physics Department, University of California
Berkeley, CA 94720-7300, USA
**E-mail: naleefer@berkeley.edu*

We propose using a Stark interference technique to measure directly the odd-parity c_{0j} components of the electron sector $c_{\mu\nu}$ tensor of the Standard-Model Extension. This technique has been shown to be a sensitive probe of parity violation in atomic dysprosium in a low-energy, tabletop experiment, and may also be straightforwardly applied to test Lorentz invariance. We estimate that such an experiment may be sensitive to c_{0j} coefficients as small as 10^{-18}.

Tests of Lorentz invariance in atomic systems typically hinge on detecting small energy shifts of bound states of elementary particles, due to the coupling of a Lorentz violating field to either the particle spin[1,2] or momentum.[3,4] Spectroscopic measurement of the energy difference between the nearly degenerate, opposite-parity states $[\mathrm{Xe}]4f^{10}5d6s$, $J = 10$ (state A) and $[\mathrm{Xe}]4f^{9}5d^{2}6s$, $J = 10$ (state B) in atomic dysprosium (Dy) has been shown to be a sensitive probe of the parity-even c_{jk} components of the Standard-Model Extension (SME) electron-sector $c_{\mu\nu}$ tensor.[4] These coefficients shift the energy of bound electrons in direct proportion to the expectation value of $\hat{p}_j\hat{p}_k$, where $\hat{p}_j$ is the sum of the j-components of their momentum. This energy shift modulates as a function of the electrons' orientation and velocity with respect to an inertial frame, and their position in an external gravitational potential.[5–7] Such experiments are best suited as probes of the rotational symmetry-breaking c_{JK} components of the $c_{\mu\nu}$ tensor in the Sun-centered celestial equatorial frame, but are also indirectly sensitive to the parity-odd c_{TJ} terms, which contribute to the terrestrial laboratory-frame observable c_{jk} in proportion to the Earth's boost velocity $\beta \approx 10^{-4}$ with respect to the Sun.[8] Currently, the most stringent bounds on the parity-odd c_{TJ} components are kinematic constraints from the observation of highly boosted laboratory[9] or astrophysical[10] sources.

We propose a low-energy atomic physics experiment that would be directly sensitive to the parity-odd components of the $c_{\mu\nu}$ tensor. The nonrelativistic quantum hamiltonian derived in Ref. 11 includes the term

$$\delta h = \left(-a_j + me_j + mc_{(0j)}\right) \frac{p_j}{m} c, \tag{1}$$

where m is the electron mass, c is the speed of light, $c_{(0j)}$ indicates the symmetric combination $c_{0j} + c_{j0}$, and $p_j = \sum_{n=1}^{N} p_j^{(n)}$ is the total electron momentum projection along the j-axis. The $a_j - me_j \equiv (a_{\text{eff}})_j$ component of this term is unobservable in nongravitational tests of flavor-conserving systems, as it can be removed from the fully relativistic theory by redefinition of the electron wave function's global phase.[12] Note however that should these terms describe a field with a nonmetric coupling to gravity, they may be observable in experiments performed within a gravitational potential.[6] This possibility will be explored elsewhere. Here, we will focus on the $c_{(0j)}$ coefficients, which give rise to observable effects in nongravitational experiments. These terms are typically ignored in spectroscopic tests of Lorentz invariance because the expectation value of the odd-parity operator $\hat{p}_j$ is zero for any bound state. These terms do, however, contribute to the hamiltonian matrix element between two states of opposite parity, and weakly drive transitions between different states. The $c_{(0j)}$ terms can be thus be measured using experimental methods developed to detect small parity-violating matrix elements induced by electron-nucleon interactions mediated by Z_0 bosons.[13,14]

The Stark interference technique used to measure parity violation in atoms relies on observing interference between the transition amplitude induced by parity-violation and that induced by an oscillating electric field.[14,15] Consider again the opposite-parity states in dysprosium. In ^{163}Dy the nuclear spin is $I = 5/2$ and the energy difference between the $F = 10.5$ hyperfine levels of state A and state B is only 3.1 MHz. The difference of Landé g-factors between states A and B allows sublevels of equal angular-momentum projection, m_F, to be brought to complete degeneracy with a modest magnetic field of less than 2 G, as shown in Fig. 1. Near these degeneracies, the hamiltonian for the two near-degenerate states, coupled to one another by an applied electric field $E(t) = E_0 \cos \omega t$, reduces to that of a two level system, and can be represented by the matrix

$$H = \begin{pmatrix} -i\gamma_A/2 & E(t)d + i(H_w - c_{(0j)}W_j) \\ E(t)d - i(H_w - c_{(0j)}W_j) & \Delta \end{pmatrix}, \tag{2}$$

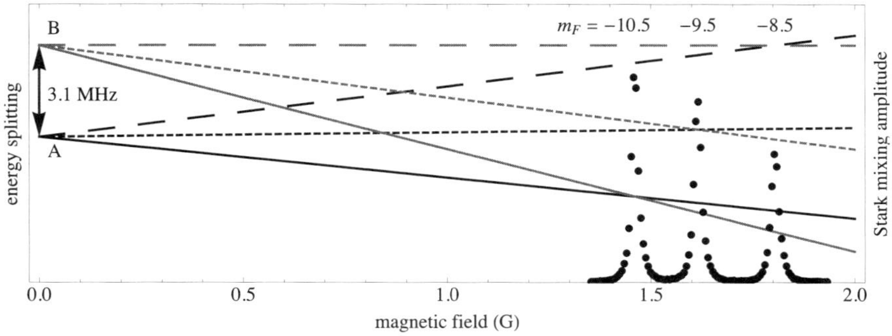

Fig. 1. An example Zeeman crossing spectrum for Dy atoms in the presence of an $\omega/(2\pi) = 10$ kHz, $E_0 = 2.5$ V/cm electric field. Solid and dashed lines show the Zeeman shifts of three magnetic sublevels of states A and B, relative to the $m_F = -8.5$ sublevel of state B, in ^{163}Dy. The filled circles show resonances in the Stark induced mixing amplitudes when sublevels of equal m_F are near degenerate.

where γ_A is the inverse lifetime of state A (B is assumed to be metastable), Δ is the residual energy difference between the states, d is the electric-dipole matrix element connecting the states, H_w is the conventional parity-violating matrix element between the two states, $W_j = \langle A, m_F | c\partial_{x_j} | B, m_F \rangle$, and sums are implied on repeated indexes. This hamiltonian is identical to that of Ref. 14, where the Lorentz-violating $c_{(0j)}W_j$ augments the parity-odd matrix element H_w. Although the odd-parity matrix element $H_w + c_{(0j)}W_j$ is too small to significantly drive transitions by itself, it can interfere with the electric field induced transition amplitude, modifying the probability of the $B \to A$ transition with a term proportional to $E_0 d(H_w - c_{(0j)}W_j)$. The detailed procedure for measuring H_w, and hence $c_{(0j)}W_j$, can be found in Ref. 14 and is not reproduced here. The only distinction is that $c_{(0j)}W_j$ can be modulated by rotation of the atoms' quantization axis or of the laboratory frame, which makes it distinguishable from conventional parity violating signals.

To constrain $c_{(0j)}$ with this method requires calculation of the W_j matrix elements, which in turn requires evaluation of the many-body electron wave function of the states A and B, as in Ref. 4. Here, we estimate the size of this matrix element by treating these states as hydrogenic wave functions of the leading order configurations for state A ([Xe]$4f^{10}5d6s$), and state B ([Xe]$4f^9 5d^2 6s$). We assume that the relevant matrix element is that between the $4f$ and $5d$ orbitals, and find that $\langle 4f | c\partial_{x_3} | 5d \rangle \sim 0.01\, c a_0^{-1} \approx 6 \times 10^{16}$ Hz, where a_0 is the Bohr radius.

The magnitude of the parity violating matrix element H_w in Dy was constrained to be $|H_w| \lesssim 5$ Hz in Ref. 14, limited by the available statistics. A search for orientation-dependent variations in the parity-violating matrix element with the same precision could constrain the $c_{(0j)}$ terms at the level of 10^{-16}, which would be competitive with astrophysical bounds.[16] Note that this estimate neglects possible enhancements due to the large nuclear charge, $Z = 66$, of Dy. Using the apparatus of Ref. 4, with minimal changes, a new experiment should shrink this statistical limit by a factor of > 50. This could allow us to detect parity-odd, $c_{(0j)}$ coefficients as small as 10^{-18} with a low-energy tabletop experiment. We caution that this is contingent on our estimate of the matrix elements $\langle A | W_j | B \rangle$ for dysprosium. A full calculation of these matrix elements will be the subject of future work.

Acknowledgments

We thank B. Altschul, D. Budker, V. Flambaum, A. Kostelecký, H. Müller, and J. Tasson for stimulating discussions. N.L. would like to thank L. Hunter and D. Phillips for their positive reception and encouragement of this idea. N.L. acknowledges the support of NSF grant PHY-1068875.

References

1. J.M. Brown *et al.*, Phys. Rev. Lett. **105**, 151604 (2010).
2. C. Gemmel *et al.*, Phys. Rev. D **82**, 111901 (2010).
3. M. Smiciklas *et al.*, Phys. Rev. Lett. **107**, 171604 (2011).
4. M.A. Hohensee *et al.*, Phys. Rev. Lett., in press.
5. C. Lane and V.A. Kostelecký, Phys. Rev. D **60**, 116010 (1999).
6. V.A. Kostelecký and J.D. Tasson, Phys. Rev. D **83**, 016013 (2010).
7. M.A. Hohensee and H. Müller, to be published.
8. V.A. Kostelecký and M. Mewes, Phys. Rev. D **66**, 056005 (2002).
9. B. Altschul, Phys. Rev. D **82**, 016002 (2010).
10. B. Altschul, Phys. Rev. Lett. **96**, 201101 (2006); Phys. Rev. D **74**, 083003 (2006).
11. V.A. Kostelecký and C. Lane, J. Math. Phys. **40**, 6245 (1999).
12. D. Colladay and V.A. Kostelecký, Phys. Rev. D **55**, 6760 (1997); Phys. Rev. D **58**,116002 (1998).
13. I.B. Khriplovich, *Parity Nonconservation in Atomic Phenomena,* Gordon and Breach, Philadelphia, 1991.
14. A.-T. Nguyen, D. Budker, D. DeMille, and M. Zolotorev, Phys. Rev. A **56**, 3453 (1997).
15. M.C. Noecker, B.P. Masterson, and C.E. Wieman, Phys. Rev. Lett. **61**, 310 (1988).
16. *Data Tables for Lorentz and CPT Violation,* V.A. Kostelecký and N. Russell, 2013 edition, arXiv:0801.0287v6.

LORENTZ AND CPT VIOLATION IN THE NEUTRINO SECTOR

JORGE S. DÍAZ
Physics Department, Indiana University
Bloomington, IN 47405, USA
E-mail: jsdiazpo@indiana.edu

Searches for Lorentz and CPT violation using neutrino oscillations and the prospects for future tests using neutrino time-of-flight measurements and beta-decay experiments are presented.

1. Introduction

Lorentz invariance is the symmetry that underlies Special Relativity. The spontaneous breakdown of this spacetime symmetry can arise in some candidate theories of quantum gravity, such as string theory.[1] In the search for physics beyond the Standard Model (SM), one path consists of the possibility that some of the observed symmetries of the SM could be broken. Following this approach, the Standard-Model Extension (SME) generalizes the SM to incorporate all the possible terms in the action that break Lorentz invariance.[2] The SME is a general framework for Lorentz violation and includes a subset of operators that also break CPT invariance. The development of the SME triggered a worldwide program searching for relativity violations in all sectors of the SM.[3]

2. Lorentz-violating neutrinos

In the SME, free neutrinos are described by a Dirac-like equation.[4] The study of Lorentz-violating operators of arbitrary dimension[5] allows a direct classification of the observable effects.

2.1. *Neutrino oscillations*

Flavor-mixing operators in the SME have been explored in the construction of realistic models for neutrino oscillations as alternatives to the conventional mass-driven description.[6] Additionally, the development of techniques to test Lorenz symmetry with neutrino-oscillation experiments[7] has led to several experimental searches. These methods have been used by Double Chooz,[8] IceCube,[9] LSND,[10] MiniBooNE,[11] MINOS,[12] and Super-Kamiokande.[13] In a recent work, the possibility of neutrino-antineutrino oscillations has been explored for the first time.[14] Using public data on neutrino interactions in MINOS, the limits on 66 different SME coefficients producing sidereal variations of the event rate were presented at this meeting. There are also 15 coefficients for Lorentz violation that produce time-independent neutrino-antineutrino oscillations. Preliminary results on the study of these effects using the antineutrino spectrum in the Double Chooz experiment were also presented at the meeting.[15]

2.2. *Neutrino time of flight*

The interferometric nature of neutrino oscillations makes them sensitive probes of new physics. Nevertheless, there exists a set of operators in the SME whose effects are unobservable in oscillations. These *oscillation-free* operators[5] affect all neutrino flavors in the same way; therefore, their experimental signatures appear in decay processes and group-velocity measurements. One method to access these operators is by comparing neutrino speed to that of the photon. Different oscillation experiments are making measurements to determine the neutrino speed. In the SME, this quantity can depend on many variables including neutrino energy, direction of propagation, and sidereal time. Additionally, SME operators that also break CPT invariance can produce differences between neutrino and antineutrino speeds. A general presentation of the theory for time-of-flight measurements can be found in Ref. 5.

2.3. *Beta decay*

The study of the experimental signatures of oscillation-free operators has led to one particular type of operator that eludes observation in time-of-flight measurements. Dimension-three operators introduce momentum-independent modifications of the neutrino energy. For this reason, they are absent in the general expression of the neutrino group velocity; hence, there are Lorentz-violating effects that remain completely unexplored to date. This feature makes beta-decay experiments an interesting probe of spacetime symmetries through the study of these *countershaded* relativity violations.[16] The unconventional energy dependence introduced by the SME operators alters the phase space of the antineutrino, leading to modifications of the beta-decay spectrum. Furthermore, the breakdown of invariance under rotations produces direction-dependent decay rates and sidereal variations. A general presentation of the theory for beta-decay measurements can be found in Ref. 16. A summary of the main results is presented below.

For neutron experiments, the simplest Lorentz-violating modification is as an isotropic distortion of the electron count rate. The modification is maximal at a well-defined energy, which allows experiments to search for this effect. The antineutrino-electron angular correlation in the decay of unpolarized neutrons gets modified by anisotropic SME operators. Unconventional effects in the experimental asymmetry include dependence on the orientation and location of the experiment and sidereal variations. Similar effects arise for the asymmetry determining the angular correlation between the antineutrino and the neutron spin in the decay of polarized neutrons.

Tritium decay experiments have been designed for direct measurements of neutrino mass. In the presence of Lorentz violation, distortions of the endpoint depend on the neutrino mass but also on the location and orientation of the experiment. Additionally, the endpoint energy can oscillate with sidereal frequency. Published data from the Mainz and Troitsk experiments implies the first limits on anisotropic effects as well as a tenfold improvement in the limit on the isotropic coefficient, previously constrained using IceCube meson thresholds.[3] These data also allow the study of an effective-dimension-two coefficient that modifies the integrated spectrum near the endpoint energy in the same way as the neutrino mass. This coefficient can mimic the mass-squared parameter that controls the shape of the spectrum near the endpoint. Since this coefficient can have any sign and also vary with sidereal time, a tachyonic-neutrino[17] behavior can appear. The first limit on this coefficient is obtained also using published data.

For two-neutrino double beta decay, the simplest Lorentz-violating modification appears as an isotropic distortion of the electron-sum spectrum. The unconventional energy dependence introduces a modification at a well defined energy that should guide future experimental searches of this effect. For neutrinoless double beta decay we find that this decay mode can occur even for massless neutrinos, in which the role of the neutrino mass is replaced by a SME Majorana coupling.

3. Future outlook

In recent years, the number of explored experimental signals of the neutrino sector of the SME has experienced a remarkable boost. Neutrino oscillations have been the main experimental technique and a considerable section of the coefficient space of the minimal SME has been constrained. Nevertheless, many effects remain to be studied. Additionally, the nonminimal sector offers new effects to be explored. Furthermore, beam experiments performing measurements of the neutrino speed and beta decay experiments studying neutron, tritium, and double beta decay can now join the worldwide program searching for violations of Lorentz and CPT symmetry.

Acknowledgments

This work was supported in part by the U.S. Department of Energy and by the Indiana University Center for Spacetime Symmetries.

References

1. V.A. Kostelecký and S. Samuel, Phys. Rev. D **39**, 683 (1989); V.A. Kostelecký and R. Potting, Nucl. Phys. B **359**, 545 (1991); Phys. Rev. D **51** 3923 (1995).
2. D. Colladay and V.A. Kostelecký, Phys. Rev. D **55**, 6760 (1997); Phys. Rev. D **58**, 116002 (1998); V.A. Kostelecký, Phys. Rev. D **69**, 105009 (2004).
3. *Data Tables for Lorentz and CPT Violation,* V.A. Kostelecký and N. Russell, 2013 edition, arXiv:0801.0287v6.
4. V.A. Kostelecký and M. Mewes, Phys. Rev. D **69**, 016005 (2004).
5. V.A. Kostelecký and M. Mewes, Phys. Rev. D **85**, 096005 (2012).
6. V.A. Kostelecký and M. Mewes, Phys. Rev. D **70**, 031902(R) (2004); T. Katori *et al.*, Phys. Rev. D **74**, 105009 (2006); V. Barger *et al.*, Phys. Lett. B **653**, 267 (2007); J.S. Díaz and V.A. Kostelecký, Phys. Lett. B **700**, 25 (2011); Phys. Rev. D **85**, 016013 (2012); V. Barger *et al.*, Phys. Rev. D **84**, 056014 (2011); S.-J. Rong and Q.-Y. Liu, Chin. Phys. Lett. **29**, 041402 (2012).
7. J.S. Díaz *et al.*, Phys. Rev. D **80**, 076007 (2009); V.A. Kostelecký and M. Mewes, Phys. Rev. D **70**, 076002 (2004).
8. Y. Abe *et al.*, Phys. Rev. D **86**, 112009 (2012).
9. R. Abbasi *et al.*, Phys. Rev. D **82**, 112003 (2010).
10. L.B. Auerbach *et al.* Phys. Rev. D **72**, 076004 (2005).
11. A.A. Aguilar-Arevalo *et al.* Phys. Lett. B **718**, 1303 (2013); T. Katori Mod. Phys. Lett. A **27**, 1230024 (2012).
12. P. Adamson *et al.* (MINOS Collaboration), Phys. Rev. Lett. **101**, 151601 (2008); Phys. Rev. Lett. **105**, 151601 (2010); Phys. Rev. D **85**, 031101 (2012).
13. T. Akiri, these proceedings.
14. B. Rebel and S. Mufson, arXiv:1301.4684; these proceedings.
15. T. Katori, these proceedings; J.S. Díaz *et al.*, IUHET 578.
16. J.S. Díaz *et al.*, arXiv:1305.4636.
17. A. Chodos *et al.*, Phys. Lett. B **150**, 431 (1985).

THE SEARCH FOR NEUTRINO-ANTINEUTRINO MIXING FROM LORENTZ-INVARIANCE VIOLATION USING NEUTRINO INTERACTIONS IN MINOS

S. MUFSON

Astronomy Department, Indiana University
Bloomington, IN 47405, USA
E-mail: mufson@astro.indiana.edu

B. REBEL

Fermi National Accelerator Laboratory
Batavia, IL 60510, USA
E-mail: brebel@fnal.gov

We searched for a sidereal modulation in the rate of neutrinos observed by the MINOS Far Detector. The detection of these signals could be a signature of neutrino-antineutrino mixing due to Lorentz and CPT violation as described by the Standard-Model Extension framework. We found no evidence for these sidereal signals and we placed limits on the coefficients in this theory describing the effect.

1. Introduction

In the Standard-Model Extension (SME), Lorentz violation (LV) and CPT violation (CPTV) could manifest themselves at observable energies through a dependence of the neutrino oscillation probability on the direction of neutrino propagation with respect to the Sun-centered inertial frame. An experiment that has both its neutrino beam and detector fixed on the Earth's surface could then observe a sidereal modulation in the number of neutrinos detected from the beam. MINOS is such an experiment.

Denote the usual neutrino survival probability in the two-flavor approximation as $P^{(0)}_{\nu_\mu \rightarrow \nu_\mu} \approx 1 - \sin^2(2\theta_{23}) \sin^2(1.27 \Delta m^2_{32} L/E)$, where θ_{23} is the angle describing mixing between the second and third mass states and Δm^2_{32} is the difference in the squares of the masses of those states. The energy of the neutrino is E and the distance it travels is L. Then LV and CPTV that would cause lepton-number violating mixing between neutrinos and antineutrinos introduces an additional perturbation term to the survival probability,[1]

$$P_{\nu_\mu \to \nu_\mu} = P^{(0)}_{\nu_\mu \to \nu_\mu} + P^{(2)}_{\nu_\mu \to \nu_\mu}, \tag{1}$$

where the perturbation term $P^{(2)}_{\nu_\mu \to \nu_\mu}$ can be written[1]

$$\begin{aligned} P^{(2)}_{\nu_\mu \to \nu_\mu} = L^2\{ & P_{\mathcal{C}} + P_{\mathcal{A}_s} \sin \omega_\oplus T_\oplus + P_{\mathcal{A}_c} \cos \omega_\oplus T_\oplus \\ & + P_{\mathcal{B}_s} \sin 2\omega_\oplus T_\oplus + P_{\mathcal{B}_c} \cos 2\omega_\oplus T_\oplus \\ & + P_{\mathcal{D}_s} \sin 3\omega_\oplus T_\oplus + P_{\mathcal{D}_c} \cos 3\omega_\oplus T_\oplus \\ & + P_{\mathcal{F}_s} \sin 4\omega_\oplus T_\oplus + P_{\mathcal{F}_c} \cos 4\omega_\oplus T_\oplus \}. \end{aligned} \tag{2}$$

Here $L = 735$ km is the distance from neutrino production in the NuMI beam to the MINOS Far Detector (FD),[2] $\omega_\oplus = 2\pi/(23^h 56^m 04.0982^s)$ is the Earth's sidereal frequency, and $T_\oplus$ is the local sidereal arrival time of the neutrino event. Equation (2) shows that harmonic variations at the sidereal frequency are visible up to $4\omega_\oplus$. The P parameters contain the LV and CPTV information on neutrino-antineutrino mixing. They depend on the SME coefficients $\widetilde{H}^{\alpha}_{a\bar{b}}$ and $\tilde{g}^{\alpha\beta}_{a\bar{b}}$, the neutrino energy, and the direction of the neutrino propagation in a coordinate system fixed on the rotating Earth.[1]

A more complete description of this work can be found in Ref. 3.

2. Data analysis

This analysis uses a dataset of neutrino interactions acquired by MINOS from May, 2005 through April, 2012. The interactions were selected using standard MINOS criteria for beam and data quality.[4] In addition, the events were required to interact within the 4.2 kiloton FD fiducial volume. This selection enables MINOS to establish each event as a charged-current (CC) ν_μ interaction by identifying the outgoing μ^-. We focused on CC events to maximize the ν_μ disappearance signal. There are a total of 2,463 CC events in this analysis.

Figure 1 shows the rate histogram in sidereal phase for the events in our analysis. The statistically significant fit to a constant rate implies there are no sidereal modulations in the data sample.

We performed a fast Fourier transform (FFT) on the rate histogram in Fig. 1 and computed the power in the harmonic terms $\omega_\oplus T_\oplus, \ldots, 4\omega_\oplus T_\oplus$ appearing in the oscillation probability, Eq. (2). The results of the FFT analysis are given in Table 1.

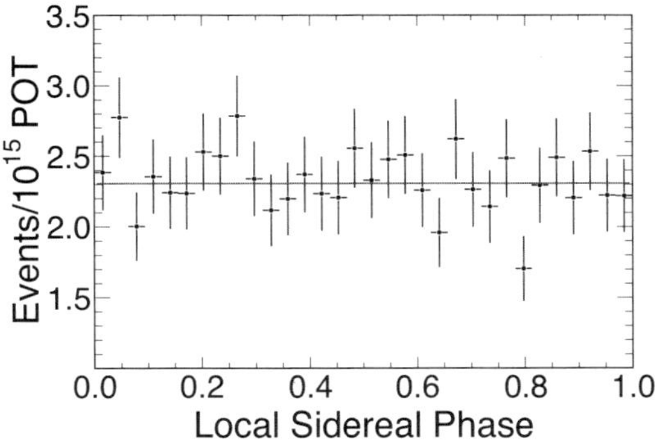

Fig. 1. The sidereal phase diagram for the CC neutrino rate for the FD data. The mean rate of 2.31 ± 0.05 events per 10^{15} protons on target is superposed. The value of χ^2 per degree of freedom is 23.7/31 for the fit.

Table 1. Results of the FFT analysis. $\mathcal{P}_\mathcal{F}$ is the probability that the power is a noise fluctuation.

	Harmonic Power (FFT)	$\mathcal{P}_\mathcal{F}$
$\omega_\oplus T_\oplus$	0.928	0.65
$2\omega_\oplus T_\oplus$	0.574	0.89
$3\omega_\oplus T_\oplus$	1.388	0.48
$4\omega_\oplus T_\oplus$	1.223	0.53

In Rebel and Mufson[3] we determine the statistical significance of the harmonic powers in Table 1 by simulation. We find that 3.45 is the 99.7% confidence level (C.L.) for the probability that a measured quadratic sum of powers for any harmonic was *not* drawn from a distribution having a sidereal signal. Since none of the harmonic powers exceed our 99.7% (FFT) detection threshold, we conclude that there is no evidence for a sidereal modulation resulting from mixing between neutrinos and antineutrinos as predicted by Eq. (2) in this neutrino dataset.

We investigated the sensitivity of our results to several sources of systematic uncertainties. We found that systematics were unimportant.

3. Limits

We determined the confidence limits for the SME coefficients by methods we developed in our previous papers.[2,5,6] In these limit calculations the odd harmonics in Eq. (2) vanish. Since we set all but one coefficient to zero in computing the limits and the odd harmonics involve products of different SME coefficients, they disappear from Eq. (2).

We start by simulating a set of experiments in which there is no sidereal modulation. Each of the simulated experiments contain the same number of interactions as the dataset. We then introduce an infinitesimal LV sidereal signal into the experiments with the one nonzero SME coefficient. We then generate the sidereal phase diagrams for these experiments and test whether there is an LV signal detectable in the phase diagrams with the same FFT analysis used to analyze the data in Fig. 1.

In our simulation, neutrinos are generated by modeling the NuMI beam line, including hadron production by the 120 GeV/c protons striking the target and the propagation of the hadrons through the focusing elements and decay pipe to the beam absorber. The simulation then calculates the probability that any neutrinos generated traverse the FD. The FD neutrino event simulation takes the neutrinos from the NuMI simulation, along with weights determined by decay kinematics, and uses this information as input into the simulation of the interactions in the FD. We inject a sidereal signal in the simulation by calculating the survival probability for each simulated neutrino based on the even harmonics in Eq. (2) using a chosen value for the magnitude of the nonzero SME coefficient, the energy of the simulated neutrino, and the distance the neutrino travels to the FD.

The 99.7% C.L. limits on the 66 $\tilde{H}_{a\bar{b}}^{\alpha}$ and $\tilde{g}_{a\bar{b}}^{\alpha\beta}$ SME coefficients are found in tables in Rebel and Mufson.[3] For these tables, we repeated the simulation for each coefficient 250 times and averaged the results.

Acknowledgments

This work was supported in part by the Indiana University Center for Spacetime Symmetries (IUCSS) and by the U.S. Department of Energy Office of Science under grant DE-FG02-91ER40661.

References

1. J.S. Díaz, V.A. Kostelecký, and M. Mewes, Phys. Rev. D **80**, 076007 (2009).
2. P. Adamson *et al.*, Phys. Rev. Lett. **101**, 151601 (2008).
3. B. Rebel and S. Mufson Astropart. Phys., in press [arXiv:1301.4684].
4. P. Adamson *et al.*, Phys. Rev. Lett. **106**, 181801 (2011).
5. P. Adamson *et al.*, Phys. Rev. Lett. **105**, 151601 (2010).
6. P. Adamson *et al.*, Phys. Rev. D **85**, 031101 (2012).

TERRESTRIAL VS. SPACEBORNE, QUANTUM VS. CLASSICAL TESTS OF THE EQUIVALENCE PRINCIPLE

M. HOHENSEE and H. MÜLLER*

Physics Department, University of California
Berkeley, CA 94720, USA
**E-mail: hm@berkeley.edu*

The equivalence principle can be tested by precision experiments based on classical and quantum systems, on the ground as well as in space. In many models, these tests are mostly equivalent in their ability to constrain physics beyond the Standard Model. We mention differences that nevertheless exist between spaceborne and quantum mechanical tests and their conventional competitors.

1. Introduction

The Einstein equivalence principle (EEP) states that gravity is equivalent to acceleration, and affects all objects in exact proportion to their mass-energy. Experimental tests[1] of the EEP are often grouped into tests of Lorentz invariance,[2] local position invariance,[3] and the weak equivalence principle (WEP),[4] but this division is largely historical. Known 'consistent' theories (e.g., energy and momentum-conserving) that violate one of these principles typically violate all. A conjecture attributed to Schiff[5] suggests that this must always be the case (although specialized counterexamples exist). Violations of the EEP are promising candidates for low-energy signals of Planck-scale physics.[6]

The gravitational Standard-Model Extension (SME)[7] provides a starting point for this discussion, though we will also go beyond it. It is constructed from the lagrangians of the Standard Model and gravity by adding new interactions that violate Lorentz invariance and the EEP. For a fermion of mass m, for example, these interactions are encoded in eight Lorentz tensors $a_\mu, b_\mu, c_{\mu\nu}, d_{\mu\nu}, e_\nu, f_\nu, g_{\lambda\mu\nu}$, and $H_{\mu\nu}$ known collectively as coefficients for Lorentz violation. Leading-order violations of the EEP arise from the six $(a^{\mathrm{p}})_0, (a^{\mathrm{n}})_0, (a^{\mathrm{e}})_0, (c^{\mathrm{p}})_{00}, (c^{\mathrm{n}})_{00}, (c^{\mathrm{e}})_{00}$, where the superscripts p, n, e denote the proton, the neutron, and the electron, respectively. Experiments with neutral matter can only measure the combination $(a^{\rm p+e})_0 \equiv (a^{\rm p})_0 + (a^{\rm e})_0$. This leaves five measurable coefficients for EEP violations.[7,8]

We will not discuss technical aspects of space tests (e.g., interrogation time, short rotational and orbital periods, vibrations) and quantum tests (e.g., control of initial conditions and systematic effects, reproducibility, or the Coriolis force[9]). A few unique signals for quantum tests have been discussed elsewhere, e.g., space-time foam[10] and dilatons.[6]

2. Unique signals for quantum tests

In any relativistic framework, the phase of a matter wave is given by $\phi = -mc^2\tau/\hbar$, where mc^2 is the rest mass-energy and $\tau = \int d\tau$ the proper time along the path.[3,8,11] Quantum tests measure the difference of that phase between two paths. In General Relativity (GR) and theories that satisfy Schiff's conjecture (e.g., the SME), changes in the proper time τ, which would be measurable by atom interferometers[3,8] and clocks, will lead to changes in the center of mass motion, which would be measurable by classical WEP tests. However, in theories that go beyond Schiff's conjecture, signals such as nonstandard gravitational redshifts might be picked up by clocks and quantum WEP tests even if they are undetectable by classical WEP tests, giving these tests a distinct power.

Classical and quantum tests differ in the species they use. Quantum tests typically use atoms with a simple electronic structure, e.g., alkalis. They are not suitable for use in classical tests, as they are soft and chemically reactive, but have a special nuclear structure. Use of such atoms is important for measuring all types of EEP violations.[8] Use of (anti-)hydrogen in quantum experiments or hydrogen-rich materials in classical ones would be of particular interest.

Spin-dependent gravitational couplings have long been studied in the context of theories of gravity with nonvanishing torsion.[12] In the SME, such effects are expected to result from the b, d, g, and H coefficients. The the e-type coefficients are not included here, as they can be lumped into the a-type coefficients. Similarly, the f-type can be lumped into the c-type coefficients.[13] These coefficients will describe how gravity might couple differently to particles exhibiting different spin-orbit couplings, i.e., having correlated external and spin degrees of freedom. Only quantum experiments will be able to study them. For example, an experiment reaching 10^{-14} or better could perform the measurement of several components of $b^{\rm p}, d^{\rm p}, g^{\rm e}, g^{\rm n}, g^{\rm p}$, and $H^{\rm p}$, for which limits do not exist at the time of this writing. The theory of these effects might start from the known relativistic hamiltonian for the gravitational SME, deriving the nonrelativistic hamiltonian by a Foldy-Wouthuysen transformation.[7] Most of the additional coefficients should lead directly to observable effects. The fluctuations with gravity might make additional ones observable.

3. Unique signals for spaceborne tests

Letting an EEP-test experiment perform measurements over a year and analyzing its signal for periodic variations at the frequencies of Earth's rotation and orbit and their combinations will allow us to separately search for effects caused by the Earth's and the Sun's gravity. Spaceborne experiments offer an additional way for such separate measurement through their orbit. Different SME parameters describing the source mass of the gravitational field can thus be constrained. Since the combinations of coefficients that produce shifts in active and passive gravitational masses are not the same,[7] and because of the different composition of the Earth and the Sun, this may result in sensitivity to additional parameter combinations. Explicit calculation has not been performed yet.

Effects of GR can be ordered by powers of the dimensionless gravitational potential U/c^2 and particle velocity $(v/c)^2$. Nontrivial higher order effects arise in GR due to the theory's nonlinearity, e.g., perihelion precession. In atom interferometry, they generate extra phase shifts[11] proportional to $(U/c^2)(v/c)^2$.

Higher-order SME terms in flat spacetime[14] are likely to cause EEP-violations at $O(1/c^3)$ and higher, though this has not been studied yet. Such effects will go beyond the traditional categorization into tests of Lorentz invariance, WEP, and local position invariance: The validity of EEP for particles at rest relative to the source mass might not imply its validity for moving objects, or its validity at one location might not imply its validity elsewhere. These effects should be studied further. An important figure of merit for experiments searching for high-order effects is the velocity the experiments attain with respect to the source mass. Spaceborne experiments clearly outperform laboratory experiments in this regard.

A chameleon is a hypothetical scalar field ϕ proposed to help explaining dark matter and the accelerated expansion of the universe.[15] Through non-linear self-coupling the chameleon is a short (millimeter)-range force close to massive objects, such as Earth. In empty space, however, it may give rise to a long-range fifth force that causes order-unity equivalence principle violations. Chameleons thus avoid detection in laboratory and solar system experiments, but might cause large EEP violations for small test particles in empty space. While a detailed analysis remains to be performed, it seems likely that models that are compatible with all previous tests could still lead to EEP violations between 10^{-19} and 10^{-11} .

Finally, we might speculate that any hints at distance-dependent gravity anomalies (perhaps varying Hubble constant[16]) are a reason to search for EEP violations at all accessible distance scales, by Earth-and space-based experiments, solar system tests, as well as astrophysics.

Acknowledgments

We thank Alan Kostelecký, Anna Nobili, Jim Phillips, Ernst Rasel, Andreas Schärer, and Nan Yu for important discussions and the David and Lucile Packard Foundation, the National Aeronautics and Space Agency, and the National Science Foundation for support.

References

1. A.M. Nobili *et al.*, Am. J. Phys. **81,** 527 (2013).
2. *Data Tables for Lorentz and CPT Violation,* V.A. Kostelecký and N. Russell, 2013 edition, arXiv:0801.0287v6.
3. R.V. Pound and G.A. Rebka Jr., Phys. Rev. Lett. **4,** 337 (1960) R.F.C. Vessot *et al.*, Phys. Rev. Lett. **45,** 2081 (1980); H. Müller *et al.*, Nature **463,** 926 (2010).
4. E.G. Adelberger *et al.*, Prog. Part. Nucl. Phys. **62,** 102 (2009); R.D. Reasenberg *et al.*, Class. Quant. Grav. **29,** 184013 (2012).
5. L.I. Schiff, Am. J. Phys. **28,** 340 (1960).
6. T. Damour, F. Piazza, and G. Veneziano, Phys. Rev. Lett. **89,** 081601 (2002); T. Damour and J.F. Donoghue, Class. Quant. Grav. **27,** 202001 (2010).
7. V.A. Kostelecký, Phys. Rev. D **69,** 105009 (2004); Q.G. Bailey and V.A. Kostelecký, Phys. Rev. D **74,** 045001 (2006); V.A. Kostelecký and J.D. Tasson, Phys. Rev. Lett. **102,** 010402 (2009); Phys. Rev. D **83,** 016013 (2011).
8. M.A. Hohensee *et al.*, Phys. Rev. Lett. **106,** 151102 (2011); M.A. Hohensee, H. Müller, and B. Wiringa, to be published.
9. S.-Y. Lan *et al.*, Phys. Rev. Lett. **108,** 090402 (2012).
10. E. Göklü and C. Lämmerzahl, Class. Quant. Grav. **25,** 105012 (2008).
11. S. Dimopoulos *et al.*, Phys. Rev. D **78,** 042003 (2008).
12. W.-T. Ni, Rep. Prog. Phys. **73,** 056901 (2010); J.C. LoPresto *et al.*, Ap. J. **376,** 757 (1991); W.G. Dixon, Nuov. Cim. XXXIV No. 2, **317,** (1964); J.E. Moody and F. Wilczek, Phys. Rev. D **30,** 130 (1984).
13. B. Altschul, J. Phys. A **39**, 13757 (2006); V.A. Kostelecký and N. Russell, Phys. Lett. B **693**, 443 (2010).
14. V.A. Kostelecký and M. Mewes, Phys. Rev. D **85,** 096005 (2012).
15. J. Khoury and A. Weltman, Phys. Rev. Lett. **93,** 171104 (2004).
16. V. Marra *et al.*, Phys. Rev. Lett. **110,** 241305 (2013).

HIGHER-ORDER LORENTZ VIOLATION

MATTHEW MEWES

Department of Physics and Astronomy, Swarthmore College
Swarthmore, PA 19081, USA

This brief review discusses Lorentz-violating operators of arbitrary dimension within the photon and neutrino sectors of the Standard-Model Extension.

1. Introduction

The Standard-Model Extension (SME) provides a general framework for theoretical and experimental studies of Lorentz and CPT violation.[1] Most early research focused on the minimal SME (mSME), which restricts attention to operators of renormalizable dimensions $d = 3$ and 4 in flat spacetime.[2] However, the full SME encompasses curved spacetime[3] and includes operators of arbitrary dimension. Here, we give a brief discussion of recent efforts to classify terms in the photon and neutrino sectors of the SME, including those of nonrenormalizable dimensions $d \geq 5$. Details can be found in Refs. 4 and 5. A summary of experimental results is given in Ref. 1.

2. Photons

The pure photon sector of the SME is given by the lagrangian[4]

$$\mathcal{L} = -\tfrac{1}{4}F_{\mu\nu}F^{\mu\nu} + \tfrac{1}{2}\epsilon^{\kappa\lambda\mu\nu}A_\lambda(\hat{k}_{AF})_\kappa F_{\mu\nu} - \tfrac{1}{4}F_{\kappa\lambda}(\hat{k}_F)^{\kappa\lambda\mu\nu}F_{\mu\nu}. \qquad (1)$$

In addition to the usual Maxwell term, there are two Lorentz-violating terms involving the CPT-odd four-vector $(\hat{k}_{AF})^\kappa$ and CPT-even tensor $(\hat{k}_F)^{\kappa\lambda\mu\nu}$. Both $(\hat{k}_{AF})^\kappa$ and $(\hat{k}_F)^{\kappa\lambda\mu\nu}$ are constants in the mSME but are momentum dependent in the full SME, where they each take the form of a power series in photon momentum. The expansion constants in the series give coefficients for Lorentz violation.

The total number of coefficients for Lorentz violation that appear in the expansions of $(\hat{k}_{AF})^\kappa$ and $(\hat{k}_F)^{\kappa\lambda\mu\nu}$ grows rapidly ($\sim d^3$) as we consider higher dimensions. To aid in classifying the numerous coefficients, a spherical-harmonic expansion of $(\hat{k}_{AF})^{\kappa}$ and $(\hat{k}_F)^{\kappa\lambda\mu\nu}$ is performed, giving sets of spherical coefficients for Lorentz violation that can be tested by different types of experiment. Several classes of experiment provide high sensitivity to various subsets of the coefficients, including searches for astrophysical birefringence and dispersion, and resonant-cavity tests. Below is a summary of coefficients that have been experimentally tested thus far.

Birefringent vacuum coefficients. The coefficients $k^{(d)}_{(E)jm}$, $k^{(d)}_{(B)jm}$, and $k^{(d)}_{(V)jm}$ lead to birefringence of light propagating *in vacuo*. The result is a change in polarization as light propagates. The effect can depend on the direction of propagation and photon energy. Polarimetry of radiation from distant astrophysical sources has led to many constraints on the mSME coefficients[6,7] and nonminimal coefficients up to dimension $d = 9$.[4,8]

Nonbirefringent vacuum coefficients. The coefficients $c^{(d)}_{(I)jm}$ affect the propagation of light in a polarization-independent way, implying no birefringence. However, they do lead to vacuum dispersion for $d \geq 6$. Time-of-flight tests involving high-energy sources, such as γ-ray bursts, search for differences in arrival times of photons at different energies. Again, this effect can depend on the photon energy and direction of propagation. Constraints on the $c^{(d)}_{(I)jm}$ coefficients for $d = 6$ and 8 have been found.[4,7,9]

Camouflage coefficients. A large class of Lorentz violation has no effect on the propagation of light. These violations are referred to as vacuum orthogonal. Among them are the so-called camouflage violations, which generically give polarization-independent effects, in addition to giving conventional light propagation. As a result, the effects of the associated camouflage coefficients $(\overline{c}^{(d)}_F)^{(0E)}_{njm}$ are particularly subtle. They can, however, be tested in resonant-cavity experiments, where they lead to tiny shifts in resonant frequencies.[4,10] Numerous cavity searches for $d = 4$ coefficients have been performed.[11] A recent experiment placed the first cavity bounds on nonminimal $d = 6$ and 8 camouflage coefficients.[12]

3. Neutrinos

Neutrinos in the SME are governed by a 6×6 effective hamiltonian that acts on the six-dimensional space that includes both neutrinos and antineutrinos. The Lorentz-violating part of the hamiltonian takes the form[5]

$$\delta h_{\text{eff}} = \frac{1}{|\vec{p}|} \begin{pmatrix} \widehat{a}_{\text{eff}} - \widehat{c}_{\text{eff}} & -\widehat{g}_{\text{eff}} + \widehat{H}_{\text{eff}} \\ -\widehat{g}^{\dagger}_{\text{eff}} + \widehat{H}^{\dagger}_{\text{eff}} & -\widehat{a}^{T}_{\text{eff}} - \widehat{c}^{T}_{\text{eff}} \end{pmatrix} . \tag{2}$$

The 3×3 matrices $\widehat{a}_{\rm eff}$, $\widehat{c}_{\rm eff}$, $\widehat{g}_{\rm eff}$, and $\widehat{H}_{\rm eff}$ are functions of the neutrino momentum vector $\vec{p}$, giving rise to unusual direction dependence and an unconventional energy-momentum relation. Note that the $\widehat{g}_{\rm eff}$, and $\widehat{H}_{\rm eff}$ terms cause mixing between neutrinos and antineutrinos. The $\widehat{c}_{\rm eff}$ and $\widehat{H}_{\rm eff}$ matrices contain the CPT-even violations, while the CPT-odd terms are in $\widehat{a}_{\rm eff}$ and $\widehat{g}_{\rm eff}$. As with photons, a spherical-harmonic expansion is used to enumerate and classify the various effects and coefficients. For example, the $\widehat{a}_{\rm eff}$ matrix has the expansion $(\widehat{a}_{\rm eff})^{ab} = \sum |\vec{p}|^{d-2}\, Y_{jm}(\hat{p})\big(a^{(d)}_{\rm eff}\big)^{ab}_{jm}$, giving spherical coefficients for Lorentz violation $\big(a^{(d)}_{\rm eff}\big)^{ab}_{jm}$. Most research in neutrinos, so far, focuses on one of the following cases.

Oscillation coefficients. Except for a small subset of flavor-diagonal coefficients, most of the spherical coefficients, $\big(a^{(d)}_{\rm eff}\big)^{ab}_{jm}$, $\big(c^{(d)}_{\rm eff}\big)^{ab}_{jm}$, $\big(g^{(d)}_{\rm eff}\big)^{ab}_{jm}$, and $\big(H^{(d)}_{\rm eff}\big)^{ab}_{jm}$, produce oscillations. Signatures of Lorentz violation in oscillations include direction dependence, unconventional energy dependence, neutrino-antineutrino oscillations, and CPT asymmetries. Neutrino oscillations are interferometric in nature, so high sensitivity to Lorentz violation is possible. While some bounds on coefficients up to $d = 10$ have been deduced from earlier analyses,[5] most of the research so far has focused on the $d = 3$ and 4 cases,[13] with many constraints on mSME coefficients.[14]

Oscillation-free coefficients. Lorentz violation also affects the kinematics of neutrino propagation. Kinematic effects are characterized, independent of oscillations, in the simple oscillation-free limit, where oscillations are neglected, and all neutrinos are treated the same. The energy in this limit is given by

$$E = |\vec{p}| + \frac{|m_l|^2}{2|\vec{p}|} + \sum |\vec{p}|^{d-3} Y_{jm}(\hat{p}) \Big[\big(a^{(d)}_{\rm of}\big)_{jm} - \big(c^{(d)}_{\rm of}\big)_{jm} \Big] , \qquad (3)$$

where $\big(a^{(d)}_{\rm of}\big)_{jm}$ and $\big(c^{(d)}_{\rm of}\big)_{jm}$ are oscillation-free coefficients. This provides a framework for a range of studies, such as time-of-flight tests, analyses of meson-decay thresholds, and Čerenkov-like decays of neutrinos, each of which has produced constraints on coefficients up to $d = 10$.[5,15]

References

1. *Data Tables for Lorentz and CPT Violation,* V.A. Kostelecký and N. Russell, 2013 edition, arXiv:0801.0287v6.
2. D. Colladay and V.A. Kostelecký, Phys. Rev. D **55**, 6760 (1997); Phys. Rev. D **58**, 116002 (1998).
3. V.A. Kostelecký, Phys. Rev. D **69**, 105009 (2004); Q.G. Bailey and V.A. Kostelecký, Phys. Rev. D **74**, 045001 (2006); V.A. Kostelecký and J.D. Tasson, Phys. Rev. Lett. **102**, 010402 (2009); Phys. Rev. D **83**, 016013 (2011).

4. V.A. Kostelecký and M. Mewes, Phys. Rev. D **80**, 015020 (2009).
5. V.A. Kostelecký and M. Mewes, Phys. Rev. D **85**, 096005 (2012).
6. S.M. Carroll *et al.*, Phys. Rev. D **41**, 1231 (1990); V.A. Kostelecký and M. Mewes, Phys. Rev. Lett. **87**, 251304 (2001); Phys. Rev. Lett. **97**, 140401 (2006); Phys. Rev. Lett. **99**, 011601 (2007); T. Kahniashvili *et al.*, Phys. Rev. D **78**, 123009 (2008); E.Y.S. Wu *et al.*, Phys. Rev. Lett. **102**, 161302 (2009); Z. Xiao and B.-Q. Ma, Phys. Rev. D **80**, 116005 (2009); G. Gubitosi *et al.*, JCAP **08**, 021 (2009); E. Komatsu *et al.*, Ap. J. Suppl. **192**, 18 (2011); Q. Exirifard, Phys. Lett. B **699**, 1 (2011).
7. V.A. Kostelecký and M. Mewes, Ap. J. **689**, L1 (2008).
8. F.W. Stecker, Astropart. Phys. **35**, 95 (2011); V.A. Kostelecký and M. Mewes, Phys. Rev. Lett. **110**, 201601 (2013).
9. V. Vasileiou, arXiv:1008.2913; V. Vasileiou *et al.*, Phys. Rev. D **87**, 122001 (2013).
10. V.A. Kostelecký and M. Mewes, Phys. Rev. D **66**, 056005 (2002); M. Mewes and A. Petroff, Phys. Rev. D **75**, 056002 (2007); M. Mewes, Phys. Rev. D **78**, 096008 (2008); Q. Exirifard, arXiv:1010.2057; M. Mewes, Phys. Rev. D **85**, 116012 (2012).
11. J. Lipa *et al.*, Phys. Rev. Lett. **90**, 060403 (2003); H. Müller *et al.*, Phys. Rev. Lett. **91**, 020401 (2003); P. Wolf *et al.*, Gen. Rel. Grav. **36**, 2351 (2004); P. Wolf *et al.*, Phys. Rev. D **70**, 051902 (2004); P. Antonini *et al.*, Phys. Rev. A **71**, 050101 (2005); Phys. Rev. A **72**, 066102 (2005); P.L. Stanwix *et al.*, Phys. Rev. Lett. **95**, 040404 (2005); S. Herrmann *et al.*, Phys. Rev. Lett. **95**, 150401 (2005); M.E. Tobar *et al.*, Phys. Rev. A **72**, 066101 (2005); P.L. Stanwix *et al.*, Phys. Rev. D **74**, 081101 (2006); H. Müller *et al.*, Phys. Rev. Lett. **99**, 050401 (2007); Ch. Eisele *et al.*, Phys. Rev. Lett. **103**, 090401 (2009); S. Herrmann *et al.*, Phys. Rev. D **80**, 105011 (2009); M.A. Hohensee *et al.*, Phys. Rev. D **82**, 076001 (2010); F. Baynes *et al.*, Phys. Rev. D **84**, 081101 (2011); F.N. Baynes *et al.*, Phys. Rev. Lett. **108**, 260801 (2012); Y. Michimura *et al.*, Phys. Rev. Lett. **110**, 200401 (2013).
12. S. Parker *et al.*, Phys. Rev. Lett. **106**, 180401 (2011).
13. V.A. Kostelecký and M. Mewes, Phys. Rev. D **69**, 016005 (2004); Phys. Rev. D **70**, 031902 (2004); Phys. Rev. D **70**, 076002 (2004); T. Katori *et al.*, Phys. Rev. D **74**, 105009 (2006); V. Barger *et al.*, Phys. Lett. B **653**, 267 (2007); J.S. Díaz *et al.*, Phys. Rev. D **80**, 076007 (2009); S. Hollenberg *et al.*, Phys. Rev. D **80**, 053010 (2009); J.S. Díaz and V.A. Kostelecký, Phys. Lett. B **700**, 25 (2011); V. Barger *et al.*, Phys. Rev. D **84**, 056014 (2011); J.S. Díaz and V.A. Kostelecký, Phys. Rev. D **85**, 016013 (2012); J.S. Díaz *et al.*, arXiv:1305.4636.
14. L.B. Auerbach *et al.*, Phys. Rev. D **72**, 076004 (2005); P. Adamson *et al.*, Phys. Rev. Lett. **101**, 151601 (2008); **105**, 151601 (2010); R. Abbasi *et al.*, Phys. Rev. D **82**, 112003 (2010); T. Katori, Mod. Phys. Lett. A **27**, 1230024 (2012); P. Adamson *et al.*, Phys. Rev. D **85**, 031101 (2012); Y. Abe *et al.*, Phys. Rev. D **86**, 112009 (2012); A.A. Aguilar-Arevalo *et al.*, Phys. Lett. B **718**, 1303 (2013); B. Rebel and S. Mufson, arXiv:1301.4684.
15. B. Altschul, Phys. Rev. D **87**, 096004 (2013).

TOWARDS A MEASUREMENT OF THE FREE FALL OF ANTIHYDROGEN WITH THE AEGIS EXPERIMENT

A. KELLERBAUER
Max Planck Institute for Nuclear Physics, Saupfercheckweg 1
69117 Heidelberg, Germany
E-mail: a.kellerbauer@cern.ch

On behalf of the AEGIS Collaboration*

After the first production of cold antihydrogen about ten years ago, second-generation experiments whose aim it is to measure the fundamental properties of this anti-atom are beginning to become operational. The AEGIS experiment will test the weak equivalence principle of General Relativity by measuring the gravitational interaction between matter and antimatter with the help of a pulsed, cold antihydrogen beam. The vertical deflection of antihydrogen atoms in the gravitational field of the Earth will be determined with a moiré deflectometer. In the present paper, the principle of the experiment will be reviewed and its present status will be presented.

1. Introduction

The weak equivalence principle (WEP) is one of the cornerstones of General Relativity (GR). It states that the trajectory of a falling test body in a gravitational field is independent of its composition. While the WEP has been very well tested with ordinary matter,[1] a verification with antimatter has never been performed. Such an experiment has only recently come within

*S. Aghion, O. Ahlén, C. Amsler, A. Ariga, T. Ariga, A.S. Belov, G. Bonomi, P. Bräunig, J. Bremer, R. S. Brusa, L. Cabaret, C. Canali, R. Caravita, F. Castelli, G. Cerchiari, S. Cialdi, D. Comparat, G. Consolati, J.H. Derking S. Di Domizio, L. Di Noto, M. Doser, A. Dudarev, A. Ereditato, R. Ferragut, A. Fontana, P. Genova, M. Giammarchi, A. Gligorova, S.N. Gninenko, S. Haider, S. Hogan, T. Huse, E. Jordan, T. Kaltenbacher, J. Kawada, A. Kellerbauer, M. Kimura, A. Knecht, D. Krasnický, V. Lagomarsino, S. Lehner, A. Magnani, C. Malbrunot, S. Mariazzi, V.A. Matveev, F. Merkt, F. Moia, G. Nebbia, P. Nédélec, M.K. Oberthaler, N. Pacifico, V. Petrácek, C. Pistillo, F. Prelz, M. Prevedelli, C. Regenfus, C. Riccardi, O. Røhne, A. Rotondi, H. Sandaker, P. Scampoli, M. Špaček, J. Storey, M.A. Subieta Vasquez, G. Testera, D. Trezzi, R. Vaccarone, E. Widmann, S. Zavatarelli, J. Zmeskal.

reach with the production of copious amounts of cold, neutral antihydrogen[2,3] at CERN's Antiproton Decelerator (AD).[4] It is the primary aim of AEGIS (Antimatter Experiment: Gravity, Interferometry, Spectroscopy) to measure the local acceleration g of antihydrogen in the gravitational field of the Earth to (initially) 1%.[5]

Gravity is the only fundamental interaction which is not currently expressed as a quantum field theory. With a view to a possible unification of GR with the Standard Model (SM), various quantum gravity theories are being discussed. In quantum gravity, the exchange of a tensor (spin-2) graviton corresponds to ordinary 'newtonian' gravity. Additional exchange bosons with different coupling constants and ranges are conceivable. For instance, the existence of vector (spin-1) and scalar (spin-0) exchange bosons is allowed by current torsion balance measurements on ordinary matter under the condition that their coupling constants and ranges match precisely.[6]

Over the years, many arguments have been brought forward against an anomalous gravitational acceleration of antimatter.[7] However, all of these arguments rely on assumptions, such as the validity of CPT or the relevance (and choice) of an absolute gravitational potential. A comprehensive review which is relevant for a 1% antimatter gravity experiment is presented in Ref. 8. Ultimately, the question of antimatter gravity can only be answered by a dedicated, direct experiment, such as AEGIS.

The AEGIS experiment involves the following main experimental steps:[9] (1) production of e^+ from a source of β-decaying radioisotope; (2) capture and accumulation of $\bar{p}$ from the AD in a Penning trap; (3) production of Ps by implanting e^+ into a nano-porous insulator material;[10,11] (4) excitation of Ps to a Rydberg state; (5) recombination of $\overline{\mathrm{H}}$ by resonant charge exchange[12] between Rydberg Ps and cold $\bar{p}$ according to the reaction $\mathrm{Ps}^* + \bar{p} \longrightarrow \overline{\mathrm{H}}^* + e^-$; (6) formation of an $\overline{\mathrm{H}}$ beam by Stark acceleration with inhomogeneous electric fields;[13] (7) measurement of g in a two-grating moiré deflectometer[14] coupled with a position-sensitive detector.

2. Antigravity within the SME

The Standard-Model Extension (SME)[15] can be used as a general framework to study the implications of Lorentz violation on physical observables and to establish quantitative bounds on Lorentz-violating terms from experiment. Within the gravitational SME,[16] terms violating Lorentz symmetry are due to the coefficients $(\bar{a}^w_{\mathrm{eff}})_\mu$ and $(\bar{c}^w)_{\mu\nu}$, where the index w designates the particle type (p, n, e^-). The former corresponds to the coupling of a particle to a field with nonmetric interaction with gravity, whereas the latter is due to a fixed background field which modifies the effective metric. If GR holds, both coefficients vanish for all particle types. Anomalous gravitational effects on antimatter may arise from the fact that the $\bar{a}$ coefficients are CPT odd, whereas the $\bar{c}$ coefficients are CPT even.

The number of free coefficients in the SME may be reduced by creating subsets with reasonable constraints on certain coefficients. To discuss the implications of antimatter gravity, a toy model termed the 'isotropic parachute model' (IPM) has been developed.[16] The IPM is constructed by considering a diagonal and traceless tensor $(\bar{c}^w)_{\mu\nu}$ and a vector $(\bar{a}^w_{\text{eff}})_\mu$ with vanishing spatial terms. In the IPM, the $\bar{a}$ and $\bar{c}$ coefficients are furthermore related by $\alpha(\bar{a}^w_{\text{eff}})_0 = \pm\frac{1}{3}m^w(\bar{c}^w)_{00}$, where α is an unobservable scaling factor and the minus sign applies for antiparticles ($\bar{p}$, $\bar{n}$, e^+). The effective classical lagrangian can be written as $L = \frac{1}{2}m^w_{i,\text{eff}}v^2 - gzm^w_{g,\text{eff}}$, where g is the local gravitational acceleration and $m^w_{i,\text{eff}}$ and $m^w_{g,\text{eff}}$ are effective inertial and gravitational masses which (for subatomic particles) are given by $m^w_{i,\text{eff}} = m^w[1 + \frac{5}{3}(\bar{c}^w)_{00}]$, $m^w_{g,\text{eff}} = m^w[1 + (\bar{c}^w)_{00}] + 2\alpha(\bar{a}^w_{\text{eff}})_0$. We observe that $m^w_{g,\text{eff}} = m^w_{i,\text{eff}}$ for ordinary matter. In the case of antimatter the effective gravitational mass, and accordingly the acceleration, are *reduced* by[17] $m^w_{g,\text{eff}}/m^w_{i,\text{eff}} \equiv 1 + \beta^w \approx [1 - \frac{4}{3}(\bar{c}^w)_{00}]$, hence the name 'parachute.'

A priori the coefficients are only constrained by the condition that they must be perturbatively small relative to m^w, and a reduction in the gravitational acceleration of antimatter of as much as 50% is conceivable.[16] However, it has recently been realized that a gravitational anomaly β^w (both for matter and for antimatter particles) contributes to the internally bound kinetic energy T_{int} of composite systems.[18] WEP tests in existing gravity experiments with ordinary matter can therefore be used to constrain the six parameters β^w or linear combinations thereof. The sensitivity of the parameters to WEP violation varies across the nuclide chart, with particular sensitivities to the antimatter anomaly parameters in the light elements.

Hohensee *et al.* have extracted bounds on linear combinations of five out of six β^w parameters from experiments involving atom interferometry and clock comparisons and were able to constrain the corresponding $\bar{a}$ and $\bar{c}$ coefficients at the 10^{-6} level.[18] It should be stressed, however, that these constraints are only valid (or meaningful) within the simple IPM, not the full gravitational SME.

3. Setup and current status

The AEGIS experiment was approved in 2008. Construction at the AD began in 2010. As of June 2013, the following main components are installed and operational: positron source and accumulator; high-field (5 T) and low-field (1 T) superconducting magnets; e^+ transfer line; $\bar{p}$ capture trap; $\overline{\mathrm{H}}$ recombination trap with Ps converter target holder; laser system for Ps excitation. A prototype of the moiré deflectometer was commissioned off-site with metastable Ar atoms. During beam times in June and December 2012, the capture and stacking of $\bar{p}$ was demonstrated, and a $\bar{p}$ lifetime in the capture trap of ≈ 10 min was observed.

During the 2013/2014 shutdown of the CERN accelerators, the design and construction of the remaining components of the AEGIS apparatus are being completed. In particular, the design of the deflectometer's $\overline{\mathrm{H}}$ detector, which will consist of a combination of nuclear emulsions[19] and Si strip detectors is being finalized, and construction will begin shortly. In parallel, a proton source is being installed, and the main components of the apparatus will be commissioned by use of the mirror reaction $\mathrm{Ps}^* + p \longrightarrow \mathrm{H}^* + e^+$.

4. Conclusions

The gravitational interaction between matter and antimatter has never been studied experimentally. The AEGIS experiment will determine g for antimatter to 1% by measuring the vertical deflection of a horizontal $\overline{\mathrm{H}}$ beam. Data taking is expected to begin in early 2015.

References

1. E.G. Adelberger *et al.*, Prog. Part. Nucl. Phys. **62**, 102 (2009).
2. M. Amoretti *et al.*, Nature **419**, 456 (2002).
3. G. Gabrielse *et al.*, Phys. Rev. Lett. **89**, 213401 (2002).
4. J.Y. Hémery and S. Maury, Nucl. Phys. A **655**, 345c (1999).
5. G. Drobychev *et al.*, CERN-SPSC-2007-017 (2007).
6. T.A. Wagner *et al.*, Class. Quantum Grav. **29**, 184002 (2012).
7. M.M. Nieto and T. Goldman, Phys. Rep. **205**, 221 (1991).
8. M. Fischler *et al.*, Fermilab report FN-0822-CD-T (2008).
9. A. Kellerbauer *et al.*, Nucl. Instr. Meth. B **266**, 351 (2008).
10. D.W. Gidley, *et al.*Annu. Rev. Mater. Res. **36**, 49 (2006).
11. G. Consolati *et al.*, Chem. Soc. Rev. **42**, 3821 (2013).
12. C.H. Storry *et al.*, Phys. Rev. Lett. **93**, 263401 (2004).
13. E. Vliegen *et al.*, Phys. Rev. A **76**, 023405 (2007).
14. M.K. Oberthaler *et al.*, Phys. Rev. A **54**, 3165 (1996).
15. D. Colladay and V.A. Kostelecký, Phys. Rev. D **55**, 6760 (1997).
16. V.A. Kostelecký and J.D. Tasson, Phys. Rev. D **83**, 016013 (2003).
17. M.A. Hohensee *et al.*, Phys. Rev. Lett. **106**, 151102 (2011).
18. M.A. Hohensee, H. Müller and R.B. Wiringa, submitted for publication.
19. S. Aghion *et al.*, J. Instrum., in press.

TESTS OF LORENTZ INVARIANCE USING HIGH-ENERGY ASTROPHYSICS OBSERVATIONS

FLOYD W. STECKER
Astrophysics Science Division, NASA Goddard Sapce Flight Center
Greenbelt, MD 20771, USA
E-mail: Floyd.W.Stecker@nasa.gov

High-energy astrophysics observations provide the best possibilities to detect a very small violation of Lorentz invariance, such as may be related to the structure of spacetime near the Planck scale. I will discuss the possible signatures of Lorentz invariance violation that can be manifested by observing the spectra, polarization, and timing of γ-rays from active galactic nuclei and γ-ray bursts. Other sensitive tests are provided by observations of the spectra of ultrahigh-energy cosmic rays and very high-energy neutrinos. I will also discuss a new time-of-flight analysis of observations of GRB 090510 by the Fermi γ-ray Space Telescope. These results, based on high-energy astrophysical observations, have fundamental implications for spacetime physics and quantum gravity models.

Everything that is not forbidden is compulsory.
— *Murray Gell-Mann*

1. Introduction

Colladay and Kostelecký[1] proposed an effective field theory framework for quantifying and cataloging the empirical effects of small violations of CPT and Lorentz invariance known as the Standard-Model Extension (SME). The SME is based on the introduction of small Lorentz and CPT violating perturbations in the individual free particle Lagrangians. Coleman and Glashow[2] have presented a simplified formalism, assuming rotational invariance, wherein particle interactions that violate Lorentz invariance can be modified in terms of the maximum attainable velocities (MAV) of the various particles involved. Thus superluminal particle velocities can be directly related to Lorentz invariance violation. This in turn can lead to the allowance or prohibition of various particle interactions.

2. Why high-energy astrophysics observations?

Lorentz invariance implies scale-free spacetime. The group of Lorentz boosts is unbounded. Very large boosts probe physics at ultrashort distance intervals λ. To probe physics at these distance intervals, particularly the nature of space and time, we need to go to ultrahigh energies $E = 1/\lambda$. Cosmic γ-rays and cosmic rays provide the highest observable energies in the Universe. Physics at the Planck scale of $\sim 10^{-35}$ m, e.g., quantum gravity, may involve a fundamental length scale. That, in itself, violates Lorentz invariance. Some frameworks for considering Lorentz invariance violation include effective field theory such as SME, deformed special relativity, stochastic spacetime foam, loop quantum gravity, string inspired models (D-branes), and emergent spacetime theories.[3] I will discuss some particular astrophysical tests of Lorentz invariance violation and limits on MAV.

3. Limits from photons and electrons

I follow the well-defined formalism for Lorentz invariance breaking discussed in Ref. 2. Within this scenario, the MAV of an electron need not equal the *in vacuo* velocity of light, i.e., $c_e \neq c_\gamma$. The physical consequences of this violation of Lorentz invariance depend on the sign of the difference. I define the parameter $\delta_{ij} \equiv \delta_i - \delta_j$ as the difference between the MAVs of particles i and j.

3.1. *Direct limits*

Direct limits on $\delta_{e\gamma}$ follow from the observation of 2 TeV cosmic-ray electrons and 50 TeV γ-rays from the Crab Nebula.[2,4] These limits are $\delta_{e\gamma} < 1.3 \times 10^{-13}$ and $-\delta_{e\gamma} < 2 \times 10^{-16}$.

3.2. *Threshold for annihilation of γ-rays*

If $\delta_{e\gamma} > 0$ the threshold energy for the pair production process $\gamma + \gamma \rightarrow e^+ + e^-$ is altered because the square of the four-momentum becomes

$$2\epsilon E_\gamma(1 - \cos\theta) - 2E_\gamma^2\delta_{e\gamma} = 4\gamma^2 m_e^2 > 4m_e^2. \tag{1}$$

The Mrk 501 γ-ray spectrum of a strong flare the nearby blazer Mrk 501 above $\sim$20 TeV can be understood as a result of expected intergalactic absorption by pair production. It follows that $\delta_{e\gamma} \leq 2(m_e/E_\gamma)^2 = 1.3 \times 10^{-15}$.[4]

3.3. *Time of flight for photons from γ-ray bursts*

The Large Area Telescope aboard the Fermi satellite has placed direct time-of-flight limits on the energy dependence of photon velocity.[5] This limit rules out the simple E/M_{Planck} dependence for photon retardation as predicted in some quantum gravity scenarios.[6]

3.4. *Vacuum birefringence*

Then, using the reported detection of polarized soft γ-ray emission from the γ-ray burst GRB041219a that is indicative of an absence of vacuum birefringence, together with a very recent improved method for estimating the redshift of the burst, I have derived strong constraints on the dimension-5 Lorentz invariance violation term

$$\Delta\mathcal{L}_\gamma = \frac{\xi}{M_{Pl}} n^a F_{ad} n \cdot \partial(n_b \tilde{F}^{bd}). \tag{2}$$

I obtain an upper limit on $|\xi|$ of 2.4×10^{-15}, corresponding to a constraint on the dimension-5 SME coefficient $k^{(5)}_{(V)00} \leq 4.2\times10^{-34}$ GeV^{-1}.[7] It also gives a directional frame independent constraint in terms of the SME coefficients of

$$|\sum_{jm} Y_{jm}(37°, 0°) k^{(5)}_{(V)jm}| \leq 1.2\times10^{-34}\ \text{GeV}^{-1}. \tag{3}$$

Other directional constraints from various γ-ray bursts have recently been derived.[8]

3.5. *Synchrotron emission from the Crab nebula*

Synchrotron γ-rays from the strong April 2011 Crab Nebula flare were observed by the Large Area Telescope on Fermi up to an energy $\sim$400 MeV.[9] This places an upper limit on $\delta_{e\gamma}$ of 6×10^{-20}.[10] This equals the limit on the SME parameter $-c^e_{TT}$.

4. Limits from neutrinos

The observation of two PeV-scale neutrino events reported by Ice Cube[11] can, in principle, allow one to place constraints on Lorentz invariance violation in the neutrino sector. This observation implies an upper limit on $\delta_{\nu e}$ of $\sim 3.1\times10^{-19}$. Combined with the best limit on $\delta_{e\gamma}$ given in Section 3.5, this gives an upper limit on $\delta_{\nu\gamma}$ of $\sim 3\times10^{-19}$.[10] This equals the limit on the SME parameter $-\mathring{c}^{(4)}$.

5. Limits from hadrons

Owing to the GZK effect,[12,13] protons with energies above 100 EeV should be attenuated from distances beyond ~100 Mpc because they interact with the CBR photons with a resonant photoproduction of pions.[14] The threshold energy for this reaction is changed by $2\delta_{\pi p}E_{\pi}^{2}$. This leads to an upper limit $\delta_{\pi p} \leq 3.23 \times 10^{-24}(\omega/\omega_0)^2$ where $\omega_0 = kT_{CMB} = 2.35 \times 10^{-4}$ eV with $T_{CMB} = 2.725 \pm 0.02$ K.[2] By comparing the observed spectrum of ultrahigh energy cosmic rays with the predicted modification from Lorentz invariance violation an extremely low upper limit on $\delta_{\pi p}$ of 4.5×10^{-23} is obtained.[15]

6. Conclusion

Presently, we have no positive evidence for modifying special relativity at even the highest energies observed. Theoretical models involving Planck scale physics and quantum gravity need to meet all of the present observational constraints.

References

1. D. Colladay and V.A. Kostelecký, Phys. Rev. D **58**, 116002 (1998).
2. S.R. Coleman and S.L. Glashow, Phys. Rev. D **59**, 116008 (1999).
3. D. Mattingly, Liv. Rev. Rel. **8**, 5 (2005).
4. F.W. Stecker and S.L. Glashow, Astropart. Phys. **16**, 97 (2001).
5. V. Vasileiou *et al.*, Phys. Rev. D **87**, 122001 (2013).
6. J. Ellis, N.E. Mavromatos, and D.V. Nanopoulos, Phys. Lett. B **665**, 412 (2008).
7. F.W. Stecker, Astropart. Phys. **35**, 95 (2011).
8. V.A. Kostelecký and M. Mewes, arXiv:1301.5367.
9. A.A. Abdo *et al.*, Science **331**, 739 (2011).
10. F.W. Stecker, arXiv:1306.6095.
11. M.G. Aartsen *et al.*, Ice Cube Collaboration, Phys. Rev. Lett. **110**, 151105 (2013).
12. K. Griesen, Phys. Rev. Lett. **16**, 748 (1966).
13. G.T. Zatsepin and V.A. Kuz'min, Zh. Eks. Teor. Fiz. Pis. Red. **4**, 144 (1966).
14. F.W. Stecker, Phys. Rev. Lett. **21**, 1016 (1968).
15. F.W. Stecker and S.T. Scully, New J. Phys. **11**, 085003 (2009).

THE KATRIN EXPERIMENT: STATUS AND OUTLOOK

D.S. PARNO

Center for Experimental Nuclear Physics and Astrophysics
and Department of Physics, University of Washington
Seattle, WA 98195, USA
E-mail: dparno@uw.edu

On behalf of the KATRIN Collaboration

The KATRIN experiment, presently under construction in Karlsruhe, Germany, will improve on previous laboratory limits on the neutrino mass by a factor of ten. KATRIN will use a high-activity, gaseous T_2 source and a very high-resolution spectrometer to measure the shape of the high-energy tail of the tritium-decay β spectrum. The shape measurement will also be sensitive to new physics, including sterile neutrinos and Lorentz violation. This report summarizes recent progress in the experiment.

Neutrino oscillation experiments show that the neutrino-mass eigenstates ν_1, ν_2, ν_3 are distinct from the weak-flavor eigenstates ν_e, ν_μ, ν_τ.[1] In the quasi-degenerate regime, where the mass splittings are small compared to the absolute mass scale, laboratory β-decay experiments can extract the antineutrino mass through a shape analysis of the high-energy tail of the β spectrum, making no model assumptions apart from basic kinematics. Previous efforts to perform such a measurement using tritium have yielded an upper limit of $\sqrt{\sum_i |U_{ei}|^2 m_{\bar{\nu}_i}^2} < 2\text{eV}$ at 90% confidence, where $m_{\bar{\nu}_i}$ is the mass of the ith neutrino-mass eigenstate and U_{ei} is an element of the neutrino mixing matrix.[1] The KArlsruhe TRItium Neutrino (KATRIN) experiment will improve this sensitivity by a factor of 10.[2]

Figure 1 shows the KATRIN beamline. The rear section (a) monitors the windowless, gaseous T_2 source (WGTS, b), in which gaseous tritium decays in a magnetic field. A two-stage transport section (c) adiabatically guides βs along magnetic field lines toward a pair of spectrometers (d, e). Each spectrometer is a magnetic adiabatic collimating/electrostatic filter.[3] As βs travel from a region of strong magnetic field to a field minimum,

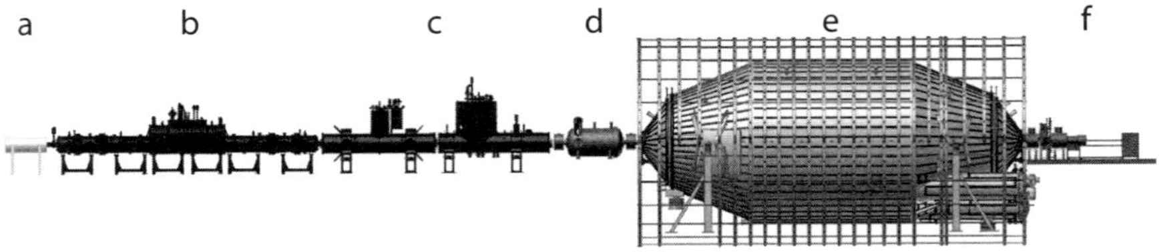

Fig. 1. The KATRIN beamline.

their magnetic orbital moments remain constant while their momenta are rotated to be nearly longitudinal. A longitudinal retarding potential thus sets a kinetic-energy threshold for βs passing through to the detector (f). By adjusting the retarding potential and thus the integration threshold, KATRIN will make a precise measurement of the tritium β spectrum near the endpoint. The fit to the spectral shape will have at least four free parameters: the antineutrino mass, the β endpoint energy, the signal rate, and the background rate.

Here, we highlight some recent achievements of the KATRIN collaboration, from research and development to construction and commissioning.

The 10-m source tube, presently under construction, will use components of the WGTS Demonstrator apparatus, a prototype used to test aspects of the source design. Using a cooling system based on neon thermosiphons, the Demonstrator was maintained for a week at the design temperature of 30 K. The demonstrated temperature stability of $\Delta T/T = 5 \times 10^{-5}\ \mathrm{h}^{-1}$ is a factor of 20 better than the design specification.[4]

The molecular composition of the source affects the source activity as well as the molecular final state after decay of a T atom. The LAser RAman (LARA) system uses Raman spectroscopy to continuously monitor the composition of the gas to be injected into the source. In a prototype of the circulation system, 0.1% precision was demonstrated in 250 s of acquisition.[5]

The commissioning of the pre-spectrometer revealed potential pitfalls in the spectrometer design. For example, ring-shaped background patterns were observed to accumulate over the course of several hours. Through analysis, simulation, and dedicated runs, these were traced to radon atoms that escaped from the getter pumps and decayed inside the spectrometer flux tube. A shakeoff or internal-conversion electron from such a decay could be stored in the spectrometer for hours, slowly losing energy via collisions with residual gas molecules. The resulting low-energy secondary electrons were then accelerated by the retarding potential; their distribution on the detector plane was ring-shaped due to the magnetron motion of the primary.[6–8] With the identification of this new background source, the main-spectrometer design was modified to include liquid-nitrogen-cooled, radon-trapping baffles between the getter pumps and the main volume.

A two-layer wire electrode system containing 23,440 wires lines the 700-m^2 interior surface of the main-spectrometer vessel, which is 23.8 m long with a maximum diameter of 9.8 m.[9] This system improves field stability and allows fine shaping of the retarding potential. Since the wire electrodes are held at a slightly more negative potential than the hull of the vessel, they also reduce background by preventing secondary electrons from entering the flux tube. This system has been installed and baked out to a temperature of 300°C.

The 0.93-eV design resolution of the main spectrometer is set by the ratio of the minimum magnetic field, in the central analyzing plane, to the maximum magnetic field at one end. In order to achieve a magnetic-field minimum of precisely 3 G, two sets of air coils have been installed around the main spectrometer.[10] The circular low-field-correction coils, which are coaxial with the main spectrometer, compensate for distortions due to residual magnetization and to fringe fields from the superconducting solenoids at either end. Compensation for the Earth's magnetic field is achieved by coils in straight sections, describing a cylindrical surface centered on the flux tube and joined by vertical endrings. The resulting magnetic field is monitored by stationary sensors and by robotic mobile sensor units that travel along the support structures for the low-field-correction coils.[11]

The focal-plane detector is a monolithic silicon PIN diode; its active area has a diameter of 90 mm. It is segmented into 148 pixels laid out in a dartboard pattern, allowing the characterization of inhomogeneities in the analyzing plane. A novel mounting structure,[12] in which electrical contacts are made by spring-loaded pogo pins, minimizes radioactivity in the region of the detector. With a built-in calibration source, we have demonstrated an energy resolution of 1.48 ± 0.01 keV (full width at half maximum) for electrons at the tritium β-spectrum endpoint energy of 18.6 keV.

In May 2013, the main spectrometer and detector sections were connected for the first time, beginning an initial commissioning period to study backgrounds and the spectrometer transmission function. After hardware upgrades, a second commissioning phase will follow in 2014, along with commissioning of the source and transport sections. We anticipate the start of KATRIN data-taking in late 2015.

KATRIN expects to achieve a sensitivity of $\sqrt{\sum_i |U_{ei}|^2 m_{\bar{\nu}_i}^2} < 0.2$ eV at the 90% confidence level in three full years of running, or approximately five calendar years. The precise spectral-shape analysis also gives potential access to new physics, including sterile neutrinos in the mass range suggested by the reactor antineutrino anomaly.[13] Endpoint measurements from previous tritium-based neutrino-mass experiments have recently been used to set limits on the coefficients of two Lorentz-violating operators in the Standard-Model Extension, which KATRIN could improve by two orders of magnitude.[14] The KATRIN measurement plan, which involves cycling through all retarding-potential measurements with a period of about 15 minutes,[15] will give access to the complex coefficient of an additional Lorentz-violating operator that produces oscillations of the endpoint with a period of one sidereal day.[14]

Acknowledgments

The KATRIN experiment is funded by the Helmholtz Gemeinschaft, the Bundesministerium für Bildung und Forschung, and the US Department of Energy. The author is funded by grant number DE-FG02-97ER41020 from the US Department of Energy.

References

1. Particle Data Group, J. Beringer *et al.*, Phys. Rev. D **86**, 010001 (2012).
2. J. Angrik *et al.*, *KATRIN Design Report 2004*, preprint FZKA-7090 (2005).
3. G. Beamson, H.Q. Porter, and D.W. Turner, J. Phys. E: Sci. Instrum. **13**, 64 (1980).
4. S. Grohmann *et al.*, Cryogenics **55-56**, 5 (2013).
5. S. Fischer *et al.*, Fusion Sci. Technol. **60**, 925 (2011).
6. F.M. Fränkle *et al.*, Astropart. Phys. **35**, 128 (2011).
7. N. Wandkowsky *et al.*, arXiv:1304.1375.
8. N. Wandkowsky *et al.*, J. Phys. G **40**, 085102 (2013).
9. K. Valerius, Prog. Part. Nucl. Phys. **64**, 291 (2010).
10. F. Glück *et al.*, arXiv:1304.6569.
11. A. Osipowicz *et al.*, J. Inst. **7**, T06002 (2012).
12. B.A. VanDevender *et al.*, Nucl. Inst. Meth. Phys. Res. A **673**, 46 (2012).
13. J. Formaggio and J. Barrett, Phys. Lett. B **706**, 68 (2011).
14. J.S. Díaz, V.A. Kostelecký, and R. Lehnert, arXiv:1305.4636.
15. A. Sejersen Riis, S. Hannestad, and C. Weinheimer, Phys. Rev. C **84**, 045503 (2011).

STATUS OF MATTER-GRAVITY COUPLINGS IN THE SME

JAY D. TASSON

Physics and Astronomy Department, Carleton College
Northfield, MN 55057, USA
E-mail: jtasson@carleton.edu

Constraints on Lorentz violation in matter-gravity couplings are summarized along with existing proposals to obtain sensitivities that exceed current limits by up to 11 orders of magnitude.

1. Introduction

The phenomenology of matter-gravity couplings and Lorentz violation in the Standard-Model Extension (SME)[1] was developed in Refs. 2,3. The first portion of Ref. 3 built upon existing analysis of the pure-gravity sector of the SME[4] to develop the necessary theoretical tools for the experimental analysis. These theoretical developments are summarized in Ref. 5. The second portion of Ref. 3 generates explicit predictions for the detection of matter-sector coefficients for Lorentz violation in a large number of gravitational tests. Note that 'explicit prediction' here means that the experimental observable has been calculated and decomposed by signal frequency such that it is ready to fit with experimental data. Of special interest in this context are sensitivities to the 12 coefficient components of $\alpha(\overline{a}^w_{\rm eff})_\mu$, where w runs over species, proton, neutron, and electron, which are unobservable in the absence of gravity.[2] Experimental implications of the remaining spin-independent coefficient, $\overline{c}_{\mu\nu}$ were also considered. Consideration of spin-dependence in matter-gravity couplings is now underway.[6,7]

Predictions were made for the following tests: laboratory tests such as gravimeter experiments and tests of the Weak Equivalence Principle (WEP) with ordinary neutral matter; versions of these experiments with electrically charged matter, higher-generation matter, and antimatter; WEP tests in space; solar-system tests such as lunar laser ranging and precession of the parihelion of various bodies; and light-travel tests such as time-delay, Doppler shift, redshift, and null-redshift tests. These predictions were then used to place several constrains on the relevant coefficients. Following the publication of Refs. 2,3, analysis of additional tests has been performed obtaining additional constraints. Section 2 of this proceedings contribution summarizes these existing limits. The many proposed investigations that remain to be completed, which could extend existing limits, are summarized in Sec. 3. Unless otherwise stated, bounds on combinations of coefficient components assume all other SME coefficients are zero.

2. Constraints

Concurrent with the development of experimental and observational predictions, several constraints were placed using the published results of experiments. Constraints were placed on 4 combinations of the 12 $\alpha(\overline{a}^w_{\text{eff}})_\mu$ coefficient components, and 4 constraints were placed on previously unconstrained combinations of $(\overline{c}^w)_{\mu\nu}$.[3] Four of these constraints are from precession of the perihelion of bodes based on existing data,[8]

$$\begin{aligned}
&|-0.97(\overline{a}^{\odot}_{\text{eff}})_X + 0.15(\overline{a}^{\odot}_{\text{eff}})_Y + 0.18(\overline{a}^{\odot}_{\text{eff}})_Z| \lesssim 10^{-6}\ \text{GeV}\ \text{(Mercury)},\\
&|-0.97(\overline{c}^n)_{TX} + 0.15(\overline{c}^n)_{TY} + 0.18(\overline{c}^n)_{TZ}| \lesssim 10^{-5}\qquad \text{(Mercury)},\\
&|-0.97(\overline{a}^{\odot}_{\text{eff}})_X - 0.21(\overline{a}^{\odot}_{\text{eff}})_Y - 0.10(\overline{a}^{\odot}_{\text{eff}})_Z| \lesssim 10^{-6}\ \text{GeV}\ \text{(Earth)},\\
&|-0.97(\overline{c}^n)_{TX} - 0.21(\overline{c}^n)_{TY} - 0.10(\overline{c}^n)_{TZ}| \lesssim 10^{-5}\qquad \text{(Earth)},
\end{aligned} \tag{1}$$

where $(\overline{a}^{\odot}_{\text{eff}})_J = \alpha[(\overline{a}^e_{\text{eff}})_J + (\overline{a}^p_{\text{eff}})_J + 0.1(\overline{a}^n_{\text{eff}})_J]$. Note that the $(\overline{c}^n)_{\Xi\Sigma}$ constraints above are simplified using existing constraints[9] on other combinations of $(\overline{c}^w)_{\Xi\Sigma}$. The other 2 constraints on $\alpha(\overline{a}^w_{\text{eff}})_\Xi$ are from torsion pendulum measurements of WEP based on data from Ref. 10,

$$|\alpha(\overline{a}^n_{\text{eff}})_T| \lesssim 10^{-10}\ \text{GeV},\qquad |\alpha(\overline{a}^e_{\text{eff}})_T + \alpha(\overline{a}^p_{\text{eff}})_T| \lesssim 10^{-10}\ \text{GeV}, \tag{2}$$

and the final constraints are based on combined results from torsion pendulum WEP measurements and falling corner-cube WEP measurements[11]

$$|(\overline{c}^n)_Q| \lesssim 10^{-8},\qquad |(\overline{c}^e)_{TT} + (\overline{c}^p)_{TT} - (\overline{c}^n)_{TT}| \lesssim 10^{-8}, \tag{3}$$

though other WEP tests could be used if sufficient sensitivity is available.

Following Refs. 2,3, weak constraints have been achieved on 4 additional combinations of $\alpha(\overline{a}^w_{\text{eff}})_\mu$ coefficients:

$$\begin{aligned}
|\alpha(\overline{a}^n_{\text{eff}})_X + 0.83\alpha\,[(\overline{a}^p_{\text{eff}})_X + (\overline{a}^e_{\text{eff}})_X]\,| &\le 0.2\ \text{GeV},\\
|\alpha(\overline{a}^n_{\text{eff}})_Y + 0.83\alpha\,[(\overline{a}^p_{\text{eff}})_Y + (\overline{a}^e_{\text{eff}})_Y]\,| &\le 0.2\ \text{GeV},
\end{aligned} \tag{4}$$

using a torsion-strip balance,[12] and

$$\begin{aligned}
|\alpha\,[(\overline{a}^e_{\text{eff}})_X + (\overline{a}^p_{\text{eff}})_X + (\overline{a}^e_{\text{eff}})_X]\,| &= 0.44 \pm 0.28\ \text{GeV},\\
|\alpha\,[(\overline{a}^e_{\text{eff}})_Y + (\overline{a}^p_{\text{eff}})_Y + (\overline{a}^e_{\text{eff}})_Y]\,| &= 0.04 \pm 0.24\ \text{GeV},
\end{aligned} \tag{5}$$

via a reinterpretation[13] of He/K comagnetometer b_μ results.[14] Computational work on $\alpha(\overline{a}^w_{\rm eff})_J$ has also been done associated with the Cassini mission.[15] Additional work has been done associated with the separation of $\alpha(\overline{a}^w_{\rm eff})_T$ and $(\overline{c}^w)_{TT}$ as well resulting in the independent constraints[16] on $(\overline{c}^p)_{TT}$ and $(\overline{c}^n)_{TT}$ at the 10^{-6} level, $\alpha[(\overline{a}^e_{\rm eff})_T + (\overline{a}^p_{\rm eff})_T]$ and $(\overline{a}^n_{\rm eff})_T$ at the level of 10^{-6} GeV, and much stronger constraints on $(\overline{c}^e)_{TT}$ via nongravitational experiments. Ref. 9 summarizes all constraints discussed.

3. Outstanding proposed analysis

To date, 8 combinations of the 12 components of the $\alpha(\overline{a}^w_{\rm eff})_\mu$ coefficient have been constrained, and 4 of those constraints are weak. Existing experiments could improve the weak constraints by up to 6 orders of magnitude. Proposed experiments could improve these constraints by up to 11 orders of magnitude, and some could gain sensitivity to unconstrained combinations.

Analysis of existing data from the following experiments could provide up to the indicated order of magnitude improvement in sensitivities to spatial components of $\alpha(\overline{a}^w_{\rm eff})_\Xi$: torsion pendulum WEP,[10] 6 orders of magnitude; super-conducting gravimeters,[17] 6 orders; lunar laser ranging,[18] 5 orders; Cassini data,[15] 5 orders; and atom interferometry,[11,19] 4 orders. If performed, the following proposed experiments could provide even greater improvement as indicated: space-based WEP,[20] up to 11 orders; Earth-based WEP,[21,22] up to 10 orders; and gravimeters,[22] up to 9 orders.

Gravitational tests with special types of matter could obtain sensitivity to additional unconstrained combinations. The 4 unconstrained combinations of $\alpha(\overline{a}^w_{\rm eff})_\Xi$ components for ordinary matter could be accessed with charged-matter tests.[23] Tests with higher-generation matter[24] of type w would attain sensitivity to many unconstrained associated coefficient components of $\alpha(\overline{a}^w_{\rm eff})_\Xi$. Tests with antimatter[25] also have the ability to separate special combinations[3] of coefficients if sufficient sensitivity can be reached.

The above possibilities offer excellent prospects for improved tests of Lorentz symmetry, and provide the opportunity for significant progress in the ongoing search for new physics at the Planck scale.

References

1. D. Colladay and V.A. Kostelecký, Phys. Rev. D **55**, 6760 (1997); Phys. Rev. D **58**, 116002 (1998); V.A. Kostelecký, Phys. Rev. D **69**, 105009 (2004).
2. V.A. Kostelecký, and J.D. Tasson, Phys. Rev. Lett. **102**, 010402 (2009).
3. V.A. Kostelecký, and J.D. Tasson, Phys. Rev. D **83**, 016013 (2011).
4. Q.G. Bailey and V.A. Kostelecký, Phys. Rev. D **74**, 045001 (2006).

5. J.D. Tasson, in V.A. Kostelecký, ed., *CPT and Lorentz Symmetry V*, World Scientific, Singapore, 2011, arXiv:1010.3990.
6. Y. Bonder, these proceedings.
7. D. Atkinson, M. Becker, J.D. Tasson, these proceedings; in preparation.
8. C.M. Will, Living Rev. Relativity **4**, 4 (2001).
9. *Data Tables for Lorentz and CPT Violation,* V.A. Kostelecký and N. Russell, 2013 edition, arXiv:0801.0287v6.
10. S. Schlamminger, K.-Y. Choi, T.A. Wagner, J.H. Gundlach, and E.G. Adelberger, Phys. Rev. Lett. **100**, 041101 (2008); Y. Su, *et al.*, Phys. Rev. D **50**, 3614 (1994).
11. K. Kuroda and N. Mio, Phys. Rev. D **42**, 3903 (1990); T.M. Niebauer, M.P. McHugh, and J.E. Faller, Phys. Rev. Lett. **59**, 609 (1987).
12. H. Panjwani, L. Carbone, and C.C. Speake, in V.A. Kostelecký, ed., *CPT and Lorentz Symmetry V*, World Scientific, Singapore, 2011.
13. J.D. Tasson, Phys. Rev. D **86**, 124021 (2012).
14. J.M. Brown, S.J. Smullin, T.W. Kornack, and M.V. Romalis, Phys. Rev. Lett. **105**, 151604 (2010).
15. A. Hees, Ph.D. thesis, Université Catholique de Louvain, 2012; A. Hees *et al.*, arXiv:1301.1658; these proceedings.
16. M.A. Hohensee, S. Chu, A. Peters, H. Mueller, Phys. Rev. Lett. **106**, 151102 (2011); M.A. Hohensee, *et al.*, arXiv:1303.2747; these proceedings.
17. S. Shiomi, arXiv:0902.4081.
18. J.G. Williams, S.G. Turyshev, and H.D. Boggs, Phys. Rev. Lett. **93**, 261101 (2004); J.B.R. Battat, J.F. Chandler, and C.W. Stubbs, Phys. Rev. Lett. **99**, 241103 (2007).
19. K.-Y. Chung *et al.*, Phys. Rev. D **80**, 016002 (2009).
20. A.M. Nobili *et al.*, Exp. Astron. **23**, 689 (2009); these proceedings; S. Schiller, these proceedings; P. Touboul, M. Rodrigues, G. Métris, and B. Tatry, Comptes Rendus de l'Académie des Sciences, Series IV, **2**, 1271 (2001); G. Amelino-Camelia *et al.*, Exp. Astron. **23**, 549 (2009).
21. J. Philips, these proceedings; R.D. Reasenberg, in V.A. Kostelecký, ed., *CPT and Lorentz Symmetry II*, World Scientific, Singapore, 2005; A.M. Nobili, these proceedings; V. Iafolla, S. Nozzoli, E.C. Lorenzini, I.I. Shapiro, and V. Milyukov, Class. Quantum Grav. **17**, 2327 (2000); H. Dittus and C. Mehls, Class. Quantum Grav. **18**, 2417 (2001).
22. S. Dimopoulos, P.W. Graham, J.M. Hogan, and M.A. Kasevich, Phys. Rev. Lett. **98**, 111102 (2007); Phys. Rev. D **78**, 042003 (2008).
23. H. Müller, these proceedings; B. Neyenhuis, D. Christensen, and D.S. Durfee, Phys. Rev. Lett. **99**, 200401 (2007).
24. K. Kirch, arXiv:physics/0702143.
25. AEGIS Collaboration, Nat. Commun. **4**, 1785 (2013); A. Kellerbauer, these proceedings; AGE Collaboration, A.D. Cronin *et al.*, *Letter of Intent: Antimatter Gravity Experiment (AGE) at Fermilab*, February 2009; J. Walz and T.W. Hänsch, Gen. Rel. Grav. **36**, 561 (2004); P. Pérez *et al.*, *Letter of Intent to the CERN-SPSC*, November 2007; A. Voronin, P. Froelich, and V. Nesvizhevsky, Hyperfine Int. **213**, 129 (2012).

LORENTZ AND CPT VIOLATION IN THE HYDROGEN SPECTRUM

G.S. ADKINS

Department of Physics and Astronomy, Franklin & Marshall College
Lancaster, PA 17604, USA
E-mail: gadkins@fandm.edu

T.J. YODER

Department of Physics, Massachusetts Institute of Technology
Cambridge, MA 02139, USA
E-mail: tjyoder@mit.edu

We have studied the effect of hypothetical violations of Lorentz and CPT symmetry by calculating the corrections to the energy levels of hydrogen induced by the Standard-Model Extension (SME). Hydrogen studies are interesting because the energy levels of hydrogen can be measured with great precision and the theory for hydrogen based on the Standard Model (SM) is well understood. We obtained corrections through order α^2 times the SME parameters for all levels of hydrogen and applied them to determine the SME corrections to the transition frequency for the 2S-1S transition.

Hydrogen is an interesting system for the study of beyond-the-SM effects because it is well-understood theoretically,[1] high-precision experiments are available, and it is amenable to the calculation of hypothetical new effects using standard methods. The leading SME corrections — at $O(\alpha^0)$ times SME parameters — to the hydrogen energies have been known for some time.[2] We have obtained the next order of corrections — at $O(\alpha^2)$ times SME parameters.[3] There were several reasons for pursuing this calculation. One is that the lowest order SME correction depends on only a few combinations of SME coefficients. Perhaps those combinations are zero or small and the leading order effect occurs at higher order. Certainly we did find that the higher order result has more structure than the lowest order one in terms of dependence on SME parameters and on the quantum numbers. In addition, such a calculation is useful for learning more about the theory.

It is difficult to anticipate what issues or problems might arise in a higher order calculation without actually doing the calculation. Hydrogen is appropriate for the exploration of such issues due to its simplicity. We also explored the possibility of placing new constraints on the SME coefficients through use of the higher-order results.

There were three main parts in our calculation. First, we needed to find an effective hamiltonian to describe the interaction between the constituent fermions in hydrogen with the background SME vectors and tensors. Next, we needed to evaluate the energy shifts due to the new interactions in appropriate states of the hydrogen atom. Finally, we needed to calculate the energy shifts coming from SME modifications to the electromagnetic binding potential. We address these three parts in turn.

We work in the context of standard quantum electrodynamics extended by the minimal SME. The appropriate Lagrangian has the form[4]

$$\mathcal{L} = \frac{1}{2} i\bar{\psi}\Gamma^\nu \overleftrightarrow{D_\nu}\psi - \bar{\psi}M\psi, \tag{1}$$

where $D_\nu = \partial_\nu + iqA_\nu$ is the covariant derivative, $F_{\mu\nu} = \partial_\mu A_\nu - \partial_\nu A_\mu$, and Γ^ν and M contain the SME coefficients a^μ, b^μ, $c^{\mu\nu}$, etc.:

$$\begin{aligned}\Gamma^\nu &= \gamma^\nu + \Gamma_1^\nu = \gamma^\nu + c^{\mu\nu}\gamma_\mu + d^{\mu\nu}\gamma_5\gamma_\mu + e^\nu + if^\nu\gamma_5 + \frac{1}{2}g^{\lambda\mu\nu}\sigma_{\lambda\mu}, \\ M &= m + M_1 = m + a^\mu\gamma_\mu + b^\mu\gamma_5\gamma_\mu + \frac{1}{2}H^{\mu\nu}\sigma_{\mu\nu}.\end{aligned} \tag{2}$$

The effective fermion hamiltonian arising from the SME can be written as an expansion in the fermion momentum since the characteristic scale of momentum in a nonrelativistic Coulombic bound state is $p \sim m\alpha$. We will need the expansion up to terms of order $(p/m)^2$. To that order, the effective hamiltonian can be obtained by a simple nonrelativistic expansion of the expectation value of the self-energy operator $\Sigma(p)$:

$$\overline{u}_p \Sigma(p) u_p \to \psi^\dagger H_{\text{eff}}\psi, \tag{3}$$

where $\Sigma(p) \equiv M_1 - \Gamma_1^\mu p_\mu$ comes from the part of $\mathcal{L}$ that is bilinear in fermions. We use the Dirac representation for gamma matrices and the nonrelativistic spinor normalization

$$u_p = \left(\frac{\omega_p + m}{2\omega_p}\right)^{1/2} \begin{pmatrix} \psi \\ \frac{\vec{\sigma}\cdot\vec{p}}{\omega_p + m}\psi \end{pmatrix}, \tag{4}$$

where $\omega_p = (m^2 + \vec{p}^{\,2})^{1/2}$. It is a simple matter now to expand the matrix elements of various Dirac matrices. We find, for example, $\overline{u}_p\sigma^{ij}u_p \approx \epsilon_{ijk}\psi^\dagger\left\{\sigma^k - \vec{\sigma}\cdot\vec{p}p^k/(2m^2)\right\}\psi$. We use the noncovariant normalization

$u_p^\dagger u_p = 1$ so that ψ has the usual nonrelativistic normalization. For the SME contribution to the effective hamiltonian we find

$$H_{\rm eff} = \left(A + B_k\sigma^k\right) + \left(C_i + D_{ik}\sigma^k\right)\frac{p^i}{m} + \left(E_{ij} + F_{ijk}\sigma^k\right)\frac{p^ip^j}{m^2} + \cdots, \quad (5)$$

where $A = a_0 - me_0 - mc_{00}$, $B_k = -b_k + md_{k0} - \frac{1}{2}\epsilon_{kab}\left(mg_{ab0} - H_{ab}\right)$, $E_{ij} = -mc_{ij} - \frac{1}{2}mc_{00}\delta_{ij}$, and $F_{ijk} = \delta_{jk}\tilde{d}_i + \frac{1}{2}\epsilon_{jka}\left(\epsilon_{aip}b_p - 2mg_{a0i} - mg_{ai0}\right)$ with $\tilde{d}_i = md_{0i} + \frac{1}{2}md_{i0} - \frac{1}{4}\epsilon_{iab}H_{ab}$. We do not show the C^i and D^{ik} coefficients because they do not contribute to the final result. The masses and coefficients here are those appropriate to the electron. There is a similar contribution to $H_{\rm eff}$ coming from the proton, but its $O(\alpha^2 \cdot \text{SME})$ contribution is suppressed by a relative factor of the square of the electron to proton mass ratio and is not shown here. This effective hamiltonian agrees with that found earlier using the Foldy-Wouthuysen transformation.[5]

Having obtained the required effective hamiltonian, we are in a position to calculate SME-induced corrections to hydrogen energy levels. We couple angular momenta in the usual way: orbital to electron spin first ($\vec{L}+\vec{S}_e = \vec{J}$), then to proton spin ($\vec{J} + \vec{S}_p = \vec{F}$). States are labeled by quantum numbers n, ℓ, j, f, and m_f: $|n, \ell, j, f, m_f\rangle$. In the presence of the usual perturbations giving rise to fine structure, the Lamb shift, hyperfine splitting, and Zeeman splitting (in the presence of any magnetic field), these states are nondegenerate. We use standard nondegenerate perturbation theory to calculate the effect of $H_{\rm eff}$.

The correction at $O(\alpha^0 \cdot \text{SME})$ is given by $\langle\left(A + B_k\sigma^k\right)\rangle$ (we are including electron effects only). The expectation value of the electron spin matrix is conveniently worked out by use of the Wigner-Ekhart theorem, leading to

$$\Delta E^0_{\rm electron} = A + \xi_e B_3 m_f, \quad (6)$$

where ξ_e is the reduced matrix element $\xi_e = 4(j-\ell)(4j-2f+1)/((2\ell+1)(2j+1))$. (The proton effect is just as easy to calculate but we do not show it here.) This contribution was first obtained by Bluhm *et al.*[2]

There is no correction at $O(\alpha^1 \cdot \text{SME})$ because the expectation value $\langle\vec{p}\rangle$ vanishes in states of definite parity.

At $O(\alpha^2 \cdot \text{SME})$ we need the expectation values of p^ip^j and $p^ip^j\sigma^k$. We found it useful to factor the radial and angular parts: $p^ip^j = p^2\hat{p}^i\hat{p}^j$ and use the fact that the wave functions factorize into radial and angular parts as well. The radial expectation value was done by way of the virial theorem: $\langle p^2\rangle_n = -2mE_n = m^2\alpha^2/n^2$. The angular evaluation was facilitated by the separation of $\hat{p}^i\hat{p}^j$ into parts having $\ell = 2$ and $\ell = 0$: $\hat{p}^i\hat{p}^j = \left(\hat{p}^i\hat{p}^j - \delta_{ij}/3\right) +$

$\delta_{ij}/3$. After a lengthy exercise in angular momentum algebra we obtained our final result for the electron part of the energy correction at this order:

$$\Delta E^2_{\text{electron}} = \Bigg\{ -\frac{5}{6}\tilde{c}_{00} + \frac{(2j+3)(2j-1)}{16j(j+1)}\left(\frac{1}{3} - \frac{m_f^2}{f(f+1)}\right)\tilde{c}_Q + \frac{m_f}{2f+1}\left(\frac{2}{2j+1}\tilde{d}_3 - 2(j-\ell)\tilde{g}_{D3}\right)\Bigg\}\left(\frac{\alpha}{n}\right)^2, \tag{7}$$

where $\tilde{c}_{00} = m\delta_{ij}c_{ij}$, $\tilde{c}_Q = m\delta^Q_{ij}c_{ij}$ with $\delta^Q_{ij} = \delta_{i1}\delta_{j1} + \delta_{i2}\delta_{j2} - 2\delta_{i3}\delta_{j3}$, and $\tilde{g}_{D3} = -b_3 + m(g_{102} - g_{201} + g_{120})$.

Another energy correction at $O(\alpha^2 \cdot \text{SME})$ comes from the photon sector. The photon propagator is corrected by SME interactions, which lead to corrections to the potential energy between the electron and proton. For example, the correction to the potential involving k_F was found by Bailey and Kostelecký:[6] $\Delta V = (\alpha/r)(k_F)_{i0j0}\,(\hat{x}_i\hat{x}_j - \delta_{ij})$. The corresponding energy correction is

$$\Delta E^2_{\text{photon}} = \Bigg\{ -\frac{2}{3}\tilde{\kappa}_0 - \frac{(2j+3)(2j-1)}{16j(j+1)}\left(\frac{1}{3} - \frac{m_f^2}{f(f+1)}\right)\tilde{\kappa}_Q\Bigg\}\left(\frac{\alpha}{n}\right)^2, \tag{8}$$

where $\tilde{\kappa}_0 = m\delta_{ij}(k_F)_{i0j0}$ and $\tilde{\kappa}_Q = m\delta^Q_{ij}(k_F)_{i0j0}$.

An interesting application of our result is to the 2S-1S transition in hydrogen, which has been measured to extreme precision.[7] The SME correction to this energy splitting due to both electron and photon effects is

$$\Delta E(2\text{S} - 1\text{S}) = \left\{\frac{5}{2}\tilde{c}_{00} + 2\tilde{\kappa}_0 - m_f\left(\tilde{d}_3 - \tilde{g}_{D3}\right)\right\}\frac{\alpha^2}{4}. \tag{9}$$

This correction and, of course, experimental input, has been used to set a bound on the diagonal spatial components of the SME tensor $c_{\mu\nu}$.[8]

References

1. M.I. Eides, H. Grotch, and V.A. Shelyuto, *Theory of Light Hydrogenic Bound States*, Springer, Berlin, 2007.
2. R. Bluhm, V.A. Kostelecký, and N. Russell, Phys. Rev. Lett. **82**, 2254 (1999).
3. T.J. Yoder and G.S. Adkins, Phys. Rev. D **86**, 116005 (2012).
4. D. Colladay and V.A. Kostelecký, Phys. Rev. D **55**, 6760 (1997); V.A. Kostelecký, C.D. Lane, and A.G.M. Pickering, Phys. Rev. D **65**, 056006 (2002).
5. V.A. Kostelecký and C.D. Lane, J. Math. Phys. **40**, 6245 (1999).
6. Q.G. Bailey and V.A. Kostelecký, Phys. Rev. D **70**, 076006 (2004).
7. C.G. Parthey *et al.*, Phys. Rev. Lett. **107**, 203001 (2011).
8. B. Altschul, Phys. Rev. D **81**, 041701(R) (2010).

TESTING THE UNIVERSALITY OF FREE FALL WITH A DUAL-SPECIES ATOM INTERFEROMETER ON STE-QUEST

MARKUS KRUTZIK* and ACHIM PETERS

Institute of Physics, Humboldt-University of Berlin
Berlin 12489, Germany
** E-mail: markus.krutzik@physik.hu-berlin.de*

On behalf of the STE-QUEST ATI team

The Space-Time Explorer and Quantum Equivalence Space Test (STE-QUEST) satellite mission is devoted to testing several aspects of General Relativity using an atomic clock and a differential dual-species atom interferometer in space. The latter aims at performing a quantum test of the Einstein equivalence principle in the perigee phase of a highly elliptical Earth orbit by probing the universality of free fall with coherent matter waves. In this paper, we give a brief summary on the mission and the prospects for the dual-species atom interferometer.

1. Introduction

General relativity (GR) is one of the cornerstones of modern physics, a classical theory based on the foundational Einstein equivalence principle (EEP). Although being quite successful in describing phenomena in the macrocosmos, its classical and therefore fundamentally incomplete nature demands an extension to the quantum world. The search for unification of GR with quantum mechanics is an active field in physics, and several attempts or extended theories addressing this problem predict violations of the basic principles of GR. A satellite mission dedicated to conceive different aspects of EEP, observing new boundaries and limits in given frameworks (e.g., the Standard-Model Extension (SME)) or even violations will help to complete our understanding of the interplay between quantum physics and gravity — an important step towards the unification of all fundamental forces of nature.

2. STE-QUEST mission outline

The STE-QUEST mission proposal is a response to the call for medium-size (M-class) missions by ESA's Cosmic Vision plan (2015-2025). It was recommended by the ESA advisory structure and finally selected for an assessment study (phase A), which is currently ongoing. The proposed satellite will operate in an highly elliptical orbit, host an advanced cesium atomic clock (ATC) and a dual-species atom interferometer (ATI) operated with two different isotopes of rubidium. Its scientific objectives mainly focus on testing two fundamental aspects of the EEP with unprecedented precision. Using the clock on the satellite in comparison with a highly accurate ground-based network of clocks enables Earth and Sun gravitational redshift tests with fractional uncertainties of 1×10^{-7} and 2×10^{-6}, respectively.

The measurement of the differential acceleration between two different atomic isotopes (^{87}Rb and ^{87}Rb) is carried out by a dual-species atom interferometer setup. As a source for coherent matter waves, Bose-Einstein condensates will be generated in a hybrid trap setup, consisting of an atom chip and a crossed optical dipole trap, and simultaneously interrogated by a symmetric interferometer sequence. Detecting the phase shift of interfering matter waves within the gravitational field of the Earth will allow for a precision measurement of their propagation.[1,2] Here, the proposed mission duration of 5 years enables a determination of the Eötvös ratio,

$$\eta = \frac{|a_{\mathrm{Rb}_{87}} - a_{\mathrm{Rb}_{85}}|}{g}, \tag{1}$$

being the normalized differential acceleration of the two atomic species ^{87}Rb and ^{85}Rb to an accuracy of 1.5×10^{-15}. This is beyond state-of-the-art precision achieved by lunar laser ranging[3,4] and torsion balances.[5]

3. Dual-species atom interferometer payload

The dual-species atom interferometer instrument can be subdivided into a physics package, a laser system, and an electronics module. See Fig. 1. Including 20% component contingency, a total mass budget below 300 kg, a peak power level below 1 kW, and a telemetry budget of 110 kbps are allocated.

The physics package consists of the atomic source, an ultrahigh vacuum science chamber operated at a 10^{-11} mbar level, the detection unit for fluorescence and absorption imaging of the atomic clouds, the vacuum pump system and a multilayer magnetic shield. The science chamber is made from titanium, features a dodecagon design housing the three-layer atom

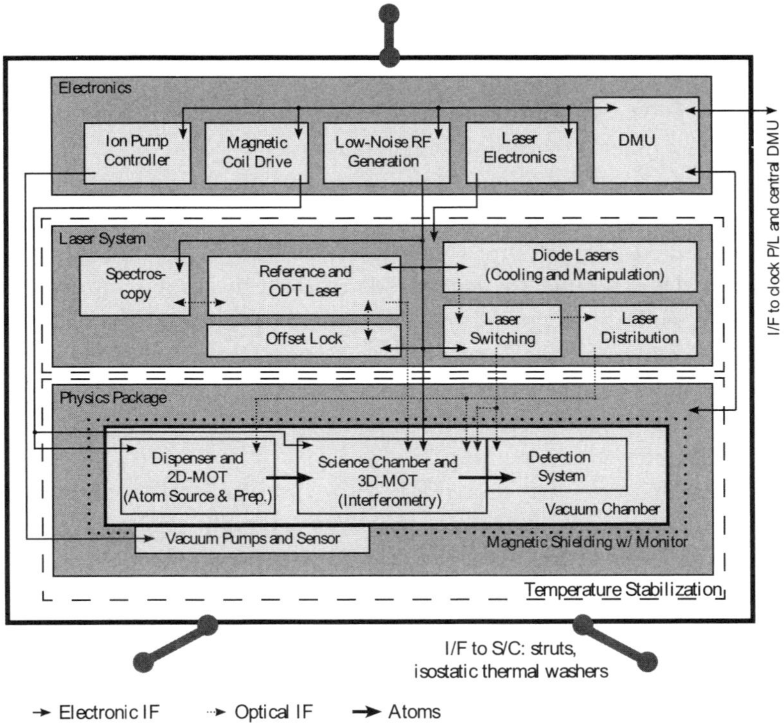

Fig. 1. Schematic of the STE-QUEST dual-species atom interferometer payload. Shown here are the main subdivisions electronics module, laser system, and physics package with the corresponding electrical and optical interfaces, respectively.

chip and allows for a 12 cm interferometer baseline. Our chip-based 3D magneto-optical trap (MOT) is loaded via a high-flux 2D MOT setup with an oven-type atom source based on a heated and controllable reservoir. It uses two counter-propagating laser beams parallel to the atom chip and two additional laser beam reflected by the chip surface with an angle of 45°. Miniaturized atom chip devices greatly simplify the cold atom apparatus and are mainly used to generate a conservative trapping potential with only modest current requirements for fast (pre-)evaporation of the atomic samples. The center of a crossed optical dipole trap is superimposed with the conservative magnetic trap provided by the atom chip, and by simply lowering the optical power of the two beams during the final evaporation step, both atomic species can be simultaneously cooled down to degeneracy. Necessary magnetic offset fields are created using three pairs of coils in Helmholtz configuration.

As external magnetic stray fields have to be suppressed by a factor $> 10,000$, a three layer mu-metal shield enclosing the physics package is foreseen. The magnetic shield needs to be capable of withstanding internal fields up to 160 G without permanent damage. Field monitoring in general is realized by precise space-borne magnetometers.

The laser system combines a telecom technology based frequency reference and optical dipole trap laser with micro-integrated, high power diode laser modules at $\lambda = 780$ nm. Laser beam manipulation and switching according to the experimental sequence is realized in Zerodur optical bench technology while fiber optical splitter systems provide the capability for precise distribution of the laser light. The main advantages here are small and robust units with well-defined interfaces. These allow for improvements within, or even replacement of, a single module whenever it is necessary without affecting adjacent units.

A digital management unit (DMU) to control all electronic devices is based on instrument control units (ICU), which additionally supports the spacecraft interface, CPU, detector capture, and housekeeping data gathering. To fulfill stringent requirements on the controllability of the magnetic fields, low noise current drivers for magnetic field generation are forseen. A low noise RF generator provides microwave signals at 3 GHz and 6.8 GHz respectively, as well as a stable 100 MHz reference. These precise signals are, for example, necessary for RF-knife techniques to lower the temperature of the atoms during pre-evaporation, to drive electro-optical components as well as beeing offset signals for laser modulation or phase comparison. All laser sources require low noise current supplies with a high bandwidth control loop and temperature controllers to maintain sufficient frequency stability.

4. Experimental sequence, sensitivity and error sources

A measurement cycle starts by simultanously generating two degenerate samples of ^{87}Rb and ^{85}Rb in a hybrid trap realized by an miniaturized atom chip and a crossed optical dipole trap, the latter used for the very last evaporation step and a controlled release of the condensates.

Each cloud of atom contains $N_{87} = N_{85} = 10^6$ atoms, whose effective temperature can be reduced to a few tens of pK with an additional delta-kick cooling step after a short free expansion time. The characteristics of a quantum gas mixture are strongly influenced by intra- and interspecies interactions, which depend on the magnitude and sign of each coupling strength or scattering length, respectively. Feshbach resonances allow the scattering length to be tuned and therefore ensure stable and miscible regimes for the experiment.

The prepared matter wave samples of both isotopes will then be simultaneoulsy interrogated by a double diffraction interferometer (DDI)[6] with a free evolution time of $T_{\mathrm{DDI}} = 5$ s between two succesive pulses, going beyond the capability of Earth-bound microgravity platforms such as zero-g airplanes[7] and drop towers.[8–10] The choice of different rubidium isotopes as test bodies allow for sufficient matching of the interferometer scaling factors and high common-mode surpression. Degenerate gases feature inherently low velocity spreads, which are required for high contrast measurements on these interrogation time scales. After closing the interferometer, both interference signals will be subsequently read out leading to a total cycle time of $T_{\mathrm{total}} = 20$ s. Anticipating shot noise limited detection, the sensitivity to differential accelerations is

$$\sigma_{\Delta a} = \frac{1}{C}\sqrt{\frac{2}{N}}\frac{1}{kT^2}. \tag{2}$$

With an effective wave vector of $k = 8\pi/780$ nm and an assumed contrast of $C = 0.6$, the resulting sensitivity is

$$\sigma_{\Delta a}^{\mathrm{cycle}} = 3 \times 10^{-12}\ \mathrm{m/s^2} \tag{3}$$

per experiment cycle, and can ideally be integrated down to

$$\sigma_{\Delta a}^{\mathrm{orbit}} = 5 \times 10^{-14}\ \mathrm{m/s^2} \tag{4}$$

per orbit. A number of $\sim$2000 orbits is then necessary to reach the targeted sensitivity at a level of 1.5×10^{-15}.

Systematic effects and error sources need to be analysed, e.g., the influence of gravity gradients, Coriolis accelerations and other inertial terms depending on position and velocity of the atoms, photon recoil effects and wave fronts of the interferometer lasers, static magnetic fields, spurious accelerations and the detection efficiency. These are currently under thorough investigation.

Acknowledgments

The STE-QUEST ATI consortium consists of European groups active in the fields of atom interferometery and space sciences. This instrument team was founded to respond to the call for proposals for an M3 mission within the Cosmic Vision program of ESA (http://sci.esa.int/ste-quest). The present work is supported by the German Space Agency DLR with funds provided by the Federal Ministry of Economics and Technology under grant numbers DLR 50OY1204 and 50OY1304.

References

1. M. Kasevich, and S. Chu, Appl. Phys. B **54**, 325375 (1992).
2. A. Peters, K.Y. Chung, and S. Chu, Nature **400**, 23655 (1999).
3. J. Müller, F. Hoffmann, and L. Biskupek, Class. Quantum Grav. **29**, 184006 (2012).
4. J.G. Williams, S.G. Turyshev, and D.H. Boggs, Class. Quantum Grav. **29**, 184004 (2012).
5. T.A. Wagner, S. Schlamminger, J.H. Gundlach, and E.G. Adelberger, Class. Quantum Grav. **29**, 184002 (2012).
6. T. Lévèque, A. Gauguet, F. Michaud, F. Pereira Dos Santos, and A. Landragin, Phys. Rev. Lett. **103**, 080405 (2009).
7. R. Geiger *et al.*, Nature Comm. **2**, 101038 (2011).
8. T. von Zoest *et al.*, Science **328**, 184006 (2010).
9. J. Rudolph *et al.*, Microgravity Sci. Technol. **3**, 12217 (2011).
10. H. Müntinga *et al.*, Phys. Rev. Lett. **110**, 093602 (2013).

TEST OF LORENTZ INVARIANCE WITH A RB-^{21}NE COMAGNETOMETER AT THE SOUTH POLE

M.A. SMICIKLAS* and M.V. ROMALIS

Physics Department, Princeton University
Princeton, NJ 08544, USA
**E-mail: msmicikl@princeton.edu*

Atomic spin comagnetometers are among the most sensitive devices for testing Lorentz symmetry of fermions. In Princeton, we have used our rotating comagnetometer to set the most stringent limits on CPT-odd and CPT-even Lorentz violating effects in neutrons. However, gyroscopic pickup of the Earth's rotation represents a significant systematic effect limiting sensitivity. To suppress this systematic, we have installed a Rb-^{21}Ne comagnetometer at the Amundsen-Scott South Pole Station with data collection being performed over the course of the austral winter.

1. Introduction

In this experiment, we use polarized atomic spins to test very sensitively Lorentz and CPT symmetry as described in the Standard-Model Extension (SME).[1] A comagnetometer consisting of a polarized alkali metal and noble gas ensemble is used to detect anomalous spin couplings while suppressing magnetic field sensitivity. The entire experimental apparatus is mounted on a rotatable platform allowing frequent reversals of the measured signal over relatively short time scales ($\sim$0.3 Hz), thereby limiting the effects of $1/f$ noise. This setup has previously been used to constrain CPT-odd and Lorentz violating vector coupling to the neutron using a ^{3}He-K comagnetometer.[2]

See Ref. 3 for a detailed description of the experimental setup and underlying physics. Briefly, alkali atoms in a small vapor cell are continuously optically pumped by a circularly polarized pump laser on resonance. Angular momentum is transfered by spin exchange collisions to polarize the nucleus of the noble gas. An externally applied magnetic compensation field cancels the magnetization of the polarized atoms and couples the spins of the two species. This coupling produces fast transient response of the atoms necessary for quick measurements, as well as suppression of magnetic field sensitivity. The alkali atom's polarization state is analyzed by observing Faraday rotation of a linearly polarized off-resonance probe laser, sent in orthogonal to the pump laser. See Fig. 1.

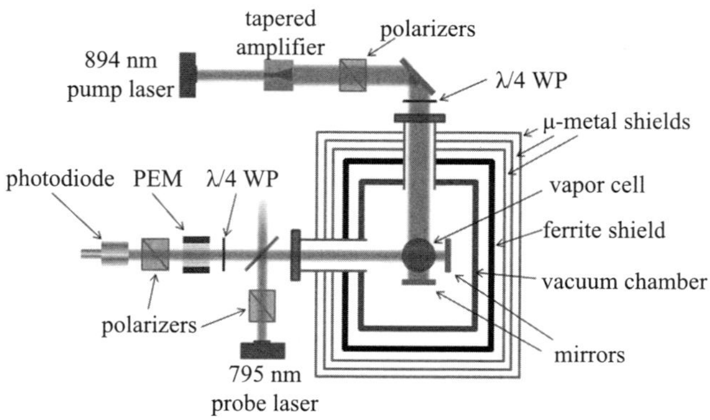

Fig. 1. Schematic diagram of the comagnetometer optical layout.

2. New tensor limit using a ^{21}Ne-Rb-K comagnetometer

An anomalous quadrupole energy shift has implications pertaining to the maximal attainable velocity (MAV) for a massive particle.[4] For this, we focus on the CPT-even c tensor components in the SME Lagrangian:

$$\mathcal{L} = -\overline{\psi} m \psi + \frac{1}{2} i \overline{\psi} (\gamma_\nu + c_{\mu\nu}\gamma^\mu) \overleftrightarrow{\partial}^\nu \psi. \quad (1)$$

Useful constraints can be determined by comparing the kinetic energy of a valence nucleon to the quadrupole energy sensitivity of the experiment. This requires a nonzero orbital angular momentum, or in other words, a nucleus with spin greater than 1/2. Previous experiments using ^{201}Hg and ^{21}Ne have been used to constrain MAV.[5,6]

The spin 1/2 ^{3}He was replaced with spin 3/2 ^{21}Ne to gain sensitivity to quadrupole energy shifts. This also has the benefit of increasing sensitivity to anomalous fields by a factor of 10 due to ^{21}Ne having a factor of 10 smaller gyromagnetic ratio than ^{3}He. For the alkali, K was replaced with Rb and operated at 10 times higher density to improve spin exchange efficiency with ^{21}Ne and overcome the extra quadrupole relaxation.[7] To polarize uniformly the optically thick Rb atoms, a hybrid pumping scheme is employed by optically pumping a much lower density of K which then transfers angular momentum to the Rb through spin exchange collisions.[8]

Data collection using ^{21}Ne-Rb-K was conducted in Princeton over the course of about two months, achieving ~1 nHz tensor frequency shift resolution. The experimental procedure was almost identical to that of ^{3}He-K, except that signals were measured at the 1st and 2nd harmonic of Earth's rotation rate. Assuming a Schmidt nucleon wave function and a kinetic energy of the valence nucleon ~5 MeV, 4 out of 5 spatial tensor components $c_{\mu\nu}$ for the neutron were measured as[9]

$$c_X = c^n_{YZ} + c^n_{ZY} = (4.8 \pm 4.4) \times 10^{-29}, \tag{2}$$

$$c_Y = c^n_{XZ} + c^n_{ZX} = -(2.8 \pm 3.4) \times 10^{-29}, \tag{3}$$

$$c_Z = c^n_{XY} + c^n_{YX} = -(1.2 \pm 1.4) \times 10^{-29}, \tag{4}$$

$$c_- = c^n_{XX} - c^n_{YY} = (1.4 \pm 1.7) \times 10^{-29}. \tag{5}$$

Improved calculations of the ^{21}Ne nuclear structure by Alex Brown *et al.* indicate enhanced sensitivity due to higher nucleon kinetic energy, as well as similar sensitivity for both the neutron and proton.[10]

3. South Pole Lorentz invariance test

Gyroscopic pick-up of Earth's rotation represents a major systematic for spatial anisotropy measurements using an atomic spin comagnetometer. In Princeton, the signal from Earth's rotation is ~10,000 times larger than the measurement resolution for the published Lorentz violation results. Calibration uncertainties and pointing accuracy limit significant improvements in precision. Careful leveling of the apparatus is also required to prevent mechanical induced signals, e.g., laser position changing due to optical breadboard strain. Operating the experiment at the South Pole effectively eliminates the large background rotation signal since Earth's axis and gravity are parallel. Two to three orders of magnitude improvement in precision is expected.

To improve performance and reliability, a number of changes were incorporated into the apparatus before moving to the South Pole, i.e., upgrading to a precision air bearing table, adding remote leveling controls for the apparatus, replacing K with Cs for hybrid pumping to reduce the light shift on Rb, installing mirrors after the cell for double pass pump and probe to suppress laser motion, among others. The apparatus was installed at the Amundsen-Scott South Pole station during the 2012-2013 austral summer. At its current location of ~230 meters from the geographic South Pole, Earth's rotation signal is suppressed to less than 1 nHz. As a preliminary test, one day of data was fit to a sidereal and semi-sidereal period which yielded a measurement resolution of ~1 nHz, roughly equal to the total resolution our previously published results. See Fig. 2.

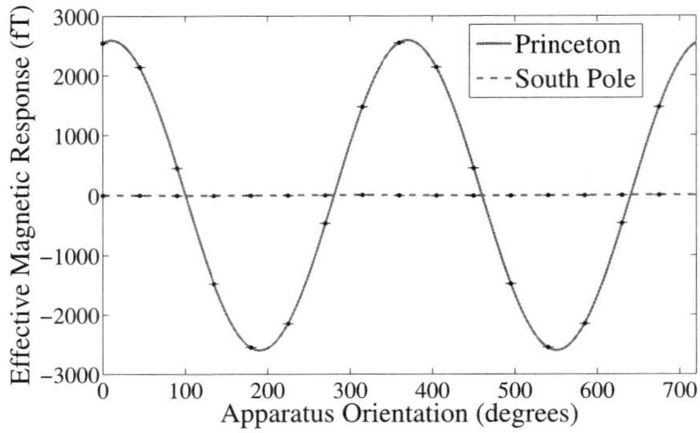

Fig. 2. Comparison of Earth rotation signal in Princeton and at the South Pole.

Data collection continues over the course of the 2013 austral winter. Future plans include returning to the South Pole during the 2013-2014 austral summer to implement improvements to automation and various environmental controls.

Acknowledgments

We thank the United States Antarctic Program and especially research associate Andrew Vernaza for their work at the South Pole. This project is funded by the NSF Division of Polar Programs under award PLR-1142032.

References

1. V.A. Kostelecký and C.D. Lane, Phys. Rev. D **60**, 116010 (1999).
2. J.M. Brown, S.J. Smullin, T.W. Kornack, and M.V. Romalis, Phys. Rev. Lett. **105**, 151604 (2010).
3. J.M. Brown, Ph.D. dissertation, Princeton University (2011).
4. S. Coleman and S.L. Glashow, Phys. Rev. D **59**, 116008 (1999).
5. S.K. Lamoreaux, J.P. Jacobs, B.R. Heckel, F.J. Raab, and E.N. Fortson, Phys. Rev. Lett. **57**, 3125 (1986).
6. T.E. Chupp *et al.*, Phys. Rev. Lett. **63**, 1541 (1989).
7. R.K. Ghosh and M.V. Romalis, Phys. Rev. A **81**, 043415 (2010).
8. M.V. Romalis, Phys. Rev. Lett. **105**, 243001 (2010).
9. M. Smiciklas, J.M. Brown, L.W. Cheuk, S.J. Smullin, and M.V. Romalis, Phys. Rev. Lett. **107**, 171604 (2011).
10. A. Brown *et al.*, in preparation.

SENSITIVITY OF ATMOSPHERIC NEUTRINOS IN SUPER-KAMIOKANDE TO LORENTZ VIOLATION

T. AKIRI

Physics Department, Duke University
Durham, NC, 27705, USA
E-mail: tarek.akiri@duke.edu

On behalf of the Super-Kamiokande Collaboration

This talk, given at CPT'13, showed Super-Kamiokande atmospheric-neutrino Monte Carlo sensitivity to Lorentz-violation effects using the perturbative model derived from the Standard-Model Extension.

1. Introduction

The Standard-Model Extension[1] (SME) is an effective field theory having all the features of the Standard Model but adding all possible Lorentz-violating terms. Recently, many experiments have been using this framework to test Lorentz invariance. Neutrino oscillations, as an interferometric effect, are a very sensitive probe for Lorentz violation (LV) effects expected to manifest around the Planck scale. Many neutrino oscillation experiments used either the short-baseline approximation or the perturbative model to search for sidereal variations in their data constraining the neutrino LV coefficients.[2]

Super-Kamiokande[3] (SK) is an underground 50 kT water Cherenkov detector located in Kamioka (Japan). Its innermost volume is instrumented with 11146 20" photomultiplier tubes (PMTs) that allow the reconstruction of neutrino interaction features based on the time and charge of the hit PMTs. In 1998, the analysis of SK atmospheric-neutrino data proved the neutrino oscillation phenomenon[4] through the disappearance of $\nu_\mu/\bar{\nu}_\mu$ and the non-appearance of $\nu_e/\bar{\nu}_e$. Using SK atmospheric-neutrino Monte Carlo (MC) and the SME perturbative model,[5] we performed a sensitivity study for isotropic Lorentz-violation effects.

2. The perturbative model

The perturbative model is derived from the SME using time-dependent perturbation theory. The LV hamiltonian is derived up to second order in the perturbative series for both $\nu \to \nu$ and $\nu \to \bar{\nu}$ oscillations. In this study, we restricted ourselves to $\nu \to \nu$ oscillations leaving two effective sets of coefficients: a_{eff} (CPT-odd) and c_{eff} (CPT-even), henceforth denoted as a and c, respectively. Furthermore, we choose to consider only the isotropic and renormalizable part that leads to spectral distortions. The perturbative hamiltonian has then the following form:

$$\delta h = \frac{1}{|p|} \begin{pmatrix} a_{ee} - c_{ee} & a_{e\mu} - c_{e\mu} & a_{e\tau} - c_{e\tau} \\ a^*_{e\mu} - c^*_{e\mu} & a_{\mu\mu} - c_{\mu\mu} & a_{\mu\tau} - c_{\mu\tau} \\ a^*_{e\tau} - c^*_{e\tau} & a^*_{\mu\tau} - c^*_{\mu\tau} & a_{\tau\tau} - c_{\tau\tau} \end{pmatrix}.$$

LV effects can be easily described by considering ν_μ disappearance at the first order in the two-flavor case:

$$P_{LV}(\nu_\mu \to \nu_\mu) = \sin(2.534 \times \frac{\Delta m^2 L}{E}) \times (\Re e(c_{\mu\tau})LE - \Re e(a_{\mu\tau})L), \quad (1)$$

with Δm^2 being the atmospheric mass splitting in eV^2, L the neutrino path-length in km and E its energy in GeV. The LV coefficients $a_{\mu\tau}$ expressed in km^{-1} and $c_{\mu\tau}$ in km^{-1}GeV^{-1} are complex.

Equation (1) shows that $a_{\mu\tau}$ and $c_{\mu\tau}$ control oscillations proportional to L and $L \times E$ respectively, each with opposite signs. In the two-flavor case, only the real parts of the LV coefficients are involved. Calculating the probabilities in the three-flavor case implies the imaginary parts as well. For a given value of the LV coefficients at first order in the perturbative series, the imaginary parts gives much smaller probabilities than the real parts. In our analysis, we extended the LV calculation up to the second order since the latter is expected to be the dominant contribution in the no-oscillation region corresponding to high energy in SK. Indeed, looking at the sine term in Eq. (1), one can see that at high energy the first order probability is suppressed in contrast to the second order (see Fig. 1). Moreover, the probability for the real and imaginary parts is similar for the second order.

3. Analysis

In its three-flavor oscillation analysis, SK divides its data into many different categories based on the event topology and the event reconstructed features.[6] The cosine of the zenith angle, related to the neutrino path-length, and the energy are used to further divide each category in zenith

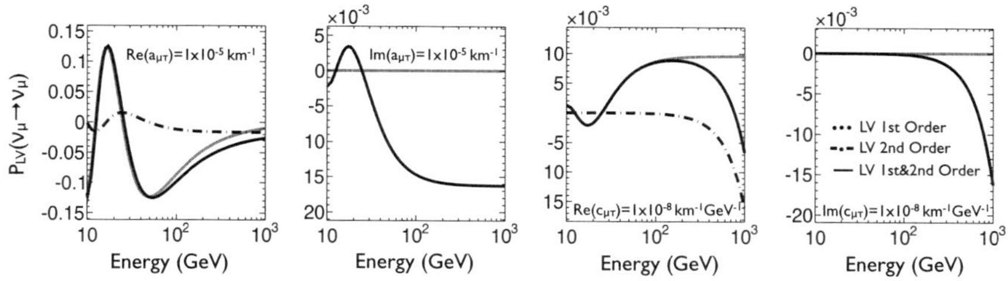

Fig. 1. Illustration of the first (dotted line), second (dashed line) and first plus second order (solid line) LV $\nu_\mu \to \nu_\mu$ oscillation probabilities for the four $\mu\tau$ coefficients taken individually as a function of energy for $L \simeq$12800 km.

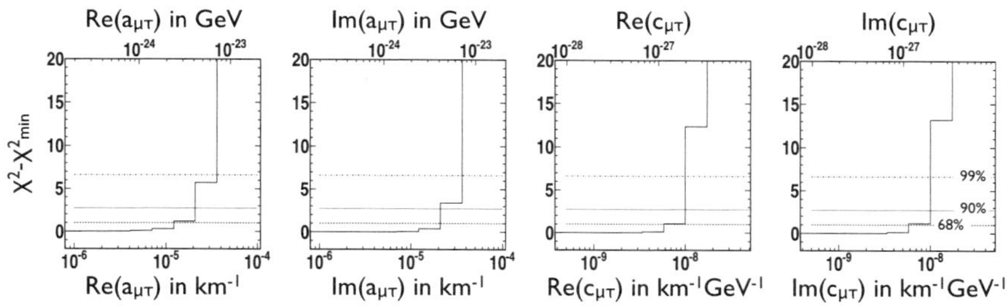

Fig. 2. $\Delta\chi^2$ as a function of the LV coefficient value tested along with the 68%, 90% and 99% C.L. lines. From left to right: Re$(a_{\mu\tau})$, Im$(a_{\mu\tau})$, Re$(c_{\mu\tau})$ and Im$(c_{\mu\tau})$.

angles and momentum bins for a total of 480 analysis bins. SK has many systematic uncertainties arising mainly from the neutrino flux and interaction as well as from detection effects. In this study, except for the CP phase δ, we also considered systematic errors associated with oscillation parameters: $\Delta m^2_{21} = 7.46 \times 10^{-5}$ eV^2, $\Delta m^2_{32} = 2.44 \times 10^{-3}$ eV^2, $\sin^2\theta_{12} = 0.32$, $\sin^2\theta_{23} = 0.5$ and $\sin^2\theta_{13} = 0.0251$ for a total of 159 systematic errors. δ has been taken as a parameter because it causes both the period and amplitude of the oscillations to change, even switching sign. We use a poissonian χ^2 that is minimized iteratively at each LV coefficient value tested. The sensitivity for each of the LV coefficients tested individually is shown in Fig. 2. The sensitivity obtained at 90% C.L. is:

- Re$(a_{\mu\tau}) < 4.1 \times 10^{-24}$ GeV,
- Im$(a_{\mu\tau}) < 5.1 \times 10^{-24}$ GeV,
- Re$(c_{\mu\tau}) < 1.7 \times 10^{-27}$,
- Im$(c_{\mu\tau}) < 1.7 \times 10^{-27}$.

Note that the sensitivity is comparable for the real and imaginary parts of both coefficients. This arises from the fact that the sensitivity comes from the highest-energy event categories where the second order LV effects are the most important. These results show that SK is extremely sensitive to LV. These sensitivities are respectively four and eight orders of magnitude better than the best limits on the isotropic coefficients a and c in the neutrino sector while comparable to the ones on the anisotropic coefficients.[7]

These results were obtained using the perturbative model that requires $|\delta h| \ll 1/L$, which we translated into $|\delta h| \leq 10\%/L$. In the sensitivity study for a and c, it appeared that, respectively, 36.1% and 1.7% of the events used in the analysis did not satisfy this perturbative criterion. These events correspond to the longest distances and highest energies, which means that cutting them will result in a loss of sensitivity. Moreover, the distance in SK is not accurately known on an event by event basis while the energies above ~10 GeV cannot be measured. It therefore appears that in contrast to beam experiments that have a fixed distance and given energy, the perturbative model is not suitable for SK atmospheric neutrinos. In the future, we intend to perform an improved analysis by using the full SME. Furthermore, neutrino oscillation experiments using either the short-baseline approximation or the perturbative model to look for LV effects should keep in mind that the perturbative criterion does not allow the calculation of LV expectation above it. In such cases, experiments should report their results in terms of a band of LV values excluded rather than simple limits.

References

1. V.A. Kostelecký and M. Mewes, Phys. Rev. D **85**, 096005 (2012).
2. LSND Collaboration, L.B. Auerbach *et al.*, Phys. Rev. D **72**, 076004 (2005); MINOS Collaboration, P. Adamson *et al.*, Phys. Rev. Lett. **101**, 151601 (2008); Phys. Rev. Lett. **105**, 151601 (2010); Phys. Rev. D **85**, 031101 (2012); MiniBooNE Collaboration, A.A. Aguilar-Arevalo *et al.*, Phys. Lett. B **718** (2013); IceCube Collaboration, R. Abbasi *et al.*, Phys. Rev. D **82**, 112003 (2010).
3. Super-Kamiokande Collaboration, S. Fukuda *et al.*, NIM A **501**, 418-462 (2003).
4. Super-Kamiokande Collaboration, Y. Fukuda *et al.*, Phys. Rev. Lett. **81**, 1562 (1998).
5. J.S. Díaz *et al.*, Phys. Rev. D **80**, 076007 (2009).
6. Super-Kamiokande Collaboration, R. Wendell *et al.*, Phys. Rev. D **81**, 092004 (2010).
7. *Data Tables for Lorentz and CPT Violation,* V.A. Kostelecký and N. Russell, 2013 edition, arXiv:0801.0287v6.

SEARCH FOR CPT AND LORENTZ SYMMETRY VIOLATION IN NEUTRAL KAONS AT KLOE/KLOE-2

ANTONIO DE SANTIS

Dip. di Fisica, 'Sapienza' Università di Roma and I.N.F.N. Sezione di Roma
P.le A. Moro, 2, I-00185 Rome, Italy
E-mail: antonio.desantis@roma1.infn.it

On behalf of the KLOE-2 Collaboration

The KLOE experiment at the DAΦNE ϕ-factory of the INFN Frascati Laboratory collected data corresponding to 2.5 fb^{-1} of integrated luminosity. A new approach to the analysis of $\phi \to K_S K_L \to \pi^+\pi^-, \pi^+\pi^-$ events has been adopted allowing us to independently measure all four CPT violating parameters Δa_μ appearing for neutral kaons in the SME. The final KLOE results on Δa_μ are presented: $\Delta a_0 = (-6.0 \pm 7.7_{stat} \pm 3.1_{syst}) \times 10^{-18}$ GeV, $\Delta a_X = (0.9 \pm 1.5_{stat} \pm 0.6_{syst}) \times 10^{-18}$ GeV, $\Delta a_Y = (-2.0 \pm 1.5_{stat} \pm 0.5_{syst}) \times 10^{-18}$ GeV, $\Delta a_Z = (3.1 \pm 1.7_{stat} \pm 0.5_{syst}) \times 10^{-18}$ GeV. The KLOE-2 experiment is going to start a new data taking campaign at DAΦNE upgraded in luminosity, allowing to reach a sensitivity of $\mathcal{O}(10^{-19})$ GeV for all Δa_μ parameters.

1. The KLOE experiment

The KLOE experiment operates at DAΦNE, the Frascati ϕ–factory. DAΦNE is an e^+e^- collider running at a center of mass energy of ~1020 MeV, the mass of the ϕ meson. Positron and electron beams collide at an angle of $\pi - 25$ mrad, producing ϕ mesons with small momentum in the orbit plane ($p_x(\phi) \sim -15$ MeV).

The KLOE detector consists of a large cylindrical drift chamber[1] surrounded by a lead-scintillating fiber electromagnetic calorimeter.[2] A superconducting coil around the EMC provides a 0.52 T axial field.

At DAΦNE the ϕ is produced almost at rest in the center of the KLOE detector and decay mostly in kaon pairs: 34% of decays are neutral kaons. The initial state of the kaon pair is produced with quantum numbers $J^{PC} = 1^{--}$. The two kaons are produced in an antisymmetric correlated state. Labeling f_1 and f_2 the decay channels for the two kaons, the time evolution of the initial state decaying into $|f_1, f_2\rangle$ final state is expressed as a function of the difference of proper decay times ($\Delta\tau = \tau_2 - \tau_1$) as:

$$I_{f_1 f_2}(\Delta\tau) \propto e^{-\Gamma|\Delta\tau|}\Big[|\eta_{f_1}|^2 e^{\frac{\Delta\Gamma}{2}\Delta\tau} + |\eta_{f_2}|^2 e^{-\frac{\Delta\Gamma}{2}\Delta\tau} - 2\Re e\Big(\eta_{f_1}\eta^*_{f_2} e^{-i\Delta m\Delta\tau}\Big)\Big], \tag{1}$$

where $\eta_{f_j} = \langle f_j|K_L\rangle/\langle f_j|K_S\rangle$, $\Gamma = \Gamma_S + \Gamma_L$ and $\Delta\Gamma = \Gamma_S - \Gamma_L$. Equation (1) shows a time interference term characteristic of the type of correlation firstly pointed out by Einstein, Podolsky and Rosen.[3]

Assuming equal final states for the two decaying kaons ($f_1 = f_2 = f$) a fully destructive interference is expected for equal decay times ($|\Delta\tau| = 0$). For $f_j = \pi^+\pi^-$ the resulting ratio of amplitudes (η_j) becomes:

$$\eta_j = \eta_{\pi^+\pi^-} = \frac{\langle\pi^+\pi^-|T|K_L\rangle}{\langle\pi^+\pi^-|T|K_S\rangle} \simeq \varepsilon_K + \varepsilon' - \delta_K. \tag{2}$$

In the Standard-Model Extension (SME) framework developed by A. Kostelecký[4] the δ_K parameter is not constant and is expected to have the following structure:

$$\delta_K \approx i\sin\phi_{SW} e^{i\phi_{SW}}\gamma_K(\Delta a_0 - \vec{\beta}_K\cdot\Delta\vec{a})/\Delta m, \tag{3}$$

where γ_K and β_K are the usual Lorentz factors for the kaon, ϕ_{SW} is the superweak phase and Δa_μ are the SME coefficients for the kaon system.

Equation (3) shows that δ_K is modulated by the kaon momentum modulus (γ_K and $|\vec{\beta}_K|$) and by its spatial direction ($\vec{\beta}_K$). In the KLOE case the two kaons are produced almost back-to-back in the ϕ decay and therefore evolve with two different δ_K ($\delta_K(\vec{P}_1) \neq \delta_K(\vec{P}_2)$). Additional angular dependence in Eqs. (1) through (3) is induced by the Earth motion (sidereal time variation) and residual ϕ momentum in the laboratory frame. The effect produced by CPT violation can be observed in the distribution of Eq. (1) provided that the two kaon final states are tagged with respect to some reference direction in the laboratory frame and taking into account the proper coordinate trasformation in the privileged reference frame where Eq. (3) holds in this form (see Eq. (14) of reference[4]).

2. Data analysis

The data reduction starts with the topological identification of the candidate signal events: two vertices with two tracks each. For each vertex the same list of kinematical selection criteria is applied in order to identify a clean and unbiased sample of $K \to \pi\pi$. The background contamination after these selections is 1.5%, as estimated from Monte Carlo. The background events are mainly due to kaon regeneration. This background is irreducible because it is made of regenerated kaons moving in the forward direction (in the limit of the angular resolution).

The data distributions have been fitted with the distribution (1) including the SME effects. The $\Delta\tau$ range is: $\Delta\tau \in [-12 : 12]\tau_S$. This choice is to limit the perturbation on the result by the regeneration on the spherical beam pipe.[5] Four sidereal time bins and two angular selection have been used for a grand total of 184 experimental points simultaneously fitted. See Fig. 1.

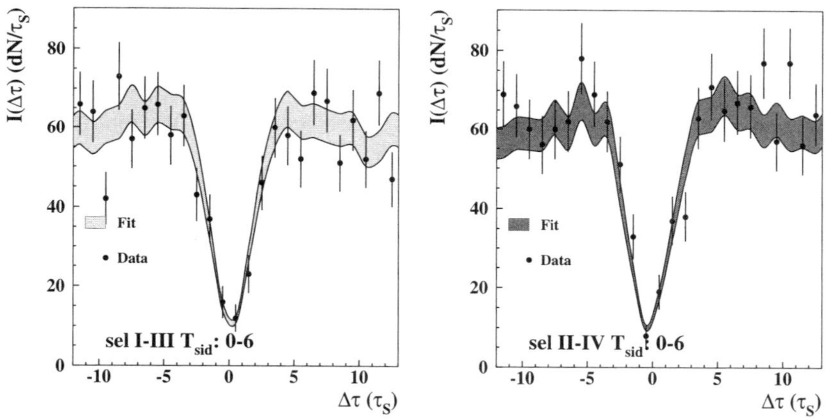

Fig. 1. Distribution of $I_{\pm}(\Delta\tau)$ and fit results for different angular selection. Black points are for data while shaded bands are the fit output. The errors on data are purely statistical, while the width of the fit result band represents the contribution to the uncertainty from Monte Carlo statistics and efficiency corrections.

The final results on the Δa_μ parameters are reported in Table 1 where errors include all the statistical fluctuations. The expected sensitivity has been reached. The total error is fully dominated by the statistical uncertainty.

Table 1. Fit results (units are 10^{-18} GeV).

χ^2/dof= 211.7/184 $\Rightarrow P(\chi^2) = 8\%$.	
$\Delta a_0 = (-6.0 \pm 7.7_{stat} \pm 3.1_{syst})$	$\Delta a_X = (\ 0.9 \pm 1.5_{stat} \pm 0.6_{syst})$
$\Delta a_Y = (-2.0 \pm 1.5_{stat} \pm 0.5_{syst})$	$\Delta a_Z = (\ 3.1 \pm 1.7_{stat} \pm 0.6_{syst})$

3. Conclusions and future plans

The continuation of the KLOE physics program with KLOE-2[6] at an improved DAΦNE machine is starting with a new beam interaction scheme[7] and with the inclusion of two pairs of electron-positron taggers for the study of the gamma-gamma physics: Low Energy Tagger[8] inside KLOE apparatus and High Energy Tagger[9] along the beam lines outside the KLOE detector. Several other upgrades for the detector are going to be installed: a pair of crystal calorimeters (CCALT)[10] near the interaction region to improve the angular acceptance for low-θ particles; a pair of tile calorimeters (QCALT),[11] covering the quadrupoles inside the KLOE detector and along the beam pipe, made of tungsten slabs and singly read-out scintillator tiles to improve the angular coverage for particles coming from the active volume of the DC; and a small and light inner tracker (IT)[12] made of four planes of cylindrical GEM to improve the resolution of the vertex reconstruction around the interaction point and to increase the low-θ acceptance for charged particles.

The results presented in Table 1 are expected to be improved with the KLOE-2 data taking campaign. The sensitivity of CPT and Lorentz invariance tests will improve both by the increased statistics and by the new interaction region with IT. The IT will improve the resolution on the vertex position and the acceptance for low-θ tracks, the IR will imply a larger beam crossing angle (from 25 to 60 mrad) that will enhance the effect of asymmetry between the two kaons. The expected sensitivity should increase to 10^{-19} GeV for all the Δa_μ parameters.

References

1. M. Adinolfi *et al.*, Nucl. Inst. Meth. A 488 (2002) 51.
2. M. Adinolfi *et al.*, Nucl. Inst. Meth. A 482 (2002) 363.
3. A. Einstein, B. Podolsky and N. Rosen, Phys. Rev. **47**, 777 (1935).
4. V.A. Kostelecký, Phys. Rev. D **64** (2001) 076001.
5. F. Ambrosino *et al.*, Phys. Lett. B **642** (2006) 315.
6. G. Amelino-Camelia *et al.*, Eur. Phys. J. C **68** (2010) 619
7. C. Milardi *et al.*, arXiv:1006.1487.
8. D. Babusci *et al.*, Nucl. Instrum. Meth. A **617**, 81 (2010).
9. F. Archilli *et al.*, Nucl. Instrum. Meth. A **617** (2010) 266.
10. F. Happacher *et al.*, Nucl. Phys. Proc. Suppl. **197**, 215 (2009).
11. M. Cordelli *et al.*, Nucl. Instrum. Meth. A **617**, 105 (2010).
12. A. Balla *et al.*, Nucl. Instrum. Meth. A **628** (2011) 194.

HOW TO TEST THE SME WITH SPACE MISSIONS?

A. HEES
Jet Propulsion Laboratory, California Institute of Technology
4800 Oak Grove Drive, Pasadena CA 91109, USA
E-mail: Aurelien.D.Hees@jpl.nasa.gov

B. LAMINE
Laboratoire Kastler Brossel, CNRS, ENS, UPMC, Campus Jussieu, F-75252 Paris
IRAP, CNRS, UPS, 14 avenue Edouard Belin, 31450 Toulouse, France

C. LE PONCIN-LAFITTE and P. WOLF
LNE-SYRTE, Observatoire de Paris, UPMC,
61 avenue de l'Observatoire, F-75014 Paris, France

In this communication, we focus on possibilities to constrain SME coefficients using Cassini and Messenger data. We present simulations of radioscience observables within the framework of the SME, identify the linear combinations of SME coefficients the observations depend on and determine the sensitivity of these measurements to the SME coefficients. We show that these datasets are very powerful for constraining SME coefficients.

1. Introduction

Since the development of General Relativity (GR), the solar system has always been a very interesting laboratory to test gravitation theory and to constrain hypothetical alternative theories of gravity. Until today, mainly two formalisms have been widely used at solar system scales to test the gravitation theory: the parametrized post-newtonian (PPN) formalism and the fifth force search.

Within the PPN formalism, the metric is phenomenologically parametrized by 10 dimensionless parameters[1] that can be constrained independently from any underlying fundamental theory. The current constraints on these PPN parameters are pretty good and can be found in Ref. 1.

The fifth force formalism consists in searching for a modification of the Newton potential of the form of a Yukawa potential parametrized by a range of interaction and an intensity.[2] The area of the parameter space excluded by experiments can be found in Ref. 3. It can be seen that very good constraints are available except for small and large interaction distances.

Even if the constraints on these two formalisms are currently very impressive, there are still theoretical motivations to improve them (for some examples, see Ref. 4). Moreover, it is also very interesting to look for deviations from GR in other frameworks than the two used so far. In particular, a consideration of a hypothetical Lorentz violation in the gravitational sector naturally leads to a parametrized expansion at the level of the action.[5] The post-newtonian metric resulting from this formalism (known as the Standard-Model Extension (SME)) is parametrized by a symmetric traceless tensor $\bar{s}_{\mu\nu}$ and differs from the PPN metric.[6] Until now, the only tracking data used to constrain these SME coefficients are the lunar laser ranging (LLR) data.[7]

In this communication, we show how spacecraft tracking data can be used to constrain SME gravity parameters. For this, we determine the incompressible signature produced by SME on tracking observations. The procedure and the software used to determine these signatures are presented in Ref. 4.

2. Simulations of tracking observations in the SME

We consider three realistic situations: a two year radioscience link between Earth and the Mercury system corresponding to Messenger data, a 32 day Doppler link between Earth and the Cassini spacecraft during its cruise between Jupiter and Saturn corresponding to the conjunction experiment,[8] and a 9 year radioscience link between Earth and the Saturn system corresponding to Cassini data. For these three situations, we determine the linear combinations of SME coefficients to which the observations are sensitive, the signatures produced by these coefficients on observations and the sensitivity of these observations to SME coefficients.

The radioscience (range and Doppler) measurements of Messenger depend on the 4 linear combinations of the 9 fundamental parameters $\bar{s}_{\mu\nu}$:

$$\bar{s}_A = \bar{s}_{XX} - 0.72\bar{s}_{YY} - 0.28\bar{s}_{ZZ}, \tag{1a}$$

$$\bar{s}_{TX}, \tag{1b}$$

$$\bar{s}_B = \bar{s}_{TY} + 0.53\bar{s}_{TZ}, \tag{1c}$$

$$\bar{s}_C = \bar{s}_{XY} + 2.954\bar{s}_{XZ} - 0.26\bar{s}_{YZ}. \tag{1d}$$

The 32 days of Doppler data from the Cassini conjunction experiment de-

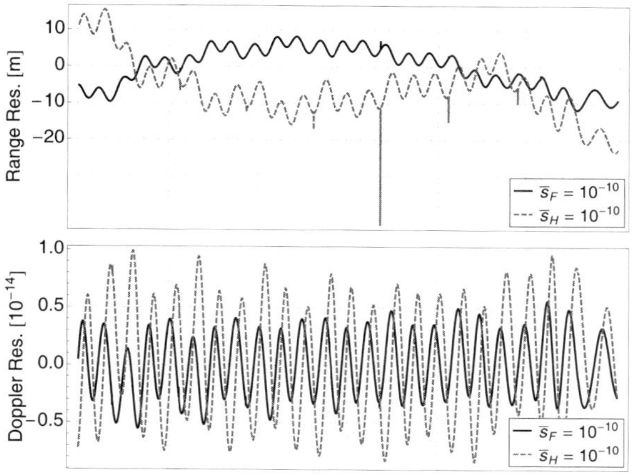

Fig. 1. Incompressible signatures produced by some of the SME linear combinations (3) on range and Doppler observations of Cassini while orbiting in the saturnian system.

pend only on two linear combinations given by

$$\bar{s}_D = \bar{s}_{XX} - 0.84\bar{s}_{YY} - 0.16\bar{s}_{ZZ} + 9.45\bar{s}_{XY} + 4.1\bar{s}_{XZ} - 0.72\bar{s}_{YZ}, \quad (2a)$$

$$\bar{s}_E = \bar{s}_{TX} + 3.69\bar{s}_{TY} + 1.55\bar{s}_{TZ}, \quad (2b)$$

while the 9 year Range and Doppler data coming from the Saturnian system depend on

$$\bar{s}_F = \bar{s}_{XX} - 0.83\bar{s}_{YY} - 0.17\bar{s}_{ZZ} - 0.76\bar{s}_{YZ}, \quad (3a)$$

$$\bar{s}_{TX}, \quad (3b)$$

$$\bar{s}_G = \bar{s}_{TY} + 0.43\bar{s}_{TZ}, \quad (3c)$$

$$\bar{s}_H = \bar{s}_{XY} + 0.56\bar{s}_{XZ}. \quad (3d)$$

Figure 1 represents the signature due to some of the SME linear combinations (3) on the Cassini radioscience measurements. These signatures correspond to residuals that would be obtained by a naive observer measuring data and analyzing them in GR (using standard procedure) while the correct gravitation theory is SME theory with the linear combinations taking the indicated values. The signatures are characteristic of the SME theory of gravity and should be searched for in the residuals of real data analysis. Similar signatures have been determined for the two other situations (Messenger and the Cassini conjunction) and for the other linear combinations but are not presented here.

Table 1. Estimated reachable uncertainties on SME coefficients

(a): Messenger

Coeff.	Uncertainties
$\bar{s}_A$	1.1×10^{-10}
$\bar{s}_{TX}$	3.1×10^{-8}
$\bar{s}_B$	1.4×10^{-8}
$\bar{s}_C$	3.2×10^{-11}

(b): Cassini conjunction

Coeff.	Uncertainties
$\bar{s}_D$	3.6×10^{-7}
$\bar{s}_E$	3.1×10^{-3}

(c): Cassini in orbit

Coeff.	Uncertainties
$\bar{s}_F$	8.6×10^{-11}
$\bar{s}_{TX}$	1.2×10^{-8}
$\bar{s}_G$	1.5×10^{-8}
$\bar{s}_H$	2.3×10^{-11}

The comparison of the amplitude of these signatures with the accuracy of the measurements gives an estimate of the uncertainties on the SME coefficients that would be reachable in a real data analysis. The estimated uncertainties on SME coefficients reachable using Messenger and Cassini data are presented in Table 1. One can see that the conjunction data are not interesting to constrain SME. On the other side, Messenger and Cassini data (while orbiting within the saturnian system) are very interesting and can improve the current LLR constraints on SME coefficients[7] by one order of magnitude. This gives a strong motivation to consider a test of SME using these datasets.

Acknowledgments

The research described in this paper was partially carried out at the Jet Propulsion Laboratory, California Institute of Technology, under contract with the National Aeronautics and Space Administration. A.H. acknowledges support from the Belgian American Educational Foundation (BAEF).

References

1. C.M. Will, *Theory and Experiment in Gravitational Physics*, Cambridge University Press, 1993; C.M. Will, Liv. Rev. Rel. **9**, (2006).
2. E. Fischbach and C.L. Talmadge, *The Search for Non-Newtonian Gravity*, Aip-Press Series, Springer, 1999; E.G. Adelberger, J.H. Gundlach, B.R. Heckel, S. Hoedl, and S. Schlamminger, Prog. Part. Nucl. Phys. **62**, 102 (2009).
3. A.S. Konopliv *et al.*, Icarus **211**, 401 (2011).
4. A. Hees *et al.*, Class. Quantum Grav. **29**, 235027 (2012).
5. V.A. Kostelecký, Phys. Rev. D **69**, 105009 (2004).
6. Q.G. Bailey and V.A. Kostelecký, Phys. Rev. D **74**, 045001 (2006).
7. J.B.R. Battat, J.F. Chandler, and C.W. Stubbs, Phys. Rev. Lett. **99**, 241103 (2007).
8. B. Bertotti, L. Iess, and P. Tortora, Nature **425**, 374 (2003).

NEUTRINO VELOCITY MEASUREMENTS WITH MINOS

IOANA ANGHEL

High Energy Physics Division, Argonne National Lab
9700 S Cass Ave, Bldg 362, F141, Argonne, IL 60439, USA
E-mail: ianghel@anl.gov

Department of Physics and Astronomy, Iowa State University
12 Physics Hall, Ames, IA 50011-3160, USA
E-mail: anghel@iastate.edu

On behalf of the MINOS Collaboration

The MINOS experiment uses two neutrino detectors separated by 735 km. Two measurements of the neutrino time of flight were conducted between the two detectors. The first analysis uses the statistics accumulated during 7 years of data taking. The second analysis was done using the data collected during 2012 making use of a new system of clocks which provided a more precise measurement of time at the primary proton target, the Near and the Far Detectors. Using the resulting neutrino time of flight between the MINOS detectors we present an improved neutrino velocity measurement.

1. Introduction

Observations of neutrinos from Supernova SN1987A[1] have set limits on the difference between the neutrino velocity and the speed of light consistent with $v = c$. Within the context of new instrumental developments, there are unprecedented opportunities to measure the neutrino velocity with high precision in long-baseline experiments.[2,3]

2. NuMI beam and MINOS experiment

The MINOS experiment has two detectors.[4] The Near Detector (ND) weights 0.98 kton, is located at the Fermi National Accelerator Facility (Fermilab), at 1.04 km downstream of the production target and 100 m underground. The Far Detector (FD) weights 5.4 kton, is located at Soudan Underground Laboratory, about 735 km downstream of the target and 705 m underground. The neutrino beam is produced at Fermilab by bombarding a graphite target with 120 GeV protons. The resulting charged pions and kaons are focused by a complex of magnetic horns and then decay within a 675 m long pipe, producing a neutrino beam with a peak energy of 3.5 GeV. During acceleration, the target protons are grouped in a $\sim$10 μs spill, containing six batches, separated by 100 ns. Each batch consists of 81 bunches spaced at 18.83 ns interval. The time resolution of the experiment is $\sim$1.5 ns, meaning that it can resolve this bunch structure. The time profile of the beam is measured using a resistive wall current monitor (RWCM)[5] situated along the beam pipe, between the Main Injector and the NuMI target. The neutrino time distribution at both MINOS detectors follows the proton beam structure.

3. Neutrino velocity measurement

Muon neutrinos are identified in the detectors throught their charged current (CC) interactions, $\nu_\mu + X \rightarrow \mu + Y$. Typically, the muon leaves a well defined linear energy deposit in the detector, crossing tens of scintillator planes. The time and position of the neutrino interaction vertex is calculated using a fit to the track hits. The neutrino velocity measurement is conceptually straightforward, consisting of a measurement of the distance between the two detectors and the time it takes for a neutrino beam to pass between them, denoted as the time of flight. The straight line distance between the front faces of the near and far detectors have been determined to 70 cm precision, due to the inertial survey of the FD location. However, the neutrino interaction cross section is very small and the process of detecting a neutrino is destructive. It is therefore necessary to consider the time distribution of the neutrino beam. The measurement of the neutrino velocity with MINOS experiment was performed in two separate phases: Phase I, where we used the entire dataset accumulated over 7 years of data taking from 2005 to 2011, and Phase II, where we used the data collected in 2012 with a new timing system installed.

3.1. *Phase I neutrino time of flight measurement*

The time of recorded neutrino interactions in the MINOS detectors is used to reject background muons from cosmic rays. The complete MINOS dataset, from seven years of running, contains about 4000 CC ν_μ events due to neutrino beam and a single background event. Such background rejection does not need the event time to be accurate to much better than a microsecond. For a better precision, we used parts of the timing system set in place at the end of 2011 intended for the Phase II measurement, to cross check some of the measured qualities used with Phase I data. In this way we found that the old GPS has a random but stable offset with respect to UTC, which changed with each power cycle. This effect determined the largest systematic uncertainty for this first measurement. For Phase I , we achieved a measurement of the difference between the time of flight of neutrino and the expectation assuming travel at the speed of light of $\delta = -15 \pm 31$ ns.

3.2. *Phase II neutrino time of flight measurement*

For Phase II data, new timing components were installed to specifically improve the timing precision. For this measurement, we considered three locations: RWCM — the place where the NuMI beamline leaves the Main Injector, ND and FD. A local Cs atomic clock was installed at each detector and a third Rb clock was installed at RWCM. In this way, the detector signals were measured against the atomic clocks, then the clock signals were transported to the surface with two-way time transfer and finally the clocks were measured with modern dual-frequency GPS receivers. This system allowed us to time the detectors against a local atomic clock, but there is an outstanding unknown latency in each neutrino detector, causing a systematic shift to the measured time of flight. The relative measurement of the latency between the FD and ND was performed with an uncertainty of 1 ns by using two mobile auxiliary muon detectors. Within the analysis, for the FD we selected 195 CC events and 177 rock and anti-fiducial (RAF) events for ND and FD.

For the ND, the number of events was sufficient to consider only the CC events. The data analysis is performed by considering the time distribution measured from the protons in the RWCM to form likelihood distribution for the arrival time of each neutrino event in the detector. The PDF is built from convoluting the time distribution with a 1.5 ns wide gaussian representing the detector timing resolution and shifted by the predicted time of flight. For each detector, we combined the event likelihoods and maximize them with respect to the time of flight from RWCM to the detector. The neutrino time of flight between the two detectors is obtained by subtracting the results of the two measurements. Figure 1[6] shows the arrival time distribution since the end of the preceding 18.83 ns long bunch. The appearance of the bunch shape in the FD demonstrates that the whole apparatus is functioning correctly and that there is a good stability and resolution of the time measurement.

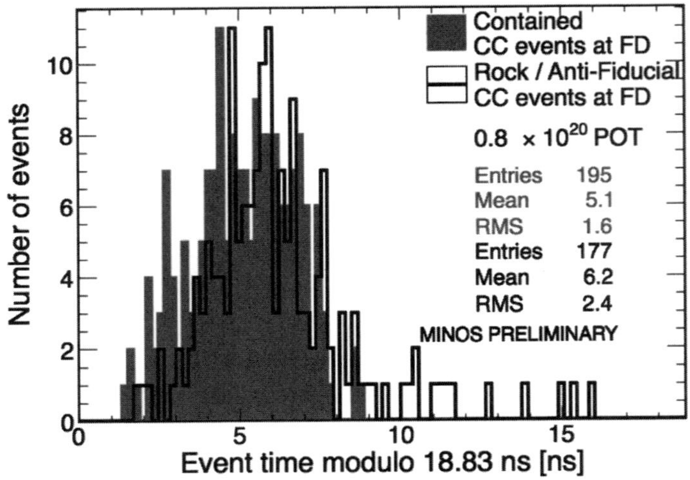

Fig. 1. Arrival time distribution at the Far Detector modulo 18.83 ns bunch separation for CC (solid histogram) and RAF (line) events. The RAF events are expected to arrive on average 1 ns later because of the extra path length traveled by the muon.[6]

The difference between the measured time it takes for the neutrinos to travel between the MINOS detectors and the time computed considering the speed of light is $\delta = -2.4 \pm 0.1(\text{stat.}) \pm 2.3(\text{syst.})$ ns, where the dominant systematic uncertainty comes from the inertial survey of the FD location. The neutrino velocity v is therefore found to be $(v/c - 1) = (1.0 \pm 1.1) \times 10^{-6}$.

References

1. R. Bionta *et al.*, Phys. Rev. Lett. **58**, 1494 (1987).
2. P. Adamson *et al.*, Phys. Rev. D **76**, 072005 (2007).
3. T. Adam *et al.*, JHEP **1210**, 093 (2012).
4. D.G. Michael *et al.*, Nucl. Instrum. Meth. A **596**, 190 (2008).
5. J. Crisp and B. Fellenz, JINST **6**, T11001 (2011).
6. P. Adamson *et al.*, in *Proceedings of the 44th Annual Precise Time and Time Interval Systems and Applications Meeting*, Institute of Navigation, 2012.

PROSPECTS FOR A GLOBAL NETWORK OF OPTICAL MAGNETOMETERS FOR EXOTIC PHYSICS (GNOME)

D.F. JACKSON KIMBALL,* S. PUSTELNY,† M. POSPELOV,‡,§
M.P. LEDBETTER,¶,|| N. LEEFER,¶ P. WLODARCZYK,** P. WCISLO,†,††
W. GAWLIK,† J. SMITH,‡‡ J. READ,‡‡ C. PANKOW,§§ and D. BUDKER¶,¶¶

**Department of Physics, California State University — East Bay, Hayward, CA 94542-3084, USA*
†Institute of Physics, Jagiellonian University, Reymonta 4, 30-059 Kraków, Poland
‡Department of Physics and Astronomy, University of Victoria Victoria, British Columbia V8P 1A1, Canada
§Perimeter Institute for Theoretical Physics, Waterloo, Ontario N2J 2W9, Canada
¶Department of Physics, University of California at Berkeley Berkeley, CA 94720-7300, USA
||AOSense, Inc., Sunnyvale, California 94085-2909, USA
***Department of Electronics, AGH University of Science and Technology Mickiewicza 30, 30-059 Kraków, Poland*
††Institute of Physics, Nicolaus Copernicus University Grudziadzka 5, 87-100 Toruń, Poland
‡‡Gravitational-Wave Physics and Astronomy Center, Department of Physics California State University Fullerton, Fullerton, CA 92831, USA
§§Center for Gravitation, Cosmology, and Astrophysics, Department of Physics University of Wisconsin-Milwaukee, Milwaukee, WI 53211, USA
¶¶Nuclear Science Division, Lawrence Berkeley National Laboratory Berkeley, CA 94720, USA

The concept and prospects of a proposed international network of geographically separated, time-synchronized ultrasensitive atomic comagnetometers to search for correlated transient signals heralding new physics is discussed. The Global Network of Optical Magnetometers for Exotic physics (GNOME) would be sensitive to nuclear and electron spin couplings to various exotic fields. To date, no such search has ever been carried out, making the GNOME a novel experimental window on new physics.

With the discovery of the Higgs boson, the last major fundamental prediction of a particle theory supported by experimental evidence has been confirmed. Nonetheless, there remain several fundamental mysteries in modern physics that defy explanation within the Standard Model, such as the source of additional CP-violation required to explain baryogenesis and the nature of dark matter and dark energy. To date no prediction of a new particle theory extending beyond the Standard Model has been confirmed experimentally, and thus, in some sense, particle physics has entered an 'era of speculation' where it is likely that many possibilities must be explored before solutions to remaining fundamental mysteries will be found. One such possibility is that the new physics required to explain baryogenesis, dark matter, or dark energy will be in the form of new fundamental forces or fields, and such new fields may interact with the intrinsic spins of elementary particles.[1,2]

There are a number of recent and ongoing experiments (many of which are described in the present volume) using atomic comagnetometers to search for heretofore undiscovered spin-dependent interactions. The basic concept of such experiments is to search for anomalous spin-dependent energy shifts caused by exotic interactions rather than electromagnetic fields. For example, there are experiments searching for anomalous spin-dependent interactions constant in time as evidence of long-range monopole-dipole[3–5] and dipole-dipole interactions[6] where the Earth is the source of mass or polarized electrons. There are also experiments searching for anomalous spin-dependent interactions periodic in time: either (1) searching for short-range monopole-dipole and dipole-dipole interactions by modulating a local mass or polarized spin sample[7–9] or (2) testing local Lorentz invariance (LLI) by moving a comagnetometer with respect to a hypothetical background field (either via a rotatable platform for the experiment[10] or through the motion of the Earth itself relative to the background field[11]).

We have recently proposed[12] to initiate a search for an entirely different class of effects: signals from *transient events*. While a single comagnetometer system could detect such transient events, it would be exceedingly difficult to distinguish confidently a true signal generated by heretofore undiscovered physics from 'false positives' induced by occasional abrupt changes of comagnetometer operational conditions (e.g., magnetic-field spikes, laser-light-mode jumps, electronic noise, etc.). Effective vetoing of false positive events requires an array of comagnetometers. Furthermore, there are key benefits in terms of noise suppression and event characterization to widely distributing the comagnetometers geographically. The Laser Interferometer Gravitational Wave Observatory (LIGO) Collaboration has developed sophisticated data analysis techniques[13] to search for similar correlated 'burst' signals from a worldwide network of gravitational wave detectors, and we have recently demonstrated that these data analysis techniques can be applied to data from synchronized comagnetometers.[14]

Our proposed comagnetometer array, the Global Network of Optical Magnetometers for Exotic physics (GNOME), would be uniquely sensitive, for example, to cosmic events generating a wave or burst of a heretofore undiscovered field,[15] or to correlated noise produced by a fluctuating[16] or oscillating[17] background field whose time-averaged value is zero. Eventually, the GNOME will consist of at least five dedicated atomic comagnetometers located at geographically separated stations. Each GNOME sensor will have an equivalent magnetometric sensitivity better than $\sim 10\ \mathrm{pG}/\sqrt{\mathrm{Hz}}$ ($1\ \mathrm{fT}/\sqrt{\mathrm{Hz}}$), achievable with state-of-the-art atomic comagnetometers.[18]

Each comagnetometer will be located within a multi-layer magnetic shield to reduce the influence of magnetic noise and perturbations. Even with shielding and comagnetometry techniques, there will inevitably be some level of transient signals and noise associated with the local environment (and possibly with global effects like the solar wind, changes to the Earth's magnetic field, etc.). Therefore, each GNOME sensor will use an auxiliary magnetometer to measure the field outside the shield and will record other relevant environmental conditions, allowing for exclusion/vetoing of data with known issues.

The signals from the GNOME sensors will be recorded with accurate timing provided by the global positioning system (GPS) and will have a characteristic temporal resolution of $\lesssim 10$ ms (determined by the comagnetometer bandwidth), enabling resolution of events that propagate at the speed of light (or slower) across the Earth ($R_e/c \approx 40$ ms). Because of the broad geographical distribution of sensors, the GNOME can achieve excellent spatial resolution, acting as an exotic physics 'telescope' with a baseline comparable to the diameter of the Earth.

Transient signals can also be generated by massive objects at rest in the galactic frame, through which the Earth travels at $\sim 10^{-3}c$ (leading to ~ 40 s delays between transient signals at different sites). The search for such massive objects will be the initial focus of the GNOME, in particular domain walls of a light pseudoscalar (axion-like) field.[12,14] Pseudoscalar domain walls are a ubiquitous feature of extensions to the Standard Model,[19] inevitably appearing in models with spontaneously broken discrete symmetries.[20] Such fields may comprise a significant fraction of cold dark matter,[21] are candidates for quintessence,[22] and are predicted by string theory[23] and other Standard-Model extensions.[2] When the earth passes through such a pseudoscalar domain wall, the gradient of the pseudoscalar field at the domain wall would exert a brief torque on atomic spins that, for reasonable parameters consistent with astrophysical and laboratory constraints, could be detected by the GNOME.[12,14]

Acknowledgments

This work has been supported by the National Science Foundation. S.P. is a scholar of the Polish Ministry of Science and Higher Education within the Mobility Plus program.

References

1. J.E. Moody and F. Wilczek, Phys. Rev. D **30**, 130 (1984).
2. B.A. Dobrescu and I. Mocioiu, JHEP **11**, 5 (2006).
3. B.J. Venema, P.K. Majumder, S.K. Lamoreaux, B.R. Heckel, and E.N. Fortson, Phys. Rev. Lett. **68**, 135 (1992).
4. B.R. Heckel, E.G. Adelberger, C.E. Cramer, T.S. Cook, S. Schlamminger, and U. Schmidt, Phys. Rev. D **78**, 092006 (2008).
5. D.F. Jackson Kimball *et al.*, Ann. Physik **525**, 514 (2013).
6. L. Hunter, J. Gordon, S. Peck, D. Ang, J.-F. Lin, Science **339**, 928 (2013).
7. A.N. Youdin, D. Krause, Jr., K. Jagannathan, L.R. Hunter, and S.K. Lamoreaux, Phys. Rev. Lett. **77**, 2170 (1996).
8. P.-H. Chu *et al.*, Phys. Rev. D **87**, 011105(R) (2013).
9. G. Vasilakis, J.M. Brown, T.W. Kornack, and M.V. Romalis, Phys. Rev. Lett. **103**, 261801 (2009).
10. M. Smiciklas, J.M. Brown, L.W. Cheuk, S.J. Smullin, and M.V. Romalis, Phys. Rev. Lett. **107**, 171604 (2011).
11. C. Gemmel *et al.*, Phys. Rev. D **82**, 111901(R) (2010).
12. M. Pospelov, S. Pustelny, M.P. Ledbetter, D.F. Jackson Kimball, W. Gawlik, and D. Budker, Phys. Rev. Lett. **110**, 021803 (2013).
13. B. Abbott *et al.*, Phys. Rev. D **69**, 102001 (2004).
14. S. Pustelny *et al.*, arXiv:1303.5524.
15. S.M. Carroll and G.B. Field, Phys. Rev. D **50**, 3867 (1994).
16. J. Ellis, N.E. Mavromatos, and M. Westmuckett, Phys. Rev. D **70**, 044036 (2004).
17. D. Budker, P.W. Graham, M. Ledbetter, S. Rajendran, and A. Sushkov, arXiv:1306.6089.
18. D. Budker and D.F. Jackson Kimball, eds., *Optical Magnetometry,* Cambridge University Press, Cambridge, 2013.
19. J. Jaeckel and A. Ringwald, Ann. Rev. Nucl. Part. Sci. **60**, 405 (2010).
20. A. Friedland, H. Murayama, and M. Perelstein, Phys. Rev. D **67**, 043519 (2003).
21. M. Pospelov, A. Ritz and M.B. Voloshin, Phys. Rev. D **78**, 115012 (2008).
22. V. Flambaum, S. Lambert and M. Pospelov, Phys. Rev. D **80**, 105021 (2009).
23. A. Arvanitaki *et al.*, Phys. Rev. D **81**, 123530 (2010).

LORENTZ-VIOLATING PHOTONS WITH A MASS TERM

MAURO CAMBIASO
Departamento de Ciencias Físicas, Universidad Andrés Bello
Santiago, Chile
E-mail: mcambiaso@unab.cl

RALF LEHNERT
Indiana University Center for Spacetime Symmetries
Bloomington, IN 47405, USA
E-mail: ralehner@indiana.edu

ROBERTUS POTTING
CENTRA, Departamento de Física, Universidade do Algarve
8005-139 Faro, Portugal
E-mail: rpotting@ualg.pt

Perturbative calculations in quantum field theory often require the regularization of infrared divergences. In quantum electrodynamics, such a regularization can for example be accomplished by a photon mass introduced via the Stueckelberg method. The present work extends this method to the QED limit of the Lorentz- and CPT-violating Standard-Model Extension.

1. Introduction

Massive vector particles are of relevance for various subfields of physics including the weak interaction, test models for photon-mass searches, and the regularization of certain infrared divergences in quantum-field calculations. In this latter context, the mass term should be introduced such that the symmetries of the original model remain unspoiled. For a U(1) Lorentz-invariant gauge theory, this can be accomplished with the Stueckelberg method.[1]

In recent years, Lorentz-violating quantum field theories have become a focus of theoretical[2] and experimental[3] inquiry. Such theories also require the regularization of infrared divergences in some circumstances, and the question arises as to whether the Stueckelberg method can be adapted to such Lorentz-breaking U(1) gauge theories. As in the Lorentz-invariant case, the adapted Stueckelberg method should be compatible with all symmetries of the theory to be regulated, so that the usual Stueckelberg approach can be employed for the Lorentz-violating extension of conventional QED. However, the Lorentz violation in this model allows a broader range of compatible Stueckelberg terms that may be introduced, and this freedom can then be used to streamline calculations in Lorentz-breaking QED.

The present work provides one possible class of extensions of the usual Stueckelberg procedure to Lorentz-violating QED. Section 2 gives a brief overview of the procedure at the lagrangian level. Some implications of the Lorentz-breaking Stueckelberg lagrangian are discussed in Sec. 3.

2. Lagrangian analysis

We begin with the usual free-photon lagrangian in the minimal Standard-Model Extension coupled to a conserved source j^μ:

$$\mathcal{L}_\gamma = -\frac{1}{4}F^2 - A \cdot j - \frac{1}{4}(k_F)^{\kappa\lambda\mu\nu} F_{\kappa\lambda} F_{\mu\nu} + \frac{1}{2}\epsilon^{\kappa\lambda\mu\nu}(k_{AF})_\kappa A_\lambda F_{\mu\nu} \,. \quad (1)$$

Here, Lorentz and CPT breakdown is controlled by the spacetime-constant backgrounds k_F and k_{AF}.

The absence of Lorentz symmetry in the above lagrangian (1) allows us to drop the requirement of Lorentz invariance for the mass-type term $\delta\mathcal{L}_m$ to be introduced for the photon. However, the inclusion of arbitrary Lorentz violation into $\delta\mathcal{L}_m$ may be problematic: consider the case in which k_F and k_{AF} are such that a subgroup of the Lorentz group remains unbroken. A regulator violating this residual symmetry may be undesirable, so that the breakdown of the remaining invariant subgroup may have to be excluded from $\delta\mathcal{L}_m$.

For our present purposes, however, we consider arbitrary Lorentz violation in $\delta\mathcal{L}_m$. This yields more general results also relevant for purposes other than infrared regularization. More specifically, we implement the Stueckelberg procedure by introducing a scalar field ϕ as follows:[4]

$$\delta\mathcal{L}_m = \frac{1}{2}(\partial_\mu\phi - mA_\mu)\hat{\eta}^{\mu\nu}(\partial_\nu\phi - mA_\nu)\,, \quad (2)$$

where m can later be identified with the photon mass and

$$\hat{\eta}^{\mu\nu} = \eta^{\mu\nu} + \hat{G}^{\mu\nu}, \quad (3)$$

with $\hat{G}^{\mu\nu}$ a Lorentz-violating operator that may contain derivatives but is otherwise spacetime constant. The small number of Lorentz-symmetric pieces still contained in an arbitrary $\hat{G}^{\mu\nu}$ can be removed if necessary.

As in the conventional Stueckelberg case, the key feature of $\mathcal{L}_\gamma + \delta\mathcal{L}_m$ is its invariance (up to total derivatives) under a local gauge transformation

$$\delta A_\mu = \partial_\mu \epsilon(x)\,, \quad \delta\phi = m\epsilon(x)\,. \tag{4}$$

With the addition of ξ-type gauge fixing $\mathcal{L}_{\text{g.f.}} = -\frac{1}{2\xi}(\partial_\mu \hat{\eta}^{\mu\nu} A_\nu + \xi m\phi)^2$ and a Faddeev-Popov contribution $\mathcal{L}_{\text{F.P.}}$, the model Lagragian $\mathcal{L} = \mathcal{L}_\gamma + \delta\mathcal{L}_m + \mathcal{L}_{\text{g.f.}} + \mathcal{L}_{\text{F.P.}}$ becomes[4]

$$\begin{aligned}\mathcal{L} = & -\frac{1}{4}F^2 - A\cdot j + \frac{1}{2}m^2 A_\mu \hat{\eta}^{\mu\nu} A_\nu - \frac{1}{2\xi}(\partial_\mu \hat{\eta}^{\mu\nu} A_\nu)^2 \\ & - \frac{1}{2}\phi(\partial_\mu \hat{\eta}^{\mu\nu}\partial_\nu + \xi m^2)\phi - \bar{c}(\partial_\mu \hat{\eta}^{\mu\nu}\partial_\nu + \xi m^2)c \\ & - \frac{1}{4}(k_F)^{\kappa\lambda\mu\nu} F_{\kappa\lambda}F_{\mu\nu} + \frac{1}{2}\epsilon^{\kappa\lambda\mu\nu}(k_{AF})_\kappa A_\lambda F_{\mu\nu}\,. \end{aligned} \tag{5}$$

Note that the scalar ϕ and the ghosts c and $\bar{c}$ are now decoupled and can be integrated out. We can therefore disregard these fields in what follows.

3. Equations of motion

The lagrangian (5) yields the following equation of motion for the photon:

$$\begin{aligned}& \left[\eta^{\mu\alpha}\eta^{\nu\beta}\partial_\mu + (k_{AF})_\mu \epsilon^{\mu\nu\alpha\beta} + (k_F)^{\mu\nu\alpha\beta}\partial_\mu\right] F_{\alpha\beta} \\ & \quad + \left[m^2 \hat{\eta}^{\mu\nu} + \frac{1}{\xi}\hat{\eta}^{\mu\alpha}\hat{\eta}^{\nu\beta}\partial_\alpha\partial_\beta\right] A_\mu = j^\nu\,. \end{aligned} \tag{6}$$

Taking the 4-divergence of this equation yields

$$\left(\hat{\eta}^{\mu\nu}\partial_\mu\partial_\nu + \xi m^2\right)\left(\partial_\alpha \hat{\eta}^{\alpha\beta} A_\beta\right) = 0\,. \tag{7}$$

Note that $(\partial_\alpha \hat{\eta}^{\alpha\beta} A_\beta)$ projects out one degree of freedom contained in A_μ. Equation (7) establishes that this degree of freedom is not excited by the source j_ν, so $(\partial_\alpha \hat{\eta}^{\alpha\beta} A_\beta)$ is an auxiliary mode. This is consistent with the expectation of three physical degrees of freedom for a massive vector field.

A plane-wave ansatz in Eq. (6) yields the model's dispersion relation; it has the following structure:[4]

$$\frac{1}{\xi}\left(1 + \frac{1}{4}\hat{G}^\alpha_{\ \alpha}\right)(\hat{\eta}^{\mu\nu} p_\mu p_\nu - \xi m^2) Q(p) = 0\,. \tag{8}$$

The first factor does not contain physical modes. The $(\hat{\eta}^{\mu\nu} p_\mu p_\nu - \xi m^2)$ piece corresponds to the auxiliary mode governed by Eq. (7). The factor $Q(p)$ is associated with the physical degrees of freedom and has the structure

$$Q = (p^2 - m^2)^3 + r_2(p^2 - m^2)^2 + r_1(p^2 - m^2) + r_0\,. \tag{9}$$

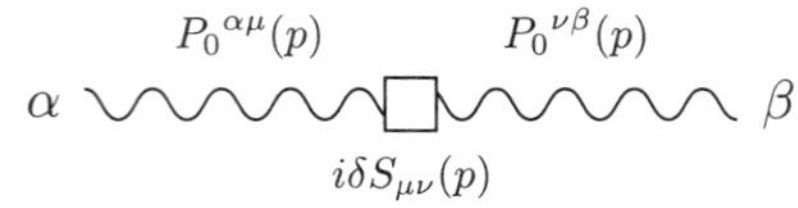

Fig. 1. Propagator insertion. The wavy lines denote the conventional Lorentz-invariant Stueckelberg propagator P_0. The square box represents the leading-order Lorentz-breaking insertion given by Eq. (10).

Here, the coefficients r_j are coordinate scalars containing the Lorentz-breaking tensors k_F and k_{AF}. They vanish in the limit $k_F, k_{AF} \to 0$. The explicit expressions for the r_j can be found in the literature.[4] Note that the physical dispersion relation (9) is consistent with the expectation of three conventional massive modes perturbed by small Lorentz violation.

The exact expression for the corresponding propagator is somewhat unwieldy.[4] But for most applications only leading-order Lorentz-violating effects need to be taken into account. The Lorentz-breaking contributions can then be incorporated via this propagator insertion (see also Fig. 1):[4]

$$\begin{aligned}\delta S(p)^{\mu\nu} &= 2i(k_{AF})_\alpha \epsilon^{\alpha\beta\mu\nu} p_\beta - 2(k_F)^{\alpha\mu\beta\nu} p_\alpha p_\beta \\ &\quad + m^2 \hat{G}^{\mu\nu} - \frac{1}{\xi}\left(\hat{G}^{\mu}_{\alpha}\, p^\alpha p^\nu + \hat{G}^{\nu}_{\alpha}\, p^\alpha p^\mu\right). \qquad (10)\end{aligned}$$

Acknowledgments

This work was supported in part by the Indiana University Center for Spacetime Symmetries, by the Universidad Andrés Bello under Grant No. UNAB DI-27-11/R, and by the Portuguese Fundação para a Ciência e a Tecnologia.

References

1. E.C.G. Stueckelberg, Helv. Phys. Acta **11**, 225 (1938); Helv. Phys. Acta **11**, 299 (1938).
2. See, e.g., D. Colladay and V.A. Kostelecký, Phys. Rev. D **58**, 116002 (1998); V.A. Kostelecký, Phys. Rev. D **69**, 105009 (2004); V.A. Kostelecký and R. Lehnert, Phys. Rev. D **63**, 065008 (2001); A.J. Hariton and R. Lehnert, Phys. Lett. A **367**, 11 (2007); R. Lehnert, Phys. Rev. D **74**, 125001 (2006).
3. See, e.g., V.A. Kostelecký and N. Russell, arXiv:0801.0287v6; M.A. Hohensee *et al.*, Phys. Rev. Lett. **102**, 170402 (2009); Phys. Rev. D **80**, 036010 (2009); J.-P. Bocquet *et al.*, Phys. Rev. Lett. **104**, 241601 (2010); J.S. Díaz *et al.*, arXiv:1305.4636.
4. M. Cambiaso *et al.*, Phys. Rev. D **85**, 085023 (2012).

TESTING THE EQUIVALENCE PRINCIPLE 10,000 TIMES BETTER ON A SOUNDING ROCKET

J.D. PHILLIPS* and R.D. REASENBERG

Harvard-Smithsonian Center for Astrophysics
Cambridge, MA 02138, USA
**E-mail: jphillips@cfa.harvard.edu*

The Sounding-Rocket Principle Of Equivalence Measurement (SR-POEM) will test the weak equivalence principle (WEP) to $2 \times 10^{-17} g$ in an experiment launched into free fall by a sounding rocket. The high sensitivity is possible in a short time because: (1) our laser distance gauges measure to 0.1 pm in 1 second; (2) the high measurement speed allows us to keep the temperature of the critical region stable to within a few μK using two cascaded thermal low-pass filters; (3) the spacing between the physics package and the test masses is kept constant by virtue of a servo (but not a drag-free satellite); (4) the test masses (TMs) are unconstrained during drops, avoiding constraint-force imperfections; and (5) the position measurement is to a plate that is almost stationary with respect to the TMs (by virtue of the position servo and the initialization of the TMs).

1. Introduction

The WEP is at the heart of gravitation theory. It has been tested with increasing accuracy for many centuries; the present best tests[1,2] yield $\sigma(\eta) < 2 \times 10^{-13}$, where $\eta = 2(a_A - a_B)/(a_A + a_B)$, and a_A and a_B are the accelerations of substances A and B solely under the influence of gravity. SR-POEM will achieve $\sigma_{SRP} < 2 \times 10^{-17}$. Theories that unify gravity with the other forces tend to predict a violation, though few predict the magnitude. The WEP has been tested for rest mass (of protons+electons and neutrons separately); for strong and electromagnetic binding energy; and for the first-order, parity conserving, weak interaction contribution.[3] Tests have not yet reached the level of the second-order, parity-violating, weak contribution. If second-order weak processes violate the WEP with unit strength,[4] the SR-POEM signal would be $\eta_{SRP} = 1.7 \times 10^{-17}$. While the presently estimated accuracy is not quite sufficient to measure this, our work on the SR-POEM error budget indicates that we may in future be able to revise our accuracy estimate, and it will be.

2. G-POEM

G-POEM is our test of the WEP on the ground. It employs repetitive free fall provided by an air bearing and a linear motor. The TM vertical separation, $\sim$0.5 m, is monitored by our Tracking Frequency laser Gauge (TFG). For G-POEM, TFG accuracy is 1 pm in 1 s: $\sigma(\eta) = 7 \times 10^{-12}$ in a single toss. The system can provide $\sim$ 50,000 tosses/day. Earth's gravity gradient, 3.1×10^{-6} s^{-2}, is a key source of systematic error. The TM separation must be known to 1.1 μm rms to reduce the error to 10% of that due to the TFG. Top-bottom swaps compensate for the large unknown offset between the TM center of mass and optical reference point.

Our first use of G-POEM will be a test of coefficients in the Standard-Model Extension resulting in a WEP violation varying once per sidereal day, 1/solar day, etc.[5] This test does not need top-bottom swaps. The principal systematic error is due to the arrival of 10^4 kg of people each morning, two floors above our experiment. All G-POEM subsystems have been tested individually. In summer 2013 we will refine the motion system. Then, the laser gauge will be upgraded based on work already done for SR-POEM. These are the last major prerequisites for a science run. To test the WEP with G-POEM to high sensitivity, we will increase the symmetry of the TMs to improve rejection of gradient due to local mass. Using SR-POEM's new 3-block TM will improve G-POEM testing of SR-POEM technology.

3. SR-POEM

SR-POEM employs two TMs, each having three blocks connected by a Y. This design improves over the previous 2-block design in that the three principal moments of inertia are equal, and gradient due to nearby masses is better suppressed. The TMs reside in a housing made of ULE glass inside which are deposited capacitance gauge (CG) electrodes. The housing is connected to the TFG plate above it by a metering structure.

Each TM block has a flat mirror that is the lower end of an optical resonator whose length is measured by a TFG. The resonators' upper ends are curved mirrors mounted in the TFG plate. The TFG uses Pound-Drever-Hall locking of its laser to the resonance. The required accuracy is $\sigma \leq 0.1\,\mathrm{pm}\,(1\,\mathrm{s}/\tau)^{1/2}$ for averaging time $\tau \leq 300$ s. Laboratory tests[6] have nearly achieved this: a simulation using data from the present TFG shows that the error contribution is 0.16 to 0.84 σ_{SRP}. We expect to make modest further improvements. When the distance has changed enough that the tunable laser can no longer follow, the TFG 'hops' the laser so that it locks to a new fringe. This capability is well tested.[7] Absolute distance L is measured at no additional cost by hopping repeatedly. Then, $L = c/2\Phi$, where $\Phi \sim 600$ MHz is the fringe spacing in optical frequency.

The TMs, TM housing, metering structure, and TFG plate are maintained in a vacuum of 10^{-10} torr. Above this dropping chamber is the optics chamber, which has a modest vacuum, sufficient for thermal stability but allowing for lubricated parts and plastic insulation. This composite vacuum chamber comprises the physics package. It is mounted to the spacecraft bus with an active hexapod. The mass of the physics package creates important differences of gravitational acceleration between the two TMs, in spite of the symmetry of the latter. To keep this acceleration difference constant, the physics package position servo keeps the TM-housing distance constant. The sensors are the TFGs and CGs, and the actuator is the hexapod. During drops, the CGs operate only in the horizontal directions and at reduced excitation voltage so as to avoid spurious vertical acceleration. Averaging over one drop, the accuracy is a few nm.

The TMs, TM housing, metering structure, and TFG plate are isolated from the payload outer wall by a pair of cascaded thermal low-pass filters based on vacuum insulation and the heat capacity of the internal parts. The temperature change is $<$ a few μK in a 120 s drop. The component that is synchronous with inversions is much smaller.

After leaving the atmosphere, we shed the hot shroud, uncage the TMs, and capture them electrostatically. After uncaging, the TMs do not touch anything until re-entry. We reduce the charge on the TMs and determine their centers of mass. Above 800 km, the principal nongravitational acceleration of the spacecraft is due to Earth thermal radiation. (Flying from Wallops near midnight, not near the summer solstice, the trajectory never enters sunlight.) We test the WEP in eight drops of 120 s each. During drops, the electrostatic TM suspension is turned off. Between drops, the suspension is turned on and the entire payload is inverted, which reverses the direction to the Earth and leaves most systematic error effects unchanged. CG position data from before and after the drop and continuous star tracker data yield mean velocity during the drop to 2×10^{-13} m/s (horizontal) and 7×10^{-4} arcsec/s (rotation). These support correction for the Coriolis effect.

An error budget is in preparation, and we plan to submit it for publication during calendar 2013. We treat several error sources here. (1) A key source is gravity gradient due to the Earth, $\partial g_z/\partial z = 1.6 \times 10^{-6}$ s^{-2} at $h = 1500$ km. An unknown change in the relative TM height by 11 pm would contribute an error $0.1\,\sigma_{SRP}$. TM velocity during the drop causes a relative position shift of 17 pm rms. We correct for this *a posteriori* using the TFG data accumulated during the drop. (2) Gradient due to the manufacturing error of components within the physics package is estimated to be $\leq 1.5 \times 10^{-10}$ s^{-2}. The horizontal velocity error at setup causes a 38 nm position shift during the drop, which results in an error of $0.09\,\sigma_{SRP}$. (3) Spatial and temporal variations of surface potential result in an electric field between TMs and housing. A change with time of the rms by 0.1 mV, *in synchronism with inversions*, would cause an error of $0.1\,\sigma_{SRP}$. The best study to date[8] found a spatial variation of 1 mV rms, at the limit of their resolution. We have been working with an equipment manufacturer, McAllister Technical Services, to develop a Kelvin probe with sensitivity improved 50-fold.[9] We intend to test a surface cleaning method that has recently realized a 100-fold reduction in the heating rate of ions trapped above gold electrodes, which had been associated with the surface and had troubled the field of quantum computing for a decade.[10] (4) TFG intracavity laser power exerts a force on the TMs. If the normalized variation between drops is 2.9×10^{-4}, the error is $0.1\,\sigma_{SRP}$. The laser intensity and alignment are stabilized. Such an alignment servo has been demonstrated in transmission[11] to an accuracy of 10^{-10} Hz$^{-1/2}$ using 160 μW laser power, for frequencies $f \geq 0.1$ Hz. For SR-POEM, we only need 5×10^{-4} Hz$^{-1/2}$, although the power is 0.1 μW and $f = 0.0033$ Hz. We have shown that the scheme of Sampas and Anderson works in reflection.[12]

References

1. S. Schlamminger *et al.*, Phys. Rev. Lett. **100**, 041101 (2008).
2. T.A. Wagner *et al.*, Class. Quantum Grav. **29**, 184002 (2012).
3. M.P. Haugan and C.M. Will, Phys. Rev. Lett. **37**, 1 (1976).
4. E. Fischbach *et al.*, Phys. Rev. D **52**, 5417 (1995).
5. V.A. Kostelecký and J.D. Tasson, Phys. Rev. D **83**, 016013 (2011).
6. R. Thapa, J.D. Phillips, E. Rocco, and R.D. Reasenberg, Opt. Lett. **36** 3759 (2011).
7. J.D. Phillips and R.D. Reasenberg, Rev. Sci. Instr. **76**, 064501 (2005).
8. N.A. Robertson *et al.*, Class. Quantum Grav. **23**, 2665 (2006).
9. J.D. Phillips, in preparation; R.D. Reasenberg *et al.*, Class. Quantum Grav. **30**, 125016 (2013).
10. D.A. Hite *et al.*, Phys. Rev. Lett. **109**, 103001 (2012).
11. N. Sampas and D.Z. Anderson, Appl. Opt. **29**, 394 (1990).
12. R.D. Reasenberg, Appl. Opt. **51**, 3132 (2012).

COVARIANT PHOTON QUANTIZATION IN THE SME

D. COLLADAY

New College of Florida
Sarasota, FL 34234, USA
E-mail: colladay@ncf.edu

The Gupta-Bleuler quantization procedure is applied to the SME photon sector. A direct application of the method to the massless case fails due to an unavoidable incompleteness in the polarization states. A mass term can be included into the photon lagrangian to rescue the quantization procedure and maintain covariance.

1. Introduction

The fermion sector of the SME was quantized consistently during the first stages of its theoretical development, at least in theories ith a significant nonzero mass parameter.[1] The photon sector has remained largely unaddressed due to several factors that make it more complicated to deal with. For example, there is no simple linear hamiltonian arising from the equation of motion that can be used for a complete set of orthogonal states. In addition, the modified equation of motion has implications for the gauge states that are nontrivial to incorporate. Addition of a mass term to the lagrangian makes the problem more similar to the fermion case and generates a tractable problem, so this is the approach used in this talk. An alternative, perturbative approach in the nonbirefringent case has also been implemented.[2]

2. Gupta-Bleuler method applied to the SME

The starting point is the Stuckelberg lagrangian including a CPT-conserving Lorentz-violating term as well as a mass term,

$$\mathcal{L} = -\frac{1}{4}F_{\mu\nu}F^{\mu\nu} - \frac{1}{4}k_F^{\mu\nu\alpha\beta}F_{\mu\nu}F_{\alpha\beta} + \frac{1}{2}m^2 A_\mu A^\mu - \frac{\lambda}{2}(\partial_\mu A^\mu)^2. \quad (1)$$

Note that the CPT-violating term has been omitted since it can cause instabilities even at tree level.[3] The gauge condition $\lambda = 1$ is also chosen for simplicity of the commutation relations. The starting assumptions are the standard covariant commutation relations for the field and the conjugate momenta

$$\pi^j = F^{j0} + k_F^{j0\alpha\beta} F_{\alpha\beta}, \quad \pi^0 = -\partial_\mu A^\mu. \tag{2}$$

Imposing equal-time canonical commutation rules

$$[A_\mu(t,\vec{x}), \pi^\nu(t,\vec{y})] = i\delta_\mu{}^\nu \delta^3(\vec{x}-\vec{y}), \tag{3}$$

along with

$$[A_\mu(t,\vec{x}), A_\nu(t,\vec{y})] = [\pi^\mu(t,\vec{x}), \pi^\nu(t,\vec{y})] = 0, \tag{4}$$

implements the standard canonical quantization in a covariant manner as is done in the conventional Gupta-Blueler method.[4] This implies that the time derivatives of the spatial components A^i satisfy the modified commutation relations

$$[\dot{A}^i(t,\vec{x}), A^j(t,\vec{y})] = -iR^{ij}\delta^3(\vec{x}-\vec{y}), \tag{5}$$

where R^{ij} is the inverse matrix of $\delta^{ij} - 2(k_F)^{0i0j}$. In any concordant frame where k_F is reasonably small, this inverse exists. The commutation relations involving $\dot{A}^0$ and $A^i$are the same as in the usual case, so it is convenient to define a covariant-looking tensor $\overline{\eta}^{\mu\nu}$ by setting $\overline{\eta}^{00} = 1$, $\overline{\eta}^{0i} = 0$, and $\overline{\eta}^{ij} = -R^{ij}$. The commutation relations are expressed as

$$[\dot{A}^\mu(t,\vec{x}), A^\nu(t,\vec{y})] = i\overline{\eta}^{\mu\nu}\delta^3(\vec{x}-\vec{y}). \tag{6}$$

This matrix is also the inverse of $\tilde{\eta}^{\mu\nu} = \eta^{\mu\nu} - 2k_F^{\mu00\nu}$ as $\overline{\eta}^{\mu\nu}\tilde{\eta}_{\nu\alpha} = \eta^\mu{}_\alpha$. Note that the time derivatives of A do not commute, rather

$$[\dot{A}^\mu(t,\vec{x}), \dot{A}^\nu(t,\vec{y})] = -2i\overline{\eta}_{\mu\alpha}(k_F^{\alpha0i\beta} + k_F^{\beta0i\alpha})\overline{\eta}_{\beta\nu}\frac{\partial}{\partial x^i}\delta^3(\vec{x}-\vec{y}), \tag{7}$$

involving the spatial derivatives of the delta function.

The equation of motion in momentum space is

$$(p^2 - m^2)\epsilon^\mu + 2(k_F)^{\mu\alpha\nu\beta}p_\alpha p_\beta \epsilon_\nu = 0, \tag{8}$$

where ϵ^μ is the polarization vector. One implication of this equation is found by dotting with p_μ yielding the condition

$$p^2 = m^2 \quad \text{or} \quad \epsilon \cdot p = 0, \tag{9}$$

A key observation is the modified orthogonality relation for the polarization vectors that follows from the equation of motion

$$\epsilon_\mu^{(\lambda')}(\vec{p})\left[\left(p_0^{(\lambda)} + p_0^{(\lambda')}\right)\tilde{\eta}^{\mu\nu} - 2\left(k_F^{\mu0i\nu} + k_F^{\nu0i\mu}\right)p_i\right]\epsilon_\nu^{(\lambda)}(\vec{p}) = 2p_0^{(\lambda)}\eta^{\lambda\lambda'}, \tag{10}$$

which holds whenever $p_0^{(\lambda)} \neq p_0^{(\lambda')}$. The normalization is chosen so that the $\lambda = 0$ polarization vector is timelike while the others are spacelike. This is possible due to the presence of a sufficiently large mass term which generically protects the normalization of the polarization vectors from vanishing. One of the issues of taking the $m \to 0$ limit is that the above orthogonality condition can fail due to some polarization vectors becoming lightlike.

3. Momentum-space expansion

The fields can be expanded in a standard Fourier expansion using

$$A_\mu(x) = \int \frac{d^3\vec{p}}{(2\pi)^3} \sum_\lambda \frac{1}{2p_0^{(\lambda)}} \left(a^\lambda(\vec{p})\epsilon_\mu^{(\lambda)}(\vec{p})e^{-ip\cdot x} + a^{\lambda\dagger}(\vec{p})\epsilon_\mu^{(\lambda)}(\vec{p})e^{ip\cdot x} \right). \tag{11}$$

The modified orthogonality relation for the polarization vectors can be used to invert this transform and solve for the raising and lowering operators. A straightforward computation then yields the standard relations

$$[a^\lambda(\vec{p}), a^{\lambda'\dagger}(\vec{q})] = -(2\pi)^3 2p_0^{(\lambda)} \eta^{\lambda\lambda'} \delta^3(\vec{p} - \vec{q}), \tag{12}$$

as well as

$$[a^\lambda(\vec{p}), a^{\lambda'}(\vec{q})] = 0, \tag{13}$$

demonstrating that the raising and lowering operators obey conventional statistical relations. There are subtle issues associated with the above expansion. Although the mass term is not explicitly present it turns out to be crucial for generating a complete set of polarization states required for the quantization procedure. When the mass is set to zero it turns out that the conjugate momentum π^0 is identically zero for most directions in momentum space. This creates a serious problem for the Gupta-Bleuler procedure as the gauge term initially added into the lagrangian is not sufficient to produce a generically nontrivial conjugate momenta for A^0. This indicates that the standard Gupta-Bleuler method in fact fails in the massless case, at least when there is birefringence present.

4. Explicit example

As an explicit example of the issue with $m \to 0$, consider the single parameter model $k_F^{0103} = k/2$ (along with required nonzero symmetric components to make it anti-self-dual and therefor pure birefringent). When $p_1 = p_2 = 0$, the matrix for K in the equation of motion is

$$K^{\mu\nu}(p) = k \begin{pmatrix} 0 & p_0 p_3 & 0 & 0 \\ p_0 p_3 & 0 & 0 & -p_0^2 \\ 0 & 0 & 0 & 0 \\ 0 & -p_0^2 & 0 & 0 \end{pmatrix} . \tag{14}$$

Searching for zero eigenvalues (with $p^2 = 0$, candidates for the gauge modes...) yields two eigenvectors, one with the polarization vector proportional to the momentum and another with

$$\epsilon = \begin{pmatrix} 0 \\ 0 \\ 1 \\ 0 \end{pmatrix} . \tag{15}$$

Both of these modes satisfy $\epsilon \cdot p = 0$ indicating that there is in fact no nontrivial mode corresponding to $\pi^0 = -\partial \cdot A$. Making the momentum more general does not help as the rank of the K matrix is generally increased to three indicating the same fundamental problem.

5. Summary

The standard Gupta-Bleuler method seems to work well when there is a mass term present in the lagrangian, but there are serious impediments to implementing this method when the mass is identically zero. The most serious issue appears to be the vanishing of π^0 implied by the equation of motion, something that is not an issue in the conventional case. In addition, certain directions in momentum space yield a set of polarization vectors that is strictly less than four-dimensional.

Acknowledgments

I would like to thank New College of Florida for summer funding that aided in the completion of this project.

References

1. D. Colladay and V.A. Kostelecký, Phys. Rev. D **55** 6760 (1997); Phys. Rev. D **58** 116002 (1998).
2. M. Hohensee, D. Phillips, and R. Walsworth, arXiv:1210.2683.
3. S. Carroll, G. Field, and R. Jackiw, Phys. Rev. D **41**, 1231(1990).
4. S. Gupta, Proc. Phys. Soc. **A63** 681 (1950); K. Bleuler, Helv. Phys. Acta **23** 567 (1950).

USING MICROWAVE CAVITIES TO TEST LORENTZ INVARIANCE

S.R. PARKER,* P.L. STANWIX, E.N. IVANOV and M.E. TOBAR

School of Physics, The University of Western Australia
Crawley, WA 6009, Australia
**E-mail: stephen.parker@uwa.edu.au*

M. NAGEL, E.V. KOVALCHUK and A. PETERS

Institut für Physik, Humboldt-Universität zu Berlin
Berlin, 12489, Germany

Microwave cavities can be used to generate signals that exhibit fractional frequency stabilities on the order of parts in 10^{-16}. Such signals provide an excellent tool for constraining potential violations of local Lorentz invariance in the photon sector of the Standard-Model Extension. In this work we describe the current methods employed to design and construct such an experiment.

1. Introduction

A cryogenic sapphire oscillator[1,2] (CSO) exploits the low loss properties of sapphire at cryogenic temperatures to generate microwave signals with short term fractional frequency stability below 10^{-15}. If two CSOs are aligned orthogonally the beat frequency between the oscillators can be used to search for orientation dependent deviations in the speed of light. The sensitivity of such an experiment is dictated by the stability of the beat frequency. Active rotation of the experiment can increase the effective sensitivity although it can also induce extra systematic noise. Microwave oscillators have been used in some of the most sensitive tests of Lorentz invariance[3,4] and we will provide a brief outline of the design and principles behind such an experiment.

2. Cryogenic sapphire oscillators

Cryogenic sapphire oscillators use a sapphire loaded cavity as the frequency discriminating element of a Pound-Drever-Hall phase locked loop. The fractional frequency stability of a CSO output at 11.2 GHz is state-of-the-art for integration times up to 10^3 seconds and is rivaled only by microwave signals generated through optical-comb division for very short integration times (up to 5 seconds). A HEMEX grade cylindrical sapphire crystal is mounted in a silver plated copper cavity and cooled to cryogenic temperatures. Magnetic loop probes are used to excite resonant whispering gallery modes in the sapphire where the majority of the electromagnetic energy is confined. At cryogenic temperatures sapphire has a low loss tangent that gives rise to unloaded electrical quality factors on the order of 10^9 for whispering gallery modes. In this regime the sapphire crystal also exhibits a frequency-temperature turning point. A sensor and heater mounted on the copper post that holds the cavity is used to keep the temperature constant at this turning point, making the resonator frequency first order insensitive to temperature fluctuations.

Fluctuations in the power incident on the sapphire loaded cavity give rise to frequency fluctuations due to changes in temperature and radiation pressure in the sapphire. Recent work[5] has indicated that this source of frequency instability limits CSO performance over time scales most pertinent to actively rotated Lorentz invariance experiments. This is a Fourier frequency independent limit and represents the current best achievable frequency stability for a CSO.

3. Rotation

It is beneficial to actively rotate the experiment in the laboratory instead of relying on passive rotation provided by the Earth. By using a rotation period that coincides with the lowest level of beat frequency instability one can maximize the sensitivity of the experiment. Rotation can however induce extra sources of noise as demonstrated in Fig. 1. Air bearing rotation tables use optical encoders to keep a constant rotation speed. Long term tilt variations can be compensated for by a tilt control system that heats and cools two of the three legs the experiment sits on. Frequency instability can also arise from magnetic interference and changes in the pressure of the helium gas in the return line that is boiled off from the experiment.

4. Data analysis

Experimental sensitivity is calculated using the field equations describing the electric and magnetic fields in the cavities. An Earth based experiment as described here is most sensitive to rotation violations and has a suppressed sensitivity to boost violations.

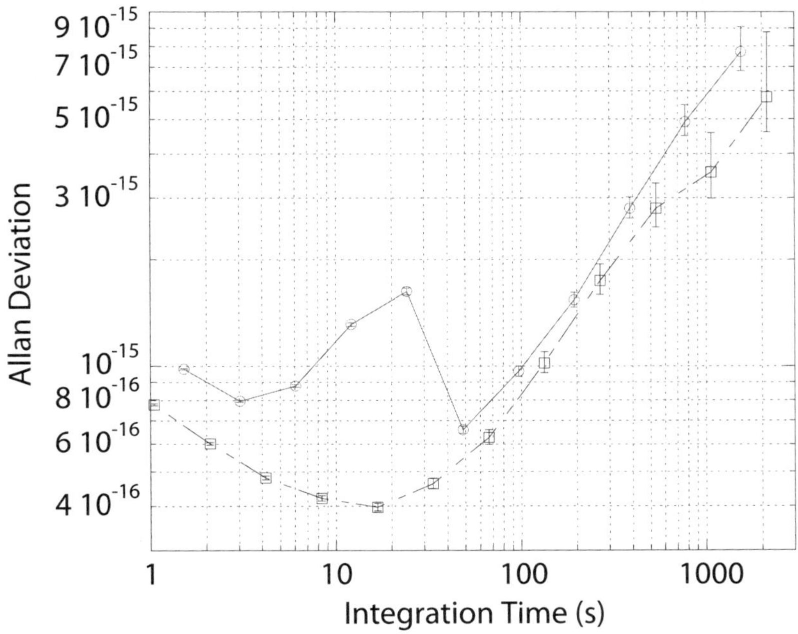

Fig. 1. Allan deviation of fractional frequency stability as a function of integration time for the beat frequency comparison of two orthogonally aligned cryogenic sapphire oscillators. Square data points are for a stationary comparison and circular data points are for a rotating measurement. Error bars are statistical.

The collected data is used to place bounds on coefficients for Lorentz violation in the photon sector of the Standard-Model Extension.[6] Analysis is performed in the Sun-centered celestial equatorial frame to allow for consistency and comparisons between different experiments. The data is demodulated around the frequency of active laboratory rotation and then a weighted least squares regression is used to fit to different frequencies of interest. These frequencies cover Earth's rotational and orbital period and relevant harmonics. In some situations it can be beneficial to fit to the time derivative of the beat frequency data as this reduces the impact of sudden systematic jumps in the beat data.[7]

5. Conclusion

The excellent fractional frequency stability performance of microwave oscillators has allowed for some of the most sensitive laboratory tests of Lorentz invariance in the photon sector of the SME. Work is still ongoing to further improve oscillator performance in the regime most relevant to Lorentz invariance experiments.

Acknowledgments

This work was partially funded by Australian Research Council Grant numbers DP1092690 and DP130100205 with support from the DAAD/Go8 Joint Research Co-operation Scheme.

References

1. C.R. Locke, E.N. Ivanov, J.G. Hartnett, P.L. Stanwix and M.E. Tobar, Rev. Sci. Instrum. **79**, 051301 (2008).
2. J.G. Hartnett, N.R. Nand and C. Lu, Appl. Phys. Lett. **100**, 183501 (2012).
3. P.L. Stanwix *et al.*, Phys. Rev. Lett. **95**, 040404 (2005).
4. *Data Tables for Lorentz and CPT Violation,* V.A. Kostelecký and N. Russell, 2013 edition, arXiv:0801.0287v6.
5. N.R. Nand, S.R. Parker, E.N. Ivanov, J.M. le-Floch, J.G. Hartnett and M.E. Tobar, Appl. Phys. Lett., in press.
6. V.A. Kostelecký and M. Mewes, Phys. Rev. D **66**, 056005 (2002).
7. M.A. Hohensee, P.L. Stanwix, M.E. Tobar, S.R. Parker, D.F. Phillips and R.L. Walsworth, Phys. Rev. D **82**, 076001 (2010).

USING QUANTUM ENTANGLEMENT TO STUDY CP AND CPT VIOLATIONS

YU SHI
Department of Physics, Fudan University
Shanghai, 200433, China
E-mail: yushi@fudan.edu.cn

We report some general phenomenological results concerning CP and CPT violations in joint decays of entangled pseudoscalar neutral mesons.

1. Introduction

Quantum entanglement refers to the situation that the quantum state of a composite system is not a direct product of its subsystems. Its significance was discovered by Einstein, Podolsky and Rosen.[1] Schrödinger coined the term and regarded it as 'the characteristic trait of quantum mechanics, the one that enforces its entire departure from classical lines of thought.'[2] Pseudoscalar neutral mesons are copiously generated as pairs entangled in flavor space.[3–9] CP and CPT violating parameters can be measured by studying the joint decays of entangled meson pairs.[3,8,10–15] This provides a venue for studying the Standard-Model Extension.[11]

2. Single meson states

$|M^0\rangle$ and $|\bar{M}^0\rangle$ are eigenstates of parity P both with eigenvalue -1, and of a characteristic flavor $\mathcal{F}$ with eigenvalues ± 1. We have $C|M^0\rangle = -|\bar{M}^0\rangle$, $C|\bar{M}^0\rangle = -|M^0\rangle$. Hence the eigenstates of CP are $|M_\pm\rangle = (|M^0\rangle \pm |\bar{M}^0\rangle)/\sqrt{2}$, with eigenvalues ± 1. In the Wigner-Weisskopf approximation, the weak decay of a meson can be described by $i\frac{\partial}{\partial t}|M(t)\rangle = H|M(t)\rangle$, where $H = \begin{pmatrix} H_{00} & H_{0\bar{0}} \\ H_{\bar{0}0} & H_{\bar{0}\bar{0}} \end{pmatrix}$. It is defined that $(1-\epsilon_M)/(1+\epsilon_M) \equiv \sqrt{H_{\bar{0}0}/H_{0\bar{0}}} \equiv q/p$, $\delta_M \equiv (H_{\bar{0}\bar{0}} - H_{00})/\sqrt{H_{0\bar{0}}H_{\bar{0}0}}$. If CP or T is conserved indirectly, then $\epsilon_M = 0$. If CPT or CP is conserved indirectly, then $\delta_M = 0$.

The eigenstates $|M_S\rangle$ and $|M_L\rangle$ of H, corresponding to the eigenvalues λ_S and λ_L respectively, are found by diagonalizing H. Starting as $|M_S\rangle$, the state of a single meson evolves as $|M_S(t)\rangle = e^{-i\lambda_S t}|M_S\rangle$. Starting as $|M_L\rangle$, the state evolves as $|M_L(t)\rangle = e^{-i\lambda_L t}|M_L\rangle$. Based on this, one obtains $|M^0(t)\rangle$, which starts as $|M^0\rangle$; $|\bar{M}^0(t)\rangle$, which starts as $|\bar{M}^0\rangle$; $|M_+(t)\rangle$, which starts as $|M_+\rangle$; and $|M_-(t)\rangle$, which starts as $|M_-\rangle$.

To characterize direct violations, for decays into flavor eigenstates $|l^{\pm}\rangle$ with eigenvalue ± 1, we define decay amplitudes $R^+ \equiv \langle l^+|\mathcal{H}|M^0\rangle$, $S^+ \equiv \langle l^+|\mathcal{H}|\bar{M}^0\rangle$, $S^- \equiv \langle l^-|\mathcal{H}|M^0\rangle$, $R^- \equiv \langle l^-|\mathcal{H}|\bar{M}^0\rangle$. They can be related to quantities a, b, c, d usually defined.[3,13] If CP is conserved directly, then $R^+ = R^-$ and $S^+ = S^-$. If CPT is conserved directly, then $(R^+)^* = R^-$ and $(S^+)^* = S^-$. If the $\Delta\mathcal{F} = \Delta Q$ rule is respected, then $S^{\pm} = 0$.

For decays into CP eigenstates $|h^{\pm}\rangle$ with eigenvalue ± 1, we define decay amplitudes $Q^+ \equiv \langle h^+|\mathcal{H}|M_+\rangle$, $X^+ \equiv \langle h^-|\mathcal{H}|M_+\rangle$, $X^- \equiv \langle h^+|\mathcal{H}|M_-\rangle$, $Q^- \equiv \langle h^-|\mathcal{H}|M_-\rangle$. These newly defined quantities are convenient. If CP is conserved directly, then $X^{\pm} = 0$. If CPT is conserved directly, then $X^{\pm}$ is purely imaginary.

3. Entangled states

The $C = -1$ entangled state of a pair of pseudoscalar mesons is $|\Psi_-\rangle = (|M^0\rangle_a|\bar{M}^0\rangle_b - |\bar{M}^0\rangle_a|M^0\rangle_b)/\sqrt{2} = (|M_-\rangle_a|M_+\rangle_b - |M_+\rangle_a|M_-\rangle_b)/\sqrt{2}$, which is produced for kaons at the ϕ resonance,[3] and for B mesons at the $\Upsilon(4s)$ resonance[4,5] and the Υ(5S) resonance with a large branching ratio.[6–8] The $C = +1$ entangled state is $|\Psi_+\rangle = (|M_0\rangle|\bar{M}_0\rangle + |\bar{M}_0\rangle|M_0\rangle)/\sqrt{2} = (|M_+\rangle_a|M_+\rangle_b - |M_-\rangle_a|M_-\rangle_b)/\sqrt{2}$, which is produced for B mesons at the Υ(5S) resonance with some branching ratio[6–8] and above the $\Upsilon(4s)$ resonance.[9]

Although physically a single meson cannot be in a CP eigenstate $|M_{\pm}\rangle$ because of CP violation, the entangled states can be *exactly* written in terms of $|M_{\pm}\rangle$. This ma8kes the expression $|M_{\pm}(t)\rangle$ and the decay amplitudes $Q^{\pm}$ and $X^{\pm}$ meaningful and useful.

Starting from $|\Psi_{\mp}\rangle$, the entangled meson pair decays to certain products at t_a and t_b. Hence the time-dependent states are $|\Psi_{\mp}(t_a, t_b)\rangle = (|M^0(t_a)\rangle_a|\bar{M}^0(t_b)\rangle_b \mp |\bar{M}^0(t_a)\rangle_a|M^0(t_b)\rangle_b)/\sqrt{2} = (|M_{\mp}(t_a)\rangle_a|M_+(t_b)\rangle_b - |M_{\pm}(t_a)\rangle_a|M_-(t_b)\rangle_b)/\sqrt{2}$.

4. Joint decay rates

For state $|\Psi(t_a, t_b)\rangle$, the joint rate that particle a decays to f at t_a while particle b decays to g at t_b is $I(f, t_a; g, t_b) = |\langle f, g|\mathcal{H}_a\mathcal{H}_b|\Psi(t_a, t_b)\rangle|^2$, where $\mathcal{H}_a$ and $\mathcal{H}_b$ represent the hamiltonians governing the weak decays of a and b, respectively. In experiments, it is more convenient to use the integrated rate $I(f,g,\Delta t) = \int_0^\infty I(f,t_a;g,t_a+\Delta t)dt_a$. Then one can find the asymmetry between the joint decays to f and g and the joint decays to f' and g',

$$A(fg,f'g',\Delta t) \equiv \frac{I[f,t_a;g,t_a+\Delta t]-I[f',t_a;g',t_a+\Delta t]}{I[f,t_a;g,t_a+\Delta t]+I[f',t_a;g',t_a+\Delta t]} = \frac{I[f,g,\Delta t]-I[f',g',\Delta t]}{I[f,g,\Delta t]+I[f',g',\Delta t]}. \tag{1}$$

We have considered the following cases of the final states: (1) the decay products are flavor eigenstates $|l^\pm\rangle$, with the equal-flavor asymmetry $A(l^+l^+,l^-l^-,\Delta t)$ and the unequal-flavor asymmetry $A(l^+l^-,l^-l^+,\Delta t)$; (2) the decay products are CP eigenstates $|h^\pm\rangle$, with the equal-CP asymmetry $A(h^+h^+,h^-h^-,\Delta t)$ and the unequal-CP asymmetry $A(h^+h^-,h^-h^+,\Delta t)$; (3) the decay products $|h_1\rangle$ and $|h_2\rangle$ are CP conjugates.

5. General results on joint decays of $|\Psi_-\rangle$[13]

Theorem 1. If the equal-flavor asymmetry is nonzero, then there exists one or two of the following violations: (1) CP is violated indirectly, (2) both CP and CPT are violated directly.

Theorem 2. If the equal-flavor asymmetry is nonzero while CPT is assumed to be conserved both directly and indirectly, then in addition to indirect CP violation, we can draw the following conclusions: (1) $|q/p| \neq 1$, i.e., T must also be violated indirectly; (2) $|\langle l^+|\mathcal{H}|\bar{M}_0\rangle| \neq |\langle l^+|\mathcal{H}|M^0\rangle|$, $|\langle l^-|\mathcal{H}|M^0\rangle| \neq |\langle l^-|\mathcal{H}|\bar{M}_0\rangle|$, despite $\langle l^+|\mathcal{H}|M^0\rangle = \langle l^-|\mathcal{H}|\bar{M}_0\rangle^*$ and $\langle l^+|\mathcal{H}|\bar{M}_0\rangle = \langle l^-|\mathcal{H}|M^0\rangle^*$.

Theorem 3. If the unequal-flavor asymmetry is nonzero, then CP must be violated, directly or indirectly or both.

Theorem 4. If the unequal-flavor asymmetry is nonzero for $\Delta t \neq 0$ while CPT is assumed to be conserved both directly and indirectly, then we can draw the following conclusions: (1) $|\langle l^+|\mathcal{H}|\bar{M}_0\rangle| = |\langle l^-|\mathcal{H}|M^0\rangle| \neq |\langle l^+|\mathcal{H}|M^0\rangle| = |\langle l^-|\mathcal{H}|\bar{M}_0\rangle|$; (2) $\langle l^-|\mathcal{H}|M^0\rangle = \langle l^+|\mathcal{H}|\bar{M}^0\rangle^* \neq 0$, which means the $\Delta\mathcal{F} = \Delta Q$ rule must be violated.

Theorem 5. The equal-CP asymmetry is constant independent of Δt.

Theorem 6. For $\Delta t = 0$, the unequal-CP asymmetry vanishes, independent of whether CP or CPT is violated.

Theorem 7. If any equal-CP joint decay rate is nonzero, then CP must be violated, directly or indirectly or both.

6. General results on joint decays of $|\Psi_+\rangle$[14]

Theorem 8. If the unequal-flavor asymmetry is nonzero, then CP must be violated indirectly.

Theorem 9. If $I(l^+, l^-, \Delta t)$ and $I(l^-, l^+, \Delta t)$ depend on the first order of ϵ_M, then CP must also be violated directly.

Theorem 10. If the unequal-flavor asymmetry depends on the first or second order of ϵ_M, then the $\Delta\mathcal{F} = \Delta Q$ rule is violated.

Theorem 11. Suppose $|h_1\rangle$ and $|h_2\rangle$ are CP conjugates. If $I(h_1, h_2; \Delta t)$ and $I(h_2, h_1; \Delta t)$ depend on the first order of ϵ_M, then CP is violated directly.

In addition, we have derived various quantitative relations of the indirect violating parameters with the decay asymmetries of $|\Psi_-\rangle$,[12,13] with those of $|\Psi_+\rangle$[14] and with four asymmetries defined for some time-ordered integrated rates of $|\Psi_-\rangle$ and $|\Psi_+\rangle$.[15]

Acknowledgments

This work is supported by the National Science Foundation of China under grant number 10875028.

References

1. A. Einstein, B. Podolsky and N. Rosen, Phys. Rev. **47**, 777 (1935).
2. E. Schrödinger, Proc. Camb. Phil. Soc. **31**, 555 (1936).
3. A . Di Domenico, Found. Phys. **40**, 852 (2010).
4. B. Aubert *et al.*, Phys. Rev. Lett. **88**, 221802 (2002).
5. A. Go *et al.*, Phys. Rev. Lett. 99, 131802 (2007).
6. O. Aquines *et al.*, Phys. Rev. Lett. 96, 152001 (2006).
7. R. Louvot *et al.*, Phys. Rev. Lett. 102, 021801 (2009).
8. D. Atwood and A. Soni, Phys. Rev. D **82**, 036003 (2010).
9. D.S. Skerib *et al.*, Phys. Rev. Lett. 67, 1692 (1991).
10. J. Bernabéu *et al.*, Phys. Lett. B **211**, 226 (1980); I. Dunietz *et al.*, Phys. Rev. D **35**, 2166 (1987); C.D. Buchanan *et al.*, Phys. Rev. D **45**, 4088 (1992); G. D'Ambrosio *et al.*, in L. Maiani *et al.*, eds., *The Second DAΦNE Physics Handbook*, SIS-Publicazioni, Frascati, 1995.
11. V.A. Kostelecký, Phys. Rev. Lett. **80**, 1818 (1998); Phys. Rev. D **61**, 016002 (1999); Phys. Rev. D **64**, 076001 (2001).
12. Y. Shi, Eur. Phys. J. C **72**, 1907 (2012).
13. Y. Shi, arXiv:1306.2676.
14. Z. Huang and Y. Shi, arXiv:1307.4459.
15. Z. Huang and Y. Shi, Eur. Phys. J. C **72**, 1900 (2012).

A CPT-EVEN LORENTZ-VIOLATING NONMINIMAL COUPLING BETWEEN FERMIONS AND PHOTONS

R. CASANA, M.M. FERREIRA JR.,* F.E.P. DOS SANTOS, E.O. SILVA

Departamento de Física, Universidade Federal do Maranhão
São Luís, Maranhão 65080-805, Brazil
**E-mail: manojr07@ibest.com.br*

E. PASSOS

Departamento de Física, Universidade Federal de Campina Grande
Caixa Postal 10071, 58109-970 Campina Grande, PB, Brazil

We propose a CPT-even and Lorentz-violating dimension-five nonminimal coupling between fermionic and gauge fields, involving the CPT-even and Lorentz-violating gauge tensor of the Standard-Model Extension. This nonminimal coupling modifies the nonrelativistic regime of Dirac particles, inducing new effects such as an electric-Zeeman-like spectrum splitting and an anomalous-like contribution to the electron magnetic moment. These new effects allow to constrain the magnitude of this nonminimal coupling in 1 part in 10^{16}.

The Standard-Model Extension (SME)[1] is the natural framework for studying properties of Lorentz violation (LV) in all sectors of the Standard Model.[2] Some time ago, a Lorentz-violating and CPT-odd nonminimal coupling between fermions and the gauge field was proposed[3] of the form $D_\mu = \partial_\mu + ieA_\mu + ig\epsilon_{\mu\lambda\alpha\beta}(k_{AF})^\lambda F^{\alpha\beta}/2$, in the context of the Dirac equation, where $(k_{AF})^\mu = (\mathrm{v}_0, \mathbf{v})$ is the Carroll-Field-Jackiw 4-vector and g is the coupling constant. The analysis of the nonrelativistic limit revealed that this coupling provides a magnetic moment $(g\mathbf{v})$ for uncharged particles,[3] and corrections to the hydrogen spectrum.[4] Contributions to the dynamics of the Aharonov-Bohm-Casher problem[5] and to the fermion-fermion ultrarelativistic scattering[6] were also analyzed. In the present work, we propose a new CPT-even,[7] dimension-five, nonminimal coupling linking the fermionic and gauge fields in the context of the Dirac equation. The starting point is a nonminimal coupling involving fundamental Dirac fermions and the electromagnetic field in the context of the Dirac equation, $(i\gamma^\mu D_\mu - m_e)\Psi^{(e)} = 0$, where $\Psi^{(e)}$ is the electron spinor wave function, and the covariant derivative with nonminimal coupling is

$$D_\mu = \partial_\mu + ieA_\mu + \frac{\lambda^{(e)}}{2}\left(K_F\right)_{\mu\nu\alpha\beta}\gamma^\nu F^{\alpha\beta}, \tag{1}$$

where $\lambda^{(e)}$ is the electron nonminimal coupling constant, and $\left(K_F\right)_{\mu\nu\alpha\beta}$ is the CPT-even LV tensor of the photon sector of SME, which can be written in terms of four 3×3 matrices $\kappa_{DE}, \kappa_{DB}, \kappa_{HE}, \kappa_{HB}$: $\left(\kappa_{DE}\right)_{jk} = -2\left(K_F\right)_{0j0k}$, $\left(\kappa_{HB}\right)_{jk} = \epsilon_{jpq}\epsilon_{klm}\left(K_F\right)_{pqlm}/2$, $\left(\kappa_{DB}\right)_{jk} = -\left(\kappa_{HE}\right)_{kj} = \epsilon_{kpq}\left(K_F\right)_{0jpq}$. The Dirac equation can be explicitly written as

$$\left[i\gamma^\mu\partial_\mu - e\gamma^\mu A_\mu + \frac{\lambda^{(e)}}{2}\left(K_F\right)_{\mu\nu\alpha\beta}\sigma^{\mu\nu}F^{\alpha\beta} - m_e\right]\Psi^{(e)} = 0, \tag{2}$$

and $\sigma^{\mu\nu} = i(\gamma^\mu\gamma^\nu - \gamma^\nu\gamma^\mu)/2$. Thus, the relevant electron lagrangian is

$$\mathcal{L}_{(e)} = \bar{\Psi}^{(e)}(i\partial\!\!\!/ - eA\!\!\!/ - m_e + \frac{\lambda^{(e)}}{2}\left(K_F\right)_{\mu\nu\alpha\beta}\sigma^{\mu\nu}F^{\alpha\beta})\Psi^{(e)}, \tag{3}$$

with $\sigma = (\sigma_x, \sigma_y, \sigma_z)$ being the Pauli matrices. In the momentum coordinates, $i\partial_\mu \to p_\mu$, the corresponding Dirac equation is

$$\begin{aligned} i\partial_t\Psi^{(e)} = \Big[&\alpha\cdot(\mathbf{p} - e\mathbf{A}) + eA_0 + m_e\gamma^0 - \lambda^{(e)} i\gamma^j\left(\mathbb{E}^j + \mathbb{B}^j\right) \\ &+ \lambda^{(e)}\gamma^0\Sigma^k\left(\tilde{\mathbb{E}}^k + \tilde{\mathbb{B}}^k\right)\Big]\Psi^{(e)}. \end{aligned} \tag{4}$$

where we have used $F_{0j} = E^j, F_{mn} = \epsilon_{mnp}B^p$, $\sigma^{0j} = i\alpha^j$, $\sigma^{ij} = -\epsilon_{ijk}\Sigma^k$,

$$\alpha^i = \begin{pmatrix} 0 & \sigma^i \\ \sigma^i & 0 \end{pmatrix}, \qquad \Sigma^k = \begin{pmatrix} \sigma^k & 0 \\ 0 & \sigma^k \end{pmatrix}. \tag{5}$$

and introduced the following definitions:

$$\begin{aligned} \mathbb{E}^k &= (\kappa_{DE})_{kj}E^j, \qquad & \mathbb{B}^k &= (\kappa_{DB})_{kj}B^j, \\ \tilde{\mathbb{E}}^k &= (\kappa_{HE})_{kq}E^q, \qquad & \tilde{\mathbb{B}}^k &= (\kappa_{HB})_{kp}B^p. \end{aligned} \tag{6}$$

In order to investigate the role played by this nonminimal coupling, we should evaluate the nonrelativistic limit of the Dirac equation. At first order in the Lorentz violating parameters, the following hamiltonian is achieved:

$$\begin{aligned} H^{(e)} = &\frac{1}{2m_e}\left[(\boldsymbol{p} - e\boldsymbol{A})^2 - e\left(\boldsymbol{\sigma}\cdot\boldsymbol{B}\right)\right] + eA_0 + \lambda^{(e)}\,\boldsymbol{\sigma}\cdot\left(\tilde{\mathbb{E}} + \tilde{\mathbb{B}}\right) \\ &- \frac{\lambda^{(e)}}{m_e}\left(\mathbb{E} + \mathbb{B}\right)\cdot(\boldsymbol{\sigma}\times\mathbf{p}) + \frac{e\lambda^{(e)}}{m_e}\left(\mathbb{E} + \mathbb{B}\right)\cdot(\sigma\times\mathbf{A}), \end{aligned} \tag{7}$$

here written for uniform fields. This hamiltonian induces new effects to be here discussed. The term $\mathbb{E}\cdot(\boldsymbol{\sigma}\times\mathbf{p})$ is a generalization of the Rashba coupling term, $\mathbf{E}\cdot(\boldsymbol{\sigma}\times\mathbf{p})$, while $\lambda^{(e)}(\boldsymbol{\sigma}\cdot\tilde{\mathbb{B}})$ implies a straightforward tree-level contribution to the anomalous magnetic moment of the electron, and the term $\lambda^{(e)}(\boldsymbol{\sigma}\cdot\tilde{\mathbb{E}})$ leads to a kind of electric Zeeman effect. We begin discussing the correction induced on the atomic spectrum of hydrogen by the term $\boldsymbol{\sigma}\cdot\tilde{\mathbb{E}}$ involving the spin operator. In this case, the correction energy is given by $\Delta_E=\lambda^{(e)}\langle nljm_jm_s|\,\boldsymbol{\sigma}\cdot\tilde{\mathbb{E}}|nljm_jm_s\rangle$. Now, we adopt a polarized spin configuration, $\boldsymbol{\sigma}=\sigma_z\hat{z}$, such that

$$\Delta_E=\pm\lambda^{(e)}\left(\kappa_{HE}\right)_{3j}E_j\frac{m_j}{2l+1},\tag{8}$$

with E_j being one of the components of the electric field, and $(\kappa_{HE})_{3j}$ a non-null element of the matrix (κ_{HE}). The positive and negative signs correspond to $j=l+1/2$ and $j=l-1/2$, respectively. We also used $\langle nljm_jm_s|\sigma_z|nljm_jm_s\rangle=m_j\hbar/(2l+1)$, $\langle nljm_jm_s|\sigma_x|nljm_jm_s\rangle=\langle nljm_jm_s|\sigma_y|nljm_jm_s\rangle=0$. The dependence on m_j leads to a spectrum splitting in $(2j+1)$ lines, representing an electric Zeeman-like effect (due to the presence of an electric field, that can be external or the atomic one). Regarding the possibility of measuring spectrum shifts as small as 10^{-10} eV, and working with a typical atomic electric field for the hydrogen fundamental level ($a_0\simeq 0.529$ Å), whose magnitude is $E\simeq 5.1\times10^{11}N/C\simeq 1.2\times10^{6}\,(\mathrm{eV})^2$, the Zeeman-like splitting will be undetectable if $|\lambda^{(e)}\left(\kappa_{HE}\right)_{3j}|E_j<10^{-10}$ eV. It leads to the following upper bound:

$$\left|\lambda^{(e)}\left(\kappa_{HE}\right)_{3j}\right|<8\times10^{-17}\,(\mathrm{eV})^{-1}.\tag{9}$$

Another effect associated with the hamiltonian (7) is concerned with the anomalous magnetic moment of the electron. The electron magnetic moment is $\boldsymbol{\mu}=-\mu\boldsymbol{\sigma}$, with $\mu=e/2m_e$, and $g=2$ the gyromagnetic factor. The anomalous magnetic moment of the electron is given by $g=2(1+a)$, with $a=\alpha/2\pi+...=0.00115965218279$ representing the deviation (value in the year 2008) in relation to the usual case. Hence, the magnetic interaction is $\boldsymbol{H}'=\mu(1{+}a)\boldsymbol{\sigma}\cdot\mathbf{B}$. Very precise measurements and QED calculations[8] yield $\Delta a\leq 3\times10^{-11}$. The term $\lambda^{(e)}\,\boldsymbol{\sigma}\cdot\tilde{\mathbb{B}}$ provides tree-level LV contributions to the magnetic interaction, $\mu\left[1+2m_e\lambda^{(e)}\left(\kappa_{HB}\right)_{33}/e\right](\boldsymbol{\sigma}_zB_0)$, for the magnetic field along the z-axis, $\mathbf{B}=B_0\hat{z}$, and a spin-polarized configuration in the z-axis. Such correction can not be larger than a, so that

$$\left|\lambda^{(e)}\left(\kappa_{HB}\right)_{33}\right|\leq 9.7\times10^{-11}\;(\mathrm{eV})^{-1},\tag{10}$$

where we have used $m_e=5.11\times10^{5}$ eV, $e=\sqrt{1/137}$. It is important to mention that the bounds here found should not be confused with the upper bounds on the $(K_F)_{\mu\nu\alpha\beta}$ components already known in the literature, as in the present case the constraint is on the magnitude of the CPT-even parameters as nonminimally coupled. A complete investigation of the corrections on the hydrogen spectrum implied by the hamiltonian (7) can be carefully carried out having as counterpart the procedures known in the literature.[9] The contributions implied by this nonminimal interaction to the photon self energy were already evaluated.[10] Also, the contributions to the fermion self-energy (searching for possible modifications in the fermionic dispersion relations) and to the vertex function are under analysis. These calculations are relevant for establishing the 1-loop consistency of this model.

Acknowledgments

The authors thank CAPES, CNPq, and FAPEMA for financial support.

References

1. D. Colladay and V.A. Kostelecký, Phys. Rev. D **55**, 6760 (1997); Phys. Rev. D **58**, 116002 (1998); S.R. Coleman and S.L. Glashow, Phys. Rev. D **59**, 116008 (1999); V.A. Kostelecký and S. Samuel, Phys. RevLett. **63**, 224 (1989); Phys. Rev. Lett. **66**, 1811 (1991); Phys. Rev. D **39**, 683 (1989); Phys. Rev. D **40**, 1886 (1989).
2. G.M. Shore, Nucl. Phys. B **717**, 86 (2005); D. Colladay and V.A. Kostelecký, Phys. Lett. B **511**, 209 (2001); R. Lehnert, Phys. Rev. D **68**, 085003 (2003); V.A. Kostelecký and C. D. Lane, J. Math. Phys. **40**, 6245 (1999); R. Lehnert, J. Math. Phys. **45**, 3399 (2004); V.A. Kostelecký and R. Lehnert, Phys. Rev. D **63,** 065008 (2001); V.A. Kostelecký and M. Mewes, Phys. Rev. Lett. **87**, 251304 (2001); Phys. Rev. D **66**, 056005 (2002).
3. H. Belich, T. Costa-Soares, M.M. Ferreira Jr., and J.A. Helayël-Neto, Eur. Phys. J. C **41,** 421 (2005).
4. H. Belich, T. Costa-Soares, M.M. Ferreira Jr., J.A. Helayël-Neto, and F.M.O. Moucherek, Phys. Rev. D **74**, 065009 (2006).
5. H. Belich, E.O. Silva, M.M. Ferreira Jr., and M.T.D. Orlando, Phys. Rev. D **83**, 125025 (2011).
6. B. Charneski, M. Gomes, R.V. Maluf, A.J. da Silva, Phys. Rev. D **86**, 045003 (2012).
7. R. Casana, M.M. Ferreira, Jr., R.V. Maluf, and F.E.P. dos Santos, Phys. Rev. D **86**, 125033 (2012); R. Casana, M.M. Ferreira, Jr., E. Passos, F.E.P. dos Santos, and E.O. Silva, Phys. Rev. D **87**, 047701 (2013).
8. G. Gabrielse, D. Hanneke, T. Kinoshita, M. Nio, and B. Odom, Phys. Rev. Lett. **97**, 030802 (2006).
9. T.J. Yoder and G.S. Adkins, Phys. Rev. D **86**, 116005 (2012).
10. R. Casana, M.M. Ferreira Jr, R.V. Maluf, and F.E.P. dos Santos, arXiv:1302.2375.

ULTRA-STABLE CRYOGENIC OPTICAL RESONATORS FOR TESTS OF FUNDAMENTAL PHYSICS

M. NAGEL,* K. MÖHLE, K. DÖRINGSHOFF, S. SCHIKORA, E.V. KOVALCHUK, and A. PETERS

Institut für Physik, Humboldt-Universität zu Berlin
Newtonstr. 15, 12489 Berlin, Germany
**E-mail: moritz.nagel@physik.hu-berlin.de*

We present the design and first measurement results for an ultra-stable cryogenically cooled optical sapphire resonator system with a potential relative frequency stability better than 3×10^{-17}. This level of oscillator stability allows for more precise tests of Einstein's theories of relativity and thus could help to find first hints of new physics. We will give some details on a projected experiment to test Lorentz invariance that will utilize these cavities.

1. Introduction

Time and frequency are the physical quantities which can be measured with by far the highest precision in modern physics and thus clock-comparison experiments are a class of exceptionally sensitive experiments that can be performed in the laboratory to test the foundations of modern physics, e.g., Lorentz invariance.

One particular type of 'clock-comparison' experiment uses optical cavities to search for possible Lorentz invariance violating anisotropies.[1] These types of experiments are mostly referred to as modern Michelson-Morley experiments because of their similarity to the classic Michelson-Morley experiment[2] performed in 1887. The basic idea is to compare the resonance frequencies $\nu = qc/2nL$ of two or more linear optical Fabry-Pérot resonators (q is an integer, c is the speed of light in vacuum, n the index of refraction of the medium, if present, parallel to the resonators axis, and L the length of the resonator) and to look for frequency changes $\delta\nu/\nu$ with respect to the orientation or velocity of the cavities in space, which would indicate a violation of rotation invariance or Lorentz boost invariance, respectively.

The comparison of the resonance frequencies is normally done by stabilizing lasers to the cavities using the Pound-Drever-Hall method[3] and taking a beat note measurement. Advantageous is the fact that in principle any type of Lorentz invariance violation that affects the isotropy of c, L , or n can be detected[4–6] in a cavity experiment ($\delta\nu/\nu = \delta c/c - \delta L/L - \delta n/n$) and thus a lot of Lorentz invariance violating coefficients of the SME[7] can be determined that alter the Maxwell and Dirac equations of motion.

Today's state-of-the-art modern Michelson-Morley experiments[8,9] are limited by the performance of the optical resonators employed. Improving the stability of optical resonators would improve the sensitivity to possible signatures of 'new physics.' A promising path towards better stability performances is the development of cryogenic optical resonators.

2. Cryogenic optical resonators

Nowadays, the frequency stability of all optimized room-temperature optical reference cavities is limited by the displacement noise within the resonator substrates and mirror coatings due to thermal noise.[10] For a reasonable cavity length (< 30 cm) the thermal noise limited relative frequency stability is restricted to $\geq 10^{-16}$.

The rather straightforward method to reduce simply the influence of thermal noise by cooling down the resonators to cryogenic temperatures is not well applicable for most room-temperature cavities. This is simply because they are mostly based on glass ceramic materials whose mechanical and thermal properties change unfavorably with low temperatures. In contrast, crystalline materials like sapphire or silicon that are normally not utilized in room-temperature cavities due to their comparatively high coefficients of thermal expansion at room temperature offer excellent material properties at cryogenic temperatures.

Therefore, we worked out a special design for an ultra-stable sapphire optical cavity system operating at 4 Kelvin. Two opposing requirements need to be considered when designing a cryogenic optical resonator and its mounting structure: high thermal conductivity towards the cold bath and low sensitivity of the optical path length to vibrations. Our resonator design has been developed using FEM computations to reduce the influence of vertical and horizontal vibrations upon the optical path length while the resonator still offers a large thermal contact area for mounting (see Fig. 1).

The prospective thermal noise limited frequency stability should be better than 3×10^{-17} (see Fig. 1), solely limited by the displacement noise originating from the high finesse mirror coatings based on Ta_2O_5/SiO_2. The displacement noise of the spacer and mirror substrates would easily even generate a theoretical thermal noise floor well in the low 10^{-20} regime due to the high stiffness (large Young modulus) and high mechanical Q-factor of sapphire. Furthermore, the thermal material properties (low CTE, high thermal conductivity) of sapphire at cryogenic temperatures make high performance active temperature stabilization feasible in order to minimize the influence of ambient temperature fluctuations. Thus, an exceptional high level of long term length stability and frequency stability, respectively, can also be expected.

In order to read out the ultra-stable eigenfrequencies of the cryogenic optical sapphire resonators, and thus making it available for other applications or precision tests of fundamental physics we use the Pound- Drever-Hall locking scheme, which stabilizes the frequency of a laser to one of the resonance frequencies of the cryogenic cavity. It has to be ensured that the electronics employed do not introduce additional noise on the frequency of the laser besides the fluctuations of the eigenfrequencies of the cryogenic optical resonator itself. Therefore, we will also implement techniques commonly used in ultra-stable resonator systems like residual-amplitude-modulation suppression and intensity-noise control.

3. Status

In total we had five 10 cm long sapphire cavities with a targeted finesse of 250,000 (corresponding to a resonance linewidth of ~6 kHz) custom-made. Upon inspection of the delivered cavities we noticed small debris on the highly reflective mirror coatings inside the cavity. Further investigation revealed that this pollution causes high losses in the fundamental modes of the cavities.

Thus, the measured finesses of the cavities lay between 13,000 (~115 kHz) and 52,000 (~29 kHz), which would set higher demands on the Pound-Drever-Hall locking electronics in order to not be influenced too much by parasitic resonance frequencies in the laser system. However, in the meantime the origin of the contamination within the fabrication process was identified by the vendor, who is currently cleaning and repairing two of the cavities.

We are in the process of setting up the final cryogenic cavity system, which has to provide a quiet environment for the sapphire resonators in order not to disturb them too much. At present, we are using a 20 year old cryostat system for pre-characterizations of the cavities and the laser lock and control circuits. In this setup, the laser beam is guided to the cavities through windows in the cryostat instead of optical fibers as it is planned for the final system.

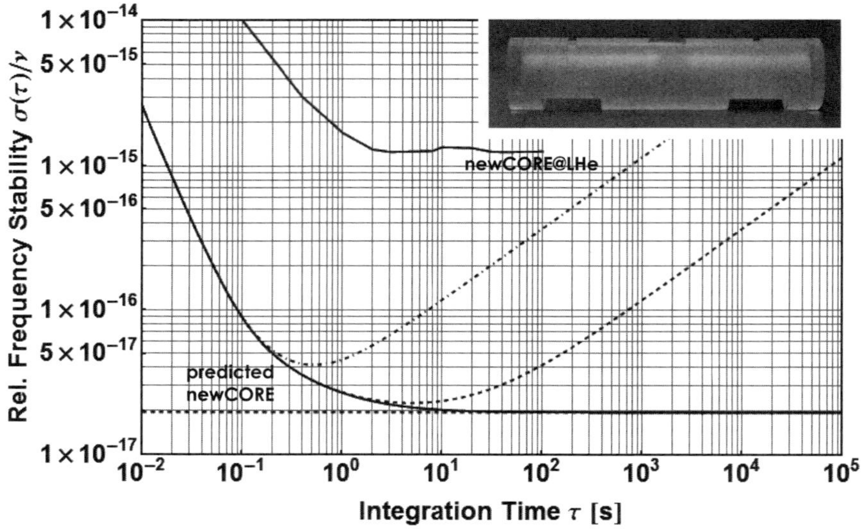

Fig. 1. Comparison of the measured and predicted relative frequency stability of the cryogenic optical resonators (CORE) including possible levels of long term random-walk noise (dashed and dot-dashed line), which cannot be predicted. The theoretical thermal noise floor with the standard Ta_2O_5/SiO_2 coating is indicated by the horizontal dashed line. The picture shows the special design of the sapphire resonators, which minimizes the influence of external forces upon the cavity length while featuring large contact areas for thermal grounding of the cavity.

Unfortunately, the old cryostat lost mechanical stiffness over the years due to transportation across Germany and a previous vacuum accident. This leads to a rather big differential movement of the sapphire resonators inside the cryostat and the laser beam system outside the cryostat. The differential movement causes comparatively large Doppler shifts of the resonance frequency of the cavities with respect to the incoming laser beam frequency, which in turn is tracked by the Pound-Drever-Hall lock. Thus, when comparing the eigenfrequencies of two of the cryogenic optical sapphire cavities by measuring the beat frequency of lasers locked to these cavities, we are at the moment mostly measuring the motion of the cavities rather than their intrinsic noise (see Fig. 1).

Therefore, we cannot yet confirm the prospected stability performance of the cryogenic cavities. Nevertheless, the electronic part of the laser stabilization system performs as intended and once the final cryogenic cavity system is set up we expect to be able to measure the predicted relative stability of $< 3 \times 10^{-17}$.

4. Intended tests of fundamental physics

Once the stability of the ultra-stable cryogenic optical sapphire resonators can be measured accurately with lasers using the Pound-Drever-Hall locking scheme, we plan to perform a modern Michelson-Morley experiment. Preparations for the needed ultra-stable rotating setup are already being done in our laboratories in parallel to the work on the cryogenic cavity system.

The measurement sensitivity of a modern Michelson-Morley experiment is limited by the provided frequency stability $\sigma(\tau)$ of the eigenfrequencies of the resonators at the integration time $\tau = T/2$, where T is the rotation period of the experimental setup. Assuming a gaussian distribution for single measurements, the measurement sensitivity can be enhanced by a factor of $1/\sqrt{N}$ by integrating over N individual measurements.

Accordingly, the optimal reachable measurement sensitivity S_{max} of a modern Michelson-Morley-type experiment spanning some time period T_{MM} and rotated at a rotation period of T can be estimated by $S_{max} = \sigma(T/2)/\sqrt{T_{MM}/(T/2)}$. Thus, the expected relative frequency stability of below 1×10^{-16} of the cryogenic optical sapphire resonators with a reasonable rotation period between 10 s and 100 s would yield a measurement sensitivity of 10^{-19} to 10^{-20} for signals of Lorentz invariance violations in a one year measurement campaign.

Actually, we plan to perform a more advanced version in which we will compare the eigenfrequencies of two rotating optical linear Fabry-Pérot resonators and of two rotating microwave whispering-gallery resonators in one setup. We will set up and perform this co-rotating experiment together with the group of Prof. M. Tobar from the University of Western Australia.[11,12]

The co-rotating setup allows testing different possible aspects of Lorentz invariance violations simultaneously by comparing optical frequencies with microwave frequencies, electromagnetic propagation in vacuum with propagation in matter, and linear modes with whispering-gallery modes. All these different aspects are considered in the full SME[13,14] and hence we can set new limits on a variety of coefficients including some with no bounds so far.

The usual right angle between the axes of the resonators in a Michelson-Morley-type experiment is insensitive[14] to some higher-order terms of the full SME. Therefore, we are either going to deviate from the usual right angle and set the resonators up in a different angle or we might compare the eigenfrequencies of the rotating resonators with stationary ones of similar performance.

5. Summary and outlook

We are developing ultra-stable cryogenic optical resonators with a prospected relative frequency stability $< 3 \times 10^{-17}$. This unprecedented stability will allow more precise tests of fundamental physics. We are planing to use these ultra-stable resonators in an advanced modern Michelson-Morley experiment in conjunction with ultra-stable cryogenic microwave whispering-gallery resonators operated by the University of Western Australia. This co-rotating experiment will allow us to set new or even first limits on several coefficients of the full SME with a sensitivity between 10^{-19} and 10^{-20}.

Moreover, we are currently investigating crystalline high finesse coatings based on $Al_{1-x}Ga_xAs$. These novel coatings could reduce the influence of thermal noise within an optical cavity by more than one order of magnitude.[15] Hence, using these coatings could boost the performance of our ultra-stable cryogenic optical sapphire resonators to a relative frequency stability around 10^{-18}. In turn, this would allow us to reach a sensitivity for Lorentz invariance violating signals in a modern Michelson-Morley-type experiment within the 10^{-21} regime.

References

1. A. Brillet and J. Hall, Phys. Rev. Lett. **42**, 549 (1979).
2. A.A. Michelson and E.W. Morley, Am. J. Sci. **34**, 333 (1887).
3. R.W.P. Drever *et al.*, Appl. Phys. B **31**, 97 (1983).
4. H. Müller *et al.*, Phys. Rev. D **68**, 116006 (2003).
5. H. Müller *et al.*, Phys. Rev. D **67**, 056006 (2003).
6. H. Müller, Phys. Rev. D **71**, 045004 (2005).
7. D. Colladay and V.A. Kostelecký, Phys. Rev. D **58**, 116002 (1998).
8. Ch. Eisele *et al.*, Phys. Rev. Lett. **103**, 090401 (2009).
9. S. Herrmann *et al.*, Phys. Rev. D **80**, 105011 (2009).
10. K. Numata *et al.*, Phys. Rev. Lett. **93**, 250602 (2004).
11. P. Stanwix *et al.*, Phys. Rev. D **74**, 081101 (2006).
12. S.R. Parker *et al.*, Phys. Rev. Lett. **106**, 180401 (2011).
13. V.A. Kostelecký and M. Mewes, Phys. Rev. D **80**, 015020 (2009).
14. M. Mewes, Phys. Rev. D **85**, 116012 (2012).
15. G. Cole *et al.*, Nature Photonics **7**, 644 (2013).

WHAT WE REALLY KNOW ABOUT NEUTRINO SPEEDS

BRETT ALTSCHUL

Department of Physics and Astronomy, University of South Carolina
Columbia, SC 29208, USA
E-mail: baltschu@physics.sc.edu

OPERA's claim to have seen faster-than-light neutrinos made a big splash in 2011. However, indirect arguments, based on gauge invariance, phase coherence in neutrino oscillations, and observations of electrons, could have already been used to rule out the OPERA claim. In fact, indirect constraints on neutrino velocities are many orders of magnitude better than direct ones.

Experimental tests of Lorentz invariance have a long history, going all the way back to 1905, when the structure of Special Relativity was first described by Einstein. In the last fifteen years, this field has grown tremendously, and researchers have become much more systematic about exploring the full parameter space of possible Lorentz violations. However, the event that brought the most attention to this area of research was the 2011 report by the OPERA experiment that neutrinos appeared to be moving faster than light. We now know that this result was purely an experimental artifact, but the initial OPERA claim raised some interesting questions, about how well Lorentz symmetry has really been tested for neutrinos and other Standard Model (SM) particles that are comparatively difficult to observe.

The apparent Lorentz violation indicated by OPERA was quite large. The initially reported neutrino speed was $\sim 1+(2.5\times10^{-5})$.[1] This was huge in comparison with some of the constraints on Lorentz violation that have been derived for other species of particles. This made the OPERA results look ever more far-fetched; even with the possibility of Lorentz violation being seriously entertained, the discrepancy between different sectors looked highly unlikely. In fact, it turns out that using data collected in a number of different types of experiments, using neutrinos and other leptons, it is possible to place rigorous (albeit indirect) constraints on neutrino speed anomalies, which would have ruled the OPERA results in advance. There are also a number of directly observed facts about neutrinos that also show that the OPERA result had to be untenable.

It will turn out that it is possible to place very strong constraints on the neutrino speed using comparisons between neutrinos and their charged lepton partners. The $SU(2)_L$ gauge invariance of the SM restricts the forms that Lorentz violations can take. In the Lorentz-violating Standard-Model Extension (SME), the operators that can contribute directly to the limiting velocity at high energies are parameterized by tensors $c_R^{\nu\mu}$ and $c_L^{\nu\mu}$. The subscripts denote the chiralities of the particles that are affected by these forms of Lorentz violation. (An alternative notation uses tensors $c^{\nu\mu}$ and $d^{\nu\mu}$, representing sums and differences of the $c_R^{\nu\mu}$ and $c_L^{\nu\mu}$.) The tensor $c_R^{\nu\mu}$ affects the velocities of right-handed leptons (and left-handed antileptons), of which there is a single charged type for each generation. In contrast, $c_L^{\nu\mu}$ determines the speeds of left-handed leptons, both charged leptons and neutrinos; because the two types of left-chiral particles are connected by $SU(2)_L$ gauge transformations, they must correspond to the same $c_L^{\nu\mu}$ tensors. This is a powerful result.

All this holds within a single fermion generation. One $c_L^{\nu\mu}$ tensor is common for e^- and ν_e. The μ^- and ν_μ share a distinct $c_L^{\nu\mu}$; and likewise for the third generation. (There could also be coefficients that are not diagonal in flavor space. These would mix neutrinos from different generations, and they could thus be constrained by looking for Lorentz-violating behavior in neutrino oscillations. The focus here is on the flavor-diagonal coefficients that primarily affect neutrino speeds and cannot be directly probed with oscillation experiments.)

The best-known refutation of the OPERA results came from analyses of the Cerenkov-like process $\nu \rightarrow \nu + e^- + e^+$, which may be allowed for neutrinos moving faster than light. The neutrino emits a virtual Z boson, which decays into an electron-positron pair. If neutrino Lorentz violation at the 10^{-5} level really existed, this process would have caused substantial energy losses in the OPERA beam.[2,3]

Changes to neutrino velocities also affect interaction and decay processes involving neutrinos. In particular, the rate of the pion decay process $\pi^+ \rightarrow \mu^+ + \nu_\mu$ that is used to produce the OPERA beam is sensitive to the energy-momentum relations for the daughter particles. Changes in the neutrino speed would affect the phase space available for this reaction.[4,5] Again, it turns out that Lorentz violation at the 10^{-5} level would have led to significant changes in the beam structure. This has been shown to be true even if the gauge invariance conditions discussed above (which tie the muon and neutrino dispersion relations together) are ignored, and only the neutrino speed is modified. The above results are important, but they were concerned with OPERA specifically, and they do not give the best current constraints on modified neutrino speeds.

While there are, in principle, separate $c_L^{\nu\mu}$ tensors for the neutrino generations, there are actually strong bounds on the intergenerational differences between those tensors. The bounds are derived from the observed existence of long-baseline neutrino oscillations. The fact that neutrino oscillations occur at all actually tells us a great deal about the relative speeds of neutrinos. If different neutrino species had substantially different speeds, the wave packets corresponding to different flavor states would not be able to move along coherently. A wave packet with a superposition of different flavors would separate in space, the faster flavors leaving the slower ones behind. Oscillations between flavors can only occur while the packets are overlapping, so having differing velocities would kill off oscillations over sufficiently long distances. Effectively, the detection of a neutrino would constitute a measurement of its time of flight, and thus a measurement of its species.

Of course, no fall-off in neutrino oscillation rates over long distances has been observed. This leads to a very strong experimental constraint on the differences between the $c_L^{\nu\mu}$ coefficients for different generations. The resulting constraints are at roughly the 10^{-19}–10^{-21} levels, depending on the specific coefficients involved.[6] These constraints turn out to be tight enough that we may effectively neglect any intergenerational differences in what follows. The fact that neutrino wave packets still overlap even after long distances of propagation constrains the flavor dependence of the neutrino speed extremely tightly.

Combining this result with the consequences of gauge invariance, it becomes possible to constrain neutrino velocity anomalies using observations of electrons. Electrons are best for this purpose because they are stable, plentiful, and observed with the highest Lorentz boost factors of any charged leptons. It turns out that the best constraints on the $c_L^{\nu\mu}$ coefficients for electrons come from astrophysical and resonant cavity data.

High-energy astrophysical data can be used to map out the energy-momentum relation for extremely energetic electrons (and sometimes positrons). The radiation emitted by an ultrarelativistic particle depends very sensitively on its velocity. The observed processes of synchrotron and inverse Compton radiation, and the absence of vacuum Cerenkov radiation and photon decay ($\gamma \to e^- + e^+$) reveal a great deal. Observations of the radiation from electrons with Lorentz factors up to $\gamma \sim 10^8$ are common, and these provide constraints on the SME coefficients that contribute to the velocity anomaly at the $1/\gamma^2$ level.[7]

Resonant cavities can be used to test some of the same coefficients at a similar level of precision. If the electron hamiltonian is anisotropic, the dimensions of a crystal will depend on its orientation. The atomic electrons mediate the interactions between the nuclei in a material; if the electrons behave anisotropically, the mean separation between adjacent atoms may be different along different crystal axes. In a resonant cavity, the physical length can be measured precisely, and tests with multiple rotating cavities can be used to disentangle this phenomenon from photon-sector Lorentz violation.[8]

Some of the measurements just discussed are sensitive to $c_L^{\nu\mu}$ and $c_R^{\nu\mu}$ separately (for example, vacuum Cerenkov radiation would occur if either helicity state were superluminal at a physically accessible energy), and some are only sensitive to the average. However, the right-left differences can be constrained directly (and quite a bit more precisely, at the 10^{-23}–10^{-24} level) using laboratory measurements of spin-dependent forces.[9] So all of the constraints discussed in the preceding two paragraphs can effectively be interpreted as bounds on the $c_L^{\nu\mu}$.

Numerically, the actual constraints range from the 10^{-14} to 10^{-19} levels. The precise values depend on which specific linear combinations are involved. However, it is generically true that the maximum achievable speed of a left-handed lepton cannot deviate from the speed of light by more than parts in 10^{14}. This is a very strong constraint, which had already ruled out the original OPERA claim by nine orders of magnitude.

References

1. T. Adam *et al.*, arXiv:1109.4897v1.
2. X.-J. Bi, P.-F. Yin, Z.-H. Yu, and Q. Yuan, Phys. Rev. Lett. **107**, 241802 (2011).
3. A.G. Cohen and S.L. Glashow, Phys. Rev. Lett. **107**, 181803 (2011).
4. B. Altschul, Phys. Rev. D **84**, 091902(R) (2011).
5. R. Cowsik, S. Nussinov, and U. Sarkar, Phys. Rev. Lett. **107**, 251801 (2011).
6. G.F. Giudice, S. Sibiryakov, and A. Strumia, Nucl. Phys. B **861**, 1 (2012).
7. B. Altschul, Phys. Rev. D **74**, 083003 (2006).
8. H. Müller, Phys. Rev. D **71**, 045004 (2005).
9. B.R. Heckel, E.G. Adelberger, C.E. Cramer, T.S. Cook, S. Schlamminger, and U. Schmidt, Phys. Rev. D **78**, 092006 (2008).

LORENTZ VIOLATION AND GRAVITY

R. BLUHM

Department of Physics, Colby College
Waterville, ME 04901, USA
E-mail: rtbluhm@colby.edu

Gravitational theories with Lorentz violation must account for a number of possible features in order to be consistent theoretically and phenomenologically. A brief summary of these features is given here. They include evasion of a no-go theorem, connections between spontaneous Lorentz breaking and diffeomorphism breaking, the appearance of massless Nambu-Goldstone modes and massive Higgs modes, and the possibility of a Higgs mechanism in gravity.

1. Gravity and the SME

The Standard-Model Extension[1] (SME) consists of the most general observer-independent effective field theory incorporating Lorentz violation. It is routinely used by both theorists and experimentalists to study and obtain bounds on possible forms of Lorentz violation.[2,3] As an effective field theory, the SME can accommodate both explicit and spontaneous Lorentz breaking. However, there are differences in these two forms of symmetry breaking that arise in the context of gravity. This overview looks at these differences and what their primary consequences are.

In a gravitational theory with Lorentz violation it is useful to use a vierbein formalism. In this approach, both the local Lorentz frames and spacetime frames are accessible and linkage between the symmetries in these frames can be examined. The vierbein provides the connection between tensor components in local Lorentz frames and tensor components in the spacetime frame.

The Lagrangian in the SME is formed as the most general scalar function (under both local Lorentz and diffeomorphism transformations) using gravitational fields, particle fields, and Lorentz-violating SME coefficients. When the Lorentz breaking is explicit, the SME coefficients are viewed as fixed background fields. However, when the Lorentz breaking is spontaneous, the

SME coefficients are vacuum expectation values (vevs) of dynamical tensor fields.

In the gravity sector of the SME, a no-go theorem shows that with explicit Lorentz breaking an inconsistency can occur between conditions stemming from the field variations and symmetry considerations with geometrical constraints that must hold, such as the Bianchi identities.[4] In contrast, the case of spontaneous Lorentz breaking was found to evade the no-go theorem. The main difference is that in a theory with explicit breaking the SME coefficients are not associated with dynamical fields, while with spontaneous Lorentz breaking they are, which creates a difference in the conditions that must hold. An important consequence of the no-go theorem is that the gravity sector of the SME can only avoid incompatibility with conventional geometrical constraints if the symmetry breaking is spontaneous.

2. Spontaneous symmetry breaking

The fact that the SME coefficients must be associated with vevs of dynamical fields that undergo spontaneous Lorentz violation leads to a number of effects that must be accounted for in the gravity sector of the SME. For example, when Lorentz symmetry is spontaneously broken, there is also spontaneous breaking of diffeomorphism symmetry. The spontaneous Lorentz breaking occurs when a nonzero tensor-valued vacuum occurs in the local Lorentz frames, which is necessarily accompanied by a vacuum value for the vierbein. When products of the vierbein vev act on the local tensor vevs, the result is that tensor vevs also appear in the spacetime frame. These tensor vevs spontaneously break local diffeomorphisms in the spacetime frame. Conversely, if a vev in the spacetime frame spontaneously breaks diffeomorphisms, then the inverse vierbein acting on it gives rise to vevs in the local frames. Consequently, spontaneous local Lorentz breaking implies spontaneous diffeomorphism breaking and vice versa.[5]

With spontaneous symmetry breaking of both Lorentz and diffeomorphism symmetry, there are standard features in particle physics that need to be investigated. These include the possible appearance of massless Nambu-Goldstone (NG) modes and massive Higgs modes, or there is the possibility of a Higgs mechanism in which the NG modes are reinterpreted as additional degrees of freedom in a theory with massive gauge fields.

In the absence of a Higgs mechanism, there can be up to as many NG modes as there are broken spacetime symmetries. Since the maximal symmetry-breaking case would yield six broken Lorentz generators and four broken diffeomorphisms, there can be as many as ten NG modes. A natural gauge choice puts all of the NG modes into the vierbein. However, this will in general lead to the appearance of ghosts, and it is for this reason that most models involve breaking fewer than ten of the spacetime symmetries.

Spontaneous symmetry breaking is usually induced by a potential term in the Lagrangian that has a degenerate minimum space. The NG modes appear as excitations away from the vacuum that stay in the minimum space, while massive Higgs modes are excitations that go up the potential well away from the minimum. In conventional gauge theory, the potential involves only scalar fields, and the massive Higgs modes are independent of the gauge fields. However, with spontaneous Lorentz breaking, the metric typically appears in the potential along with the tensor fields, and for this reason massive Higgs modes can occur that include metric excitations. This is an effect that has no analog in the case of conventional gauge theory.

In a Higgs mechanism, the would-be NG modes become additional degrees of freedom for massive gauge fields. The gauge fields associated with diffeomorphisms are the metric excitations. However, a Higgs mechanism involving the metric has been shown not to occur.[6] This is because the mass term that is generated by covariant derivatives involves the connection, which consists of derivatives of the metric and not the metric itself. However, for the broken Lorentz symmetry, where the relevant gauge fields are the spin connection, a conventional Higgs mechanism can occur.[5] This is because the spin connection appears directly in covariant derivatives acting on local tensor components, and when the local tensors acquire a vev, quadratic mass terms for the spin connection can be generated. Note, however, a viable Higgs mechanism involving the spin connection can only occur if the spin connection is a dynamical field, which requires nonzero torsion and that the geometry be Riemann-Cartan.

References

1. V.A. Kostelecký and R. Potting, Phys. Rev. D **51**, 3923 (1995); D. Colladay and V.A. Kostelecký, Phys. Rev. D **55**, 6760 (1997); Phys. Rev. D **58**, 116002 (1998).
2. R. Bluhm, arXiv:1302.1150.
3. *Data Tables for Lorentz and CPT Violation,* V.A. Kostelecký and N. Russell, 2013 edition, arXiv:0801.0287v6.
4. V.A. Kostelecký, Phys. Rev. D **69**, 105009 (2004).
5. R. Bluhm and V.A. Kostelecký, Phys. Rev. D **71**, 065008 (2005); R. Bluhm, S.-H. Fung, and V.A. Kostelecký, Phys. Rev. D **77**, 065020 (2008).
6. V.A. Kostelecký and S. Samuel, Phys. Rev. D **40**, 1886 (1989).

LORENTZ VIOLATION IN TOP-QUARK PHYSICS

M.S. BERGER
Department of Physics, Indiana University
Bloomington, IN, 47405, USA
E-mail: berger@indiana.edu

Lorentz and CPT violation can affect the rates for t-$\bar{t}$ production and decay. The Lorentz-violating coefficients in the Standard-Model Extension responsible for modifying the top-quark events have recently been bounded by the D0 Collaboration. To extend the analysis to the LHC the calculations need to be extended to include the gluon fusion production mechanism. Some of the first results of this program were presented at the Meeting.

1. Introduction

The possible existence of Lorentz and CPT violation has been increasingly studied in recent years. Many efforts toward developing a theory of quantum gravity contain Lorentz violation as a generic physical consequence although the size of the effects are usually quite small in realistic experiments with attainable energies much smaller than the Planck scale. A systematic approach for parametrizing the possible forms of Lorentz violation in the context of an effective field theory called the Standard-Model Extension (SME)[1–3] has been developed. The Lorentz- and CPT-violating physical effects are encoded in the form of coefficients. A Lorentz-violating term is an observer scalar density formed by contracting the Lorentz-violating operator with the associated coefficient. Diverse kinds of experiments can place bounds on the same coefficients, so that the experimental sensitivities can be directly compared.[4]

The Lorentz-violating coefficients provide the possibility of preferred frames and directions in space and could in principle have an effect on collider physics experiments. If the coefficient is constant in space then one expects there to be a sidereal time variation as the Earth rotates through this constant background. Collider physics scattering events naturally have access to many Lorentz frames since produced particles and their decay products are distributed with many different boosts and orientations relative to the laboratory frame and beam direction.

2. Top quark Lorentz-violating coefficients in the SME

Top quark production and decay involves many fields so the overall rates and the distributions could be affected by many different Lorentz violating coefficients in the SME. In this discussion we will focus on coefficients which involve the top quark field t. Then the relevant coefficients arise from the matrices $(a_Q)_{\mu ij}$, $(a_U)_{\mu ij}$, $(c_Q)_{\mu\nu ij}$, $(c_U)_{\mu\nu ij}$, $(H_U)_{\mu\nu ij}$ contained in the SME, where i and j are flavor indices, and the Q and U indices correspond to the left- and right-handed fields of the Standard Model. Of these five coefficients, the first two control CPT-odd operators, while the last three control CPT-even ones. At present bounds have been placed on the c-type coefficients,[5] so we will concentrate on those in the following. The coefficients relevant to the top quark production and decay have $i = j = 3$ (in general there will be flavor mixing in the Lorentz-violation sector that does not have to align with the CKM matrix of the Standard Model), so for convenience one can define $(c_L)_{\mu\nu} = (c_Q)_{\mu\nu 33}$ and $(c_R)_{\mu\nu} = (c_U)_{\mu\nu 33}$ to simplify the notation.

The Standard-Model terms involving the t and b quark fields and their interactions with the $W_\mu^\pm$ boson are

$$\begin{aligned}\mathcal{L}_{t,b}^{\text{SM}} &= \frac{1}{2} i\bar{t}\gamma^\mu \overset{\leftrightarrow}{\partial_\mu} t - m_t \bar{t}t + \frac{1}{2} i\bar{b}\gamma^\mu \overset{\leftrightarrow}{\partial_\mu} b - m_b \bar{b}b \\ &\quad + \left(\frac{g}{\sqrt{2}} W_\mu^- \bar{b}_L \gamma^\mu t_L + \text{h.c.}\right). \qquad (1)\end{aligned}$$

The additional terms in the SME corrections involving the c-type coefficient can be written in equivalent forms,

$$\begin{aligned}\mathcal{L}_{t,b}^{\text{CPT-even}} &= \frac{1}{2} i(c_L)_{\mu\nu} \bar{t}_L \gamma^\mu \overset{\leftrightarrow}{\partial^\nu} t_L + \frac{1}{2} i(c_R)_{\mu\nu} \bar{t}_R \gamma^\mu \overset{\leftrightarrow}{\partial^\nu} t_R \\ &\quad + \frac{1}{2} i(c_L)_{\mu\nu} \bar{b}_L \gamma^\mu \overset{\leftrightarrow}{\partial^\nu} b_L + \left(\frac{g}{\sqrt{2}} (c_L)_{\mu\nu} W^{-\nu} \bar{b}_L \gamma^\mu t_L + \text{h.c.}\right) \\ &= \frac{1}{2} i c_{\mu\nu} \bar{t}\gamma^\mu \overset{\leftrightarrow}{\partial^\nu} t + \frac{1}{2} i d_{\mu\nu} \bar{t}\gamma^5 \gamma^\mu \overset{\leftrightarrow}{\partial^\nu} t \\ &\quad + \frac{1}{2} i(c_L)_{\mu\nu} \bar{b}_L \gamma^\mu \overset{\leftrightarrow}{\partial^\nu} b_L + \left(\frac{g}{\sqrt{2}} (c_L)_{\mu\nu} W^{-\nu} \bar{b}_L \gamma^\mu t_L + \text{h.c.}\right), \qquad (2)\end{aligned}$$

where $c_{\mu\nu} = (c_L)_{\mu\nu} + (c_R)_{\mu\nu}$ and $d_{\mu\nu} = (c_L)_{\mu\nu} - (c_R)_{\mu\nu}$. These terms give rise to additional Feynman rules which can be treated as insertions in diagrams. By working in leading order in the Lorentz-violating coefficients, it is a straightforward procedure to enumerate diagrams and calculate contributions to the amplitudes which depend on them. At the Tevatron top quark production is dominated by the $q\bar{q} \to t\bar{t}$ process for which the Lorentz-violating corrections are known.

3. Bounds

The Lorentz violating coefficients are defined as constant in a Sun-centered frame[6–8] (X, Y, Z, T). The coefficients in the laboratory frame (x, y, z, t) then change with sidereal time because the Earth is rotating. Nonzero Lorentz-violating coefficients thus produce sidereal time-dependent signals with frequency an integer multiple of $\omega_\oplus \simeq 2\pi/(23 \text{ h } 56 \text{ min.})$.[9] The D0 Collaboration has used data collected at the Fermilab Tevatron Collider from 5.3 fb^{-1} of integrated luminosity to place the first bounds on the top-quark Lorentz-violating coefficients:[5,10]

$$
\begin{aligned}
(c_Q)_{XX33} &= -0.12 \pm 0.11 \pm 0.02, \\
(c_Q)_{YY33} &= 0.12 \pm 0.11 \pm 0.02, \\
(c_Q)_{XY33} &= -0.04 \pm 0.11 \pm 0.01, \\
(c_Q)_{XZ33} &= 0.15 \pm 0.08 \pm 0.02, \\
(c_Q)_{YZ33} &= -0.03 \pm 0.08 \pm 0.01, \\
(c_U)_{XX33} &= 0.1 \pm 0.09 \pm 0.02, \\
(c_U)_{YY33} &= -0.1 \pm 0.09 \pm 0.02, \\
(c_U)_{XY33} &= 0.04 \pm 0.09 \pm 0.01, \\
(c_U)_{XZ33} &= -0.14 \pm 0.07 \pm 0.02, \\
(c_U)_{YZ33} &= 0.01 \pm 0.07 \pm < 0.01.
\end{aligned} \tag{3}
$$

The experimental events used in the analysis contain lepton + jets final states, and a candidate signal would be expected to exhibit sidereal or twice-sidereal variations. These bounds are at roughly the 10% level, and all coefficients are consistent with zero. It is expected that the much larger top sample at the LHC could improve these bounds by at least an order of magnitude.

4. Gluon fusion

Top-quark production is dominated by the gluon-fusion mechanism at the LHC. There are three contributing diagrams (s-channel gluon and t- and

u- channel top quark) in the Standard Model, and the cross section is well-known.[11–13] The calculation of the Lorentz-violating corrections is somewhat more involved than the one for $q\bar{q} \to t\bar{t}$ because of the number of diagrams and the presence of external gluons which must be taken to be transverse. The first results of this calculation were presented at the Meeting.

5. Summary

The first bounds on the top-quark SME coefficients have been obtained by the D0 Collaboration using data collected at the Tevatron. The measurements are consistent with zero. Top quark production at the LHC is dominated by gluon fusion, and it is expected that the bounds can be improved to the percent level.

Acknowledgments

This work is supported in part by DOE grant DE-FG02-91ER40661 and by the Indiana University Center for Spacetime Symmetries. The author thanks V.A. Kostelecký for his collaboration.

References

1. D. Colladay and V.A. Kostelecký, Phys. Rev. D **55**, 6760 (1997).
2. D. Colladay and V.A. Kostelecký, Phys. Rev. D **58**, 116002 (1998).
3. V.A. Kostelecký, Phys. Rev. D **69**, 105009 (2004).
4. V.A. Kostelecký, ed., *CPT and Lorentz Symmetry I, II, III, IV, V*, World Scientific, Singapore, 1999, 2002, 2005, 2008, 2011.
5. V.M. Abazov *et al.*, Phys. Rev. Lett. **108**, 261603 (2012).
6. V.A. Kostelecký and M. Mewes, Phys. Rev. D **66**, 056005 (2002).
7. R. Bluhm *et al.*, Phys. Rev. D **68**, 125008 (2003).
8. *Data Tables for Lorentz and CPT Violation,* V.A. Kostelecký and N. Russell, 2013 edition, arXiv:0801.0287v6.
9. V.A. Kostelecký, Phys. Rev. Lett. **80**, 1818 (1998).
10. D. Whittington, these proceedings.
11. M. Gluck, J.F. Owens and E. Reya, Phys. Rev. D **17**, 2324 (1978).
12. B.L. Combridge, Nucl. Phys. B **151**, 429 (1979).
13. R.K. Ellis, W.J. Stirling and B.R. Webber, Camb. Monogr. Part. Phys. Nucl. Phys. Cosmol. **8**, 1 (1996).

TESTING FOR CPT VIOLATION IN B_s^0 SEMILEPTONIC DECAYS

R. VAN KOOTEN

Department of Physics, Indiana University, Bloomington, IN 47405, USA
E-mail: rvankoot@indiana.edu

On behalf of the DØ Collaboration*

A DØ analysis measuring the charge asymmetry $A_{\rm sl}^b$ of like-sign dimuon events due to semileptonic b-hadron decays at the Fermilab Tevatron Collider has shown indications of possible anomalous CP violation in the mixing of neutral B mesons. This result has been used to extract the first senstivity to CPT violation in the B_s^0 system. An analysis to explore further this anomaly by specifically measuring the semileptonic charge asymmetry, $a_{\rm sl}^s$, in B_s^0 decays is described, as well as how a variant of this analysis can be used to explore a larger set of CPT-violating parameters in the B_s^0 system for the first time.

1. Introduction

The interferometric systems of the particle-antiparticle oscillations of neutral mesons are particularly sensitive to testing for CP and CPT violation. In neutral meson systems, the hamiltonian is a 2×2 matrix relating the mass and weak eigenstates. Mixing between particle and antiparticle is driven by nonzero off-diagonal matrix elements due to a box diagram between $B^0_{(d\,\rm or\,s)}$ and $\bar{B}^0_{(d\,\rm or\,s)}$. T (or CP) violation in mixing can be due to differences between these off-diagonal terms and results in the two probabilities for oscillation between particle and antiparticle not being equal, i.e., $P(B^0 \to \bar{B}^0; t) \neq P(\bar{B}^0 \to B^0; t)$. CPT and Lorentz violation involves differences between *diagonal* terms of this matrix and differences in the probabilities for: $P(B^0 \to B^0; t) \neq P(\bar{B}^0 \to \bar{B}^0; t)$ and can be expressed with the parameter[1]

$$\xi = \frac{(M_{11} - M_{22}) - \frac{i}{2}(\Gamma_{11} - \Gamma_{22})}{-\Delta m - \frac{i}{2}\Delta\Gamma} \approx \frac{\beta^\mu \Delta a_\mu}{-\Delta m - \frac{i}{2}\Delta\Gamma}, \tag{1}$$

*http://www-d0.fnal.gov

where $\beta^\mu = \gamma(1, \vec{\beta})$ is the 4-velocity of the neutral B meson, and $\Delta a_\mu = r_{q_1} a_\mu^{q_1} - r_{q_2} a_\mu^{q_2}$ with r being coefficients with q_1 and q_2 as meson valence quarks and a_μ being the constant 4-vector in the Standard-Model Extension langrangian.[2] The power of using this system as an inteferometer is that Δm, i.e., the mass difference between the heavy and light mass eigenstates, is so tiny,[3] i.e., for the B_s^0 meson, $\Delta m_s = 1.16 \times 10^{-2}$ eV compared with the typical B_s^0 meson mass of approximately 5.4 GeV, allowing sensitivities down to the order of 10^{-11} on CPT-violating parameters.

2. DØ dimuon charge asymmetry

The DØ Collaboration has measured[4,5] the raw dimuon charge asymmetry $A = (N^{\mu^+\mu^+} - N^{\mu^-\mu^-})/(N^{\mu^+\mu^+} + N^{\mu^-\mu^-})$ in $p\bar{p}$ collisions at $\sqrt{s} = 1.96$ TeV regardless of muon source. From pure physics processes at the primary interaction, one of the very few sources of same-sign dileptons in the same collision event is due to B physics. If there is a nonzero asymmetry after correcting for backgrounds, the assumption is that it is coming from neutral B-meson mixing, i.e., the dimuon charge asymmetry of semileptonic B decays $A_{\text{sl}}^b = (N_b^{\mu^+\mu^+} - N_b^{\mu^-\mu^-})/(N_b^{\mu^+\mu^+} + N_b^{\mu^-\mu^-})$. This can occur, for example, when the b quark decays semileptonically directly $\overline{B}_q^0 \to \mu^-$, but for the $\bar{b}$ quark, there is first a B-meson oscillation before the semileptonic decay, i.e., $B_q^0 \to \overline{B}_q^0 \to \mu^-$. Another way to measure this asymmetry is via inclusive 'wrong-sign' decays, i.e., $\overline{B} \to \mu^+ X$ which is only possible through flavor oscillation of B_d^0 and B_s^0. A semileptonic charge asymmetry can then be constructed:

$$a_{\text{sl}}^b = \frac{\Gamma(\overline{B} \to \mu^+) - \Gamma(B \to \mu^- X)}{\Gamma(\overline{B} \to \mu^+) + \Gamma(B \to \mu^- X)}. \tag{2}$$

Assuming CPT symmetry holds, it can be shown[6] that $A_{\text{sl}}^b = a_{\text{sl}}^b$.

From the like-sign dimuon sample, after correcting for backgrounds, the CP-violating asymmetry is found for 9.0 fb^{-1} to be[5] $A_{\text{sl}}^b = -(0.736 \pm 0.172\,(\text{stat}) \pm 0.093)\%\,(\text{syst})$, which is a 3.9σ deviation from the Standard Model prediction[7] for CPT-preserving T violation, $A_{\text{sl}}^b(\text{SM}) = (-2.3^{+0.5}_{-0.6}) \times 10^{-4}$, and represents the first evidence for anomalous CP violation in the mixing of neutral B mesons. This asymmetry has contributions from both the CP-violating semileptonic asymmetry a_{sl}^d for B_d^0 and a_{sl}^s for B_s^0 oscillations: $A_{\text{sl}}^b = C_d \cdot a_{\text{sl}}^d + C_s \cdot a_{\text{sl}}^s$. The analysis is redone for various conditions on the impact parameters of the muons resulting in different values for the coefficients C_i allowing for the extraction of the specific asymmetries $a_{\rm sl}^d$ and $a_{\rm sl}^s$, albeit with a high degree of correlation: $a_{\rm sl}^d = (-0.12 \pm 0.52)\%, a_{\rm sl}^s = (-1.81 \pm 1.06)\%, \rho_{ds} = -0.7994$.

A CPT-violating effect in B-meson mixing was predicted some time ago,[8,9] and the B_s^0-$\overline{B}_s^0$ system is of particular interest for studies of CPT violation because several complete particle-antiparticle oscillations occur within a meson lifetime.[10] A measure of CPT violation is given by the inclusive 'right-charge' muon charge asymmetry $\mathcal{A}_{\rm CPT}^b$ of semileptonic decays of b hadrons,[1]

$$\mathcal{A}_{\rm CPT}^b = \frac{\Gamma(\overline{B} \to \mu^- X) - \Gamma(B \to \mu^+ X)}{\Gamma(\overline{B} \to \mu^- X) + \Gamma(B \to \mu^+ X)} \approx a_{\rm sl}^b({\rm SM}) - A_{\rm sl}^b. \tag{3}$$

Assuming the only source of T violation is the SM contribution $a_{\rm sl}^b({\rm SM}) = A_{\rm sl}^b({\rm SM})$, we find[a] $\mathcal{A}_{\rm CPT}^b = 0.00785 \pm 0.00300$. Averaging over sidereal time and the momentum (γ) spectrum, sensitivity to the spatial components $(\Delta a^{B_s})_J$ is lost; however, assuming that the only source of CPT violation comes from B_s^0-$\overline{B}_s^0$ mixing, the bound

$$-3.8 \times 10^{-12} < (\Delta a^{B_s})_T < 1.1 \times 10^{-11} \tag{4}$$

can be set at the 95% confidence level. The value of Eq. (4), documented in Ref. 1, represents the first sensitivity to CPT violation in the B_s^0-$\overline{B}_s^0$ system.

3. DØ exclusive semileptonic B_s^0 decays

In principle, the spatial components $(\Delta a^{B_s})_J$ could be accessed by studying the variation of the dimuon like-sign charge asymmetry above as a function of sidereal time. However, there are complications due to the possibility of B_d^0 CPT asymmetry and the current large uncertainty on $\Delta\Gamma_d/\Gamma_d$.[3] Instead, it is better to examine exclusive B_s^0 decays, i.e., $B_s^0 \to D_s \mu\nu$. One could fit to a complicated time-dependent CP or CPT asymmetry. This would require 'flavor tagging', i.e., determining if the meson was a B_s^0 or $\bar{B}_s^0$ at the time of production, and this costs efficiency. One can just integrate over decay time since the B_s^0 mesons oscillate very quickly and after very small decay times and lengths are 'fully mixed', i.e., there are equal probabilities of observing a B_s^0 or $\bar{B}_s^0$ despite the flavor at the time of production.

The relevant semileptonic charge CP asymmetry is therefore:

$$a_{\rm sl}^s = \frac{N(B_s \to \bar{B}_s \to D_s^- \mu^+ X) - N(\bar{B}_s \to B_s \to D_s^+ \mu^- X)}{N(B_s \to \bar{B}_s \to D_s^- \mu^+ X) + N(\bar{B}_s \to B_s \to D_s^+ \mu^- X)}. \tag{5}$$

[a]This is the updated number using the result of Ref. 5.

The DØ Collaboration has recently measured[11] this CP asymmetry in a dataset of integrated luminosity of 10.4 fb^{-1} resulting in $a^s_{\rm sl} = [-1.12 \pm 0.74\,(\text{stat}) \pm 0.17\,(\text{syst})]\%$. This result is consisent with both the SM value as well as the value extracted from the dimuon like-sign asymmetry.

The corresponding CPT asymmetry is essentially the *same* asymmetry since the necessary requirement of $B^0_s \to B^0_s$ and $\bar{B}^0_s \to \bar{B}^0_s$ is identical if the system is fully mixed, only requiring that the observed, raw asymmetry be corrected by the fraction of non-oscillated B^0_s decays. To access all four components of $(\Delta a^{B_s})_\mu$, this asymmetry is then determined separately in an appropriate number of bins in sidereal time, in this case, optimized as eight bins. Information is lost in the binning, so an unbinned analysis is also being performed in a periodogram[12] that finds the power at a given scan frequency. Finally, plans also include determining the asymmetry in bins of B^0_s meson to probe for the expected linear variation with γ from CPT-violating effects. Work continues on this analysis.

Acknowledgments

The DØ Collaboration thanks the staffs at Fermilab and collaborating institutions, and acknowledges support from agencies including the DOE and NSF (USA).

References

1. V.A. Kostelecký and R. Van Kooten, Phys. Rev. D **82**, 101702 (2010) and references therein.
2. D. Colladay and V.A. Kostelecký, Phys. Rev. D **55**, 6760 (1997); Phys. Rev. D **58**, 116002 (1998); V.A. Kostelecký, Phys. Rev. D **69**, 105009 (2004).
3. J. Beringer *et al.* (Particle Data Group), Phys. Rev. D **86**, 010001 (2012) and 2013 partial update for the 2014 edition.
4. V.M. Abazov *et al.*, Phys. Rev. D **82**, 032001 (2010); Phys. Rev. Lett. **105**, 081801 (2010).
5. V.M. Abazov *et al.*, Phys. Rev. D **84**, 052007 (2011).
6. Y. Grossman, Y. Nir, and G. Raz, Phys. Rev. Lett. **97**, 151801 (2006).
7. A. Lenz and U. Nierste, J. High Energy Phys. **0706**, 072 (2007).
8. V.A. Kostelecký and R. Potting, Phys. Rev. D **51**, 3923 (1995).
9. V.A. Kostelecký and R. Van Kooten, Phys. Rev. D **54**, 5585 (1996).
10. V.M. Abazov *et al.*, Phys. Rev. Lett. **97**, 021802 (2006); A. Abulencia *et al.*, Phys. Rev. Lett. **97**, 242003 (2006).
11. V.M. Abazov *et al.*, Phys. Rev. Lett. **110**, 011801 (2013).
12. F.A.M. Frescura, C.A. Engelbrecht and B.S. Frank, arXiv:0706.2225.

NUCLEAR β DECAY WITH LORENTZ VIOLATION

J.P. NOORDMANS,* H.W. WILSCHUT, and R.G.E. TIMMERMANS

KVI, University of Groningen
Zernikelaan 25, NL-9747 AA Groningen, The Netherlands
**E-mail: J.P.Noordmans@rug.nl*

We consider the possibility of Lorentz-invariance violation in weak-decay processes. We present a general approach that entails modifying the W-boson propagator by adding a Lorentz-violating tensor to it. We describe the effects of Lorentz violation on nuclear β decay in this scenario. In particular we show the expression for a first-forbidden transition with a spin change of two. Using data from an old experiment on the rotational invariance of yttrium-90, we derive several bounds on the Lorentz-violating parameters of the order of 10^{-6}-10^{-8}.

1. Introduction

There has been considerable scientific interest in the possibility of the violation of Lorentz symmetry, in the context of searches for a theory of quantum gravity. Its phenomenological consequences have been studied extensively, in particular in the context of the Standard-Model Extension (SME).[1] Extensive experimental efforts have been made to bound the coefficients in all sectors of the SME.[2]

Missing in these efforts is the study of weak decay. Until recently, there were, to our knowledge, only two dedicated experiments that addressed rotational invariance violation in β decay.[3,4] This effort is now supplemented by an experiment at KVI.[5]

Theoretically, the issue of calculating cross sections and decay rates is also far from settled, although some amount of work has already gone into this.[6]

We started a joint experimental and theoretical effort at KVI, addressing the issue of Lorentz violation in nuclear β decay. We incorporate Lorentz violation in weak decay by modifying the gauge boson propagator to

$$\langle W^{\mu +}(p) W^{\nu -}(-p)\rangle = \frac{-i(g^{\mu\nu} + \chi^{\mu\nu})}{M_W^2} , \tag{1}$$

where $g^{\mu\nu}$ is the usual Minkowski metric and $\chi^{\mu\nu}$ a general complex (possibly momentum-dependent) tensor, that parametrizes the Lorentz violation. What we miss in this way are mainly kinematic properties of the external particles (see Ref. 7 for more discussion).

2. β-decay transitions and forbiddenness

β-decay transitions are classified by the spin change from parent to daughter nucleus (ΔI) and the parities of the parent and daughter nuclei, denoted by π_i and π_f, respectively.

One distinguishes different β-decay transitions also by their *forbiddenness*, or the degree of suppression of the quantum mechanical amplitude that describes the process. This suppression is determined by factors of three small quantities: R/λ, v_N and αZ, which are the ratio of the nuclear radius and the de Broglie wavelength of the leptons, the velocity of the decaying nucleon in units of c, and the fine-structure constant times the charge of the daughter nucleus, respectively. Factors of R/λ come from the leptons carrying away orbital angular momentum (corresponding to higher order terms in the multipole expansion of the lepton wavefunctions). The quantities v_N and αZ originate from relativistic and Coulomb effects, respectively.

If we assume these three quantities to vanish, we limit ourselves to so-called *allowed* β-decay transitions. The leptons do not carry away orbital angular momentum in this case. Consequently, the total angular momentum of the lepton pair can be either zero or one. In the Lorentz-symmetric case this translates directly into the selection rules for allowed β decay, which can have spin change zero or one while the relative parity must be $\pi_i\pi_f = +1$. With Lorentz violation this connection is not that direct anymore, as explained later.

When allowing for factors of R/λ, v_N and αZ, the leptons can carry off orbital angular momentum and the total angular momentum of the lepton pair can be $J = 0, 1, 2, \ldots$, which allows for a larger spin change in the transition, as well as the possibility of a relative parity of $\pi_i\pi_f = -1$. For a complete classification of transitions and their forbiddenness in the Lorentz-symmetric and Lorentz-violating cases, see Ref. 8.

3. The effects of Lorentz violation

Using the propagator in Eq. (1), we have calculated the rate of general β-decay transitions. The expression for allowed β decay can be found in Ref. 7, and the one for forbidden transitions can be found in Ref. 9.

In Refs. 8 and 9 we found that there is a relative enhancement of Lorentz-violating effects in transitions with $\Delta I \geq 2$ by a factor of $\alpha Z/pR$. From the very schematic expression

$$(1, \boldsymbol{\sigma})^{\mu}_{\text{hadron}}(g_{\mu\nu} + \chi_{\mu\nu})(1, \boldsymbol{\sigma})^{\nu}_{\text{lepton}} , \tag{2}$$

that shows the hadron and lepton current, connected by the W-boson propagator, we see that the effect of the tensor $\chi^{\mu\nu}$ is to connect the singlet and triplet parts of the lepton and hadron current in an unconventional way. For example the spacetime part of $\chi^{\mu\nu}$ connects the triplet of the hadron current to the singlet of the lepton current. This means that there can be a unit of spin change in the nucleus, while the lepton pair does not carry off the corresponding unit of angular momentum. This is the cause of the enhancement factor that occurs in some Lorentz-violating terms of the amplitude of transitions with $\Delta I \geq 2$. For a more elaborate discussion of this effect, see Refs. 8 and 9.

4. Bounds on Lorentz violation

Using our results for a general β-decay transition we calculated the Lorentz-violating transition rate of a first forbidden transition with $\Delta I = 2$ and $\pi_i \pi_f = -1$, dependent on the emission direction of the outgoing β particle. It is given by

$$\frac{d\lambda}{d\Omega dE} \propto R^2 \Bigg\{ p^2 + q^2 + \frac{\alpha Z}{pR}\left[\frac{3}{10}\frac{p^3}{E}\left(\chi_r^{ij}\hat{p}^i\hat{p}^j - \frac{1}{3}\chi_r^{00}\right) \mp \frac{1}{2}p^2\tilde{\chi}_i^l\hat{p}^l \pm p^2\chi_r^{l0}\hat{p}^l\right]\Bigg\} , \tag{3}$$

where p, q, R are the β-particle momentum, the neutrino momentum and the nuclear radius respectively, $\tilde{\chi}^i = \epsilon^{ijk}\chi^{jk}$, Latin indices run over spatial components only, the subscripts r, i denote real and imaginary parts respectively, and the upper (lower) sign corresponds to β^- (β^+) decay. We see the relative enhancement factor $\alpha Z/pR$ of the Lorentz-violating effects.

Using the expression in Eq. (3) we reanalyzed results from an experiment published in 1976 and described in Ref. 3. The experiment was done using a 10 Ci strontium-90 source, which decays to yttrium-90 and subsequently to zirconium-90. The relevant decay is the one from ^{90}Y to ^{90}Z, which is a first-forbidden transition with $\Delta I = 2$ and $\pi_i \pi_f = -1$. In the experiment the source and the detector were mounted on a turntable, which rotated with a frequency of 0.75 Hz. Meanwhile the current of electrons that originated from the source was measured. Using the data of 232 good 2-h runs, asymmetries were defined in terms of the emission direction of the electrons. These asymmetries were fitted to distributions in sidereal time, given by

$$\delta = a_0 + a_1 \sin(\omega t + \phi_1) + a_2(2\omega t + \phi_2) \ . \tag{4}$$

Bounds of the order of 10^{-8} were obtained on the constants a_0, a_1, and a_2. Using Eq. (3) we calculated the theoretical prediction for the asymmetries defined in Ref. 3 and extracted the expressions for a_0, a_1, and a_2 in terms of $\chi^{\mu\nu}$. In this way we were able to put the following bounds at a 95% confidence level:[9]

$$|2X_r^{30} - \tilde{X}_i^3| < 2 \times 10^{-8} \ , \tag{5a}$$

$$|3X_r^{33} - X_r^{00}| < 3 \times 10^{-6} \ , \tag{5b}$$

$$\left[(X_r^{12} + X_r^{21})^2 + (X_r^{22} - X_r^{11})^2\right]^{1/2} < 1 \times 10^{-6} \ , \tag{5c}$$

$$\left[(X_r^{13} + X_r^{31})^2 + (X_r^{23} + X_r^{32})^2\right]^{1/2} < 1 \times 10^{-6} \ , \tag{5d}$$

$$\left[(2X_r^{20} - \tilde{X}_i^2)^2 + (2X_r^{10} - \tilde{X}_i^1)^2\right]^{1/2} < 4 \times 10^{-8} \ , \tag{5e}$$

where the Lorentz-violating parameter $X^{\mu\nu}$ is now given in the standard Sun-centered intertial reference frame.[2] These bounds are the best direct bounds on the parameter $\chi^{\mu\nu}$ of which we are aware.

References

1. D. Colladay and V.A. Kostelecký, Phys. Rev. D **55**, 6760 (1997); D. Colladay and V.A. Kostelecký, Phys. Rev. D **58**, 116002 (1998).
2. *Data Tables for Lorentz and CPT Violation,* V.A. Kostelecký and N. Russell, 2013 edition, arXiv:0801.0287v6.
3. R. Newman and S. Wiesner, Phys. Rev. D **14**, 1 (1976).
4. J.D. Ullman, Phys. Rev. D **17**, 1750 (1978).
5. S.E. Müller, Hyperfine Int. **210**, 33 (2012); H.W. Wilschut *et al.*, arXiv:1303.6419; S.E. Müller *et al.*, in preparation.
6. D. Colladay and V.A. Kostelecký, Phys. Lett. B **511**, 209 (2001); V.A. Kostelecký and R. Lehnert, Phys. Rev. D **63**, 065008 (2001); R. Lehnert, J. Math. Phys. **45**, 3399 (2004); R. Potting, Phys. Rev. D **85**, 045033 (2012);
7. J.P. Noordmans, H.W. Wilschut, and R.G.E. Timmermans, Phys. Rev. C **87**, 055502 (2013).
8. J.P. Noordmans, H.W. Wilschut, and R.G.E. Timmermans, in preparation.
9. J.P. Noordmans, H.W. Wilschut, and R.G.E. Timmermans, arXiv:1308.5570

REMARKS ON FINSLER GEOMETRY AND LORENTZ VIOLATION

N. RUSSELL

Physics Department, Northern Michigan University
Marquette, MI 49855, USA
E-mail: nrussell@nmu.edu

The physics of classical particles in a Lorentz-breaking spacetime has numerous features resembling the properties of Finsler geometry. In particular, the Lagrange function plays a role similar to that of a Finsler structure function. A summary is presented of recent results, including new calculable Finsler structures based on Lagrange functions appearing in the Lorentz-violation framework known as the Standard-Model Extension.

In conventional classical physics, a particle follows a trajectory that minimizes the spacetime interval $\int \sqrt{-dx^\mu r_{\mu\nu} dx^\nu}$. In this expression, $r_{\mu\nu}$ is a locally Minkowski metric with signature $(-,+,+,+)$ and the first coordinate is time. For particles with mass, the parameter λ of the trajectory $x^\mu(\lambda)$ may be the proper time of the particle, and in this case the path satisfies the geodesic equation $\dot{u}^\mu = -\tilde{\gamma}^\mu_{\alpha\beta} u^\alpha u^\beta$, where $u^\alpha = \dot{x}^\alpha = dx^\alpha/d\lambda$ are the four-velocity components and $\tilde{\gamma}^\mu_{\alpha\beta}$ are the Christoffel symbols derived from $r_{\mu\nu}$. In this variational problem, the total interval along the particle path, also called the action S, can be expressed as $S = \int L(x,u) d\lambda$, with suppressed summation indices in the Lagrange function $L(x,u) = \sqrt{-uru}$.

Several properties of this system are noteworthy. (a) The Lagrange function $L(x,u)$ is real and nonnegative if we restrict attention to timelike curves. (b) Since the path depends only on the initial conditions and the manifold, the action variation δS must be independent of λ. This is true for the $L(x,u)$ here, and can be ensured in general by requiring L to be positively homogeneous of degree one in the velocity u: $L(x,ku) = kL(x,u)$, for all $k > 0$. (c) We can recover the metric from the Lagrange function by evaluating the hessian $g_{\mu\nu} := -\frac{1}{2}\partial^2 L^2/\partial u^\mu \partial u^\nu$. (d) At any point x in the manifold, there is no preferred orientation or velocity because the metric is locally Minkowski. This is the property of Lorentz symmetry, and implies that experiments at point x obtain identical results regardless of their orientation or state of uniform motion.

Lorentz violation occurs if there exist physically detectable unconventional fields, such as a one-form field $a_\mu(x)$. The geodesics would be affected if it appeared in the Lagrange function, for example, $L_a(x,u) = \sqrt{-uru} + a_\nu(x)u^\nu$. Note that property (a) is ensured if the background is small, and that the homogeneity property (b) holds. In the case of property (c), the hessian $g_{\mu\nu}(x,u)$ now depends on both position and velocity, and only yields the underlying spacetime metric $r_{\mu\nu}(x)$ in the limit $a_\mu \to 0$. This spacetime breaks Lorentz symmetry, invalidating property (d), since the background field a_μ can be detected by appropriate comparisons between rotated or boosted experiments. Other Lagrange functions with unconventional backgrounds allow other ways to break Lorentz symmetry.

The properties (a), (b), (c) resemble the requirements for a Finsler geometry,[1] where the trajectories $x(\lambda)$ in a manifold M have tangent vectors y in the tangent bundle TM, and the geometric information is contained in a Finsler structure $F(x,y)$. This function, with appropriate continuity properties, is nonnegative (like property (a)), positively homogeneous (like (b)), and has a locally euclidean hessian (unlike property (c)). An important example of Finsler geometry is the Randers space,[2] with structure $F_a(x,y)$ having the same form as L_a, and Riemann metric r.

Studies of Lorentz violation have mushroomed in the last 20 years. The general effective-field-theory framework is called the Standard-Model Extension (SME),[3] and numerous experimental limits on coefficients for Lorentz violation exist.[4] To study the classical trajectories of Lorentz-breaking matter, the dispersion relations arising from the relevant modified Dirac equation[5] can be used to deduce the Lagrange function.[6] To see how this works, first note that the homogeneity condition (b) can be expressed as $L = u^\mu \partial L/\partial u^\mu = -u^\mu p_\mu$, by Euler's theorem and the definition of the canonical momentum. The Lagrange function follows by expressing the canonical momenta p_μ in terms of the four velocity u^μ. The method involves matching the wave-packet velocity with that of the classical particle and can be challenging because the Lagrange functions are roots of a polynomial that may be of high order.

Following this procedure, and related ones,[7] several classical Lagrange functions have recently been obtained, including L_a as discussed above. Since $L_a(x,u)$ is related to the Randers structure $F_a(x,y)$ by a signature change in $r_{\mu\nu}$, the question arises whether other classical Lagrange functions also give rise to Finsler structures. This is indeed the case, and several concrete, calculable Finsler geometries have been identified,[8] including

$$F_b(x,y) = \sqrt{y^2} \pm \sqrt{b^2 y^2 - (b \cdot y)^2}\,, \tag{1}$$

where r is a Riemann metric and summations have been suppressed. This structure F_b is based on a one-form, but is distinct from the Randers Finsler structure. Another example of a calculable Finsler structure arising from Lorentz-breaking backgrounds arises from the antisymmetric SME background $H_{\mu\nu}$. In four-dimensional spacetime, it has two invariants $X = H_{\mu\nu} r^{\mu\alpha} r^{\nu\beta} H_{\alpha\beta}/4$ and $Y = \epsilon^{\mu\nu\alpha\beta} H_{\mu\nu} H_{\alpha\beta}/8$. The Lagrange function for spacetime with $Y = 0$ leads to a Finsler structure $F_H = \sqrt{y^2} \pm \sqrt{-yH^2y}$ on a Riemann base manifold of any dimension. The antisymmetry of H, and the assumption of only one invariant, implies H^2 has even rank with one distinct negative eigenvalue, and satisfies the idempotent property $H^4 = -\eta^2 H^2$. Randers space F_a, 'b-space' F_b, and 'H-space' are special cases of bipartite Finsler structures,[9] which have the form

$$F_s(x,y) := \sqrt{yry} \pm \sqrt{ysy}\,, \tag{2}$$

with an underlying Riemann metric r and a symmetric two-tensor $s(x)$ with idempotent property $s^2 = \varsigma s$ for $o < \varsigma < 1$. For example, F_b is recovered for the case of $s_{\mu\nu} = b^2 r_{\mu\nu} - b_\mu b_\nu$. This bipartite structure has been shown to yield Finsler geometries with simple expressions for the hessian, its inverse, the geodesic equation, and other geometric quantities.[9]

A complementarity between F_a and F_b, and between cases of F_s, can be demonstrated by considering s as a mapping of each tangent space into itself. Since the image and kernel of the map are orthogonal, any tangent vector can be uniquely decomposed into perpendicular components, $y = y^{\|} + y^{\perp}$, forming a pythagorean triangle. Noting that the hypotenuse has the largest norm, two nonnegative complementary Finsler structures $F = ||y|| \pm \varsigma ||y^{\|}||$ and $F^{\perp} = ||y|| \pm \varsigma ||y^{\perp}||$ follow. If the rank of $s(x)$ is one, the two structures are F_a and F_b.[8] It is surprising that F_b, with this simple geometrical relationship to the Randers structure, has remained unknown for 70 years. The structure complementary to F_H is $F_H^{\perp} = \sqrt{y^2} \pm \sqrt{\eta y^2 + yH^2 y}$, and this includes all cases of $s(x)$ with even rank. Complementary Finsler structures appear for s of odd rank, and their relationship with other SME background fields is an open question.

There are numerous other open questions pertaining to the Finsler structures linked with Lorentz violation. The geodesics in bipartite Finsler spaces have the schematic form $\dot{u}^\mu + \tilde{\gamma}^\mu_{\alpha\beta} u^\alpha u^\beta = \{\widetilde{D}\ \text{SME terms}\}^\mu$, where $\widetilde{D}$ is the r-covariant derivative. The geodesics are therefore conventional if the

Lorentz-breaking fields are r parallel, and this raises the unanswered question whether r parallel SME backgrounds can be removed by suitable redefinitions. It is known that if the one-form background in Randers space is r-parallel, then the Berwald curvature vanishes,[8] but similar theorems for r-parallel SME-related structures do not exist. Shen has shown that Randers geodesics correspond to solutions of the Zermelo navigation problem.[10] Similar physical interpretations of the geodesics for F_b, and other bipartite Finsler spaces, are not known. Another open question is how to define torsions similar to the Matumoto torsion for Randers space, that characterize F_b space, bipartite Finsler spaces, and other spaces related to Lorentz-violating background fields. The first derivative of the Finsler structure $F(x, y)$ is singular at points in the tangent spaces that lie in the kernel of s. To overcome this and similar singularity issues, it is customary to exclude slits from the tangent bundle. An open question is finding alternative ways to avoid singularity issues by introducing, for example, a spin-like variable.[8] Indeed, Finsler and pseudo-Finsler geometries are active research areas with numerous approaches to open questions.[11] Structures related to the SME backgrounds with a locally Minkowski metric are of particular interest for the study of Lorentz violation in classical systems and may lead to new insights about unifying gravity and quantum mechanics.

References

1. D. Bao, S.-S. Chern, and Z. Shen, *An Introduction to Riemann-Finsler Geometry*, Springer, New York, 2000.
2. G. Randers, Phys. Rev. **59**, 195 (1941).
3. D. Colladay and V.A. Kostelecký, Phys. Rev. D **55**, 6760 (1997); Phys. Rev. D **58**, 116002 (1998); V.A. Kostelecký, Phys. Rev. D **69**, 105009 (2004).
4. *Data Tables for Lorentz and CPT Violation,* V.A. Kostelecký and N. Russell, 2013 edition, arXiv:0801.0287v6.
5. V.A. Kostelecký and R. Lehnert, Phys. Rev. D **63**, 065008 (2001).
6. V.A. Kostelecký and N. Russell, Phys. Lett. B **693**, 443 (2010).
7. D. Colladay and P. McDonald, Phys. Rev. D **85**, 044042 (2012).
8. V.A. Kostelecký, Phys. Lett. B **701**, 137 (2011).
9. V.A. Kostelecký, N. Russell, and R. Tso, Phys. Lett. B **716**, 470 (2012).
10. Z. Shen, Canad. J. Math. **55**, 112 (2003).
11. M.A. Javaloyes and M. Sánchez, arXiv:1111.5066; C. Pfeifer and M.N.R. Wohlfarth, Phys. Rev. D **84**, 044039 (2011); E. Caponio, M.A. Javaloyes, and A. Masielo, Math. Ann. **351**, 365 (2011); R.G. Torromé, P. Piccione, and H. Vitório, J. Math. Phys. **53**, 123511 (2012); C. Lämmerzahl, V. Perlick, and W. Hasse, Phys. Rev. D **86**, 104042 (2012); A.P. Kouretsis, M. Stathakopoulis, and P.C. Stavrinos, Phys. Rev. D **86**, 124025 (2012); Z. Chang, X. Li and S. Wang, arXiv:1201.1368.

LORENTZ VIOLATION IN A UNIFORM GRAVITATIONAL FIELD

Y. BONDER

Physics Department, Indiana University
Bloomington, IN 47405, USA
E-mail: ybonder@indiana.edu

We present a method to calculate the nonrelativistic hamiltonian for the minimal Standard-Model Extension matter sector in a uniform gravitational field. The resulting hamiltonian coincides with earlier results in the corresponding limits but it also includes spin-dependent terms that were previously unknown. The phenomenology associated with this hamiltonian is briefly discussed.

Lorentz invariance lies at the core of our current conception of physics. Thus, it seems necessary to test empirically Lorentz invariance. The Standard-Model Extension (SME) is a framework that includes the Standard Model and General Relativity plus all possible Lorentz violating terms,[1,2] and it is widely used as a guide to search and parametrize Lorentz violation. As was pointed out by Kostelecký,[2] gravity couplings allow access to some SME coefficients that are otherwise unobservable. Thus, to explore all possible ways Lorentz violation could manifest, one needs to study the matter-gravity SME sector.

Kostelecký and Tasson[3] studied the matter-gravity SME sector and their analysis led to bounds on some previously unmeasured SME coefficients.[4] However, in Ref. 3 they do not consider the SME coefficients associated with spin. Our proposal is to study the SME matter-gravity sector with a different — and complementary — approach. This allows us to obtain the nonrelativistic hamiltonian for the free Dirac fermion minimal SME sector in a uniform newtonian gravitational potential, including spin effects.

We now outline our work hypothesis. The fermions are assumed to be test particles, namely, the spacetime curvature generated by these particles is neglected. Moreover, the background spacetime is chosen to be the flat (and torsion-free) spacetime but as seen by a uniformly accelerated observer. We also assume that the observer's acceleration can be identified with a uniform newtonian gravitational potential Φ. Note that, in the situation at hand, the no-go theorem precluding us from having explicit Lorentz violations in curved spacetimes[2] does not apply, and it is consistent to consider explicit Lorentz violations. This allows us to ignore the mechanisms from which the Lorentz violations would spontaneously emerge, simplifying considerably our analysis. Furthermore, we require the SME coefficients to have vanishing covariant derivatives, which is equivalent to the condition that the coefficients are constant for an inertial observer.

To construct the hamiltonian we use, as a starting point, the action of the corresponding SME sector in a curved background spacetime.[2] We work in Fermi-like coordinates associated with a uniformly-accelerated and nonrotating observer, as done in Ref. 5. In principle, the equations of motion can be used to read off a hamiltonian, defined as the generator of time translations. However, one must invert a matrix contracted with the wavefunction time derivative. To invert this matrix we use the field redefinition method described in Ref. 3 that, in addition, ensures that the resulting hamiltonian is hermitian with respect to the standard inner product of nonrelativistic quantum mechanics.

We are interested in tabletop experiments where the particles are nonrelativistic. Therefore, we seek the nonrelativistic hamiltonian to first order in the SME coefficients. To do so, we perform three Foldy-Wouthuysen transformations[6] that decouple the particle and antiparticle degrees of freedom. We also remove all the unphysical terms, namely, those that can be canceled with unitary transformations. For simplicity we only present here the nonrelativistic hamiltonian for the particles (as opposed to the antiparticles) to first order in Φ; the resulting hamiltonian is

$$\begin{aligned} H_{\rm NR} = {} & A(1+\Phi) + B_i\sigma_i(1+\Phi) + \frac{1}{m}C_i\left(p_i + \frac{1}{2}\Phi p_i + \frac{1}{2}p_i\Phi\right) \\ & + \frac{1}{m}D_{ij}\sigma_j\left(p_i + \frac{1}{2}\Phi p_i + \frac{1}{2}p_i\Phi\right) + \frac{1}{2m}E_{ij}p_i(1+\Phi)p_j \\ & + \frac{1}{2m^2}F_{ijk}\sigma_k\left(p_ip_j + \frac{3}{2}p_i\Phi p_j + \frac{3}{2}p_j\Phi p_i\right) \\ & + \frac{1}{m^2}G_{ijk}\sigma_k\left(p_ip_j + \frac{1}{2}p_i\Phi p_j + \frac{1}{2}p_j\Phi p_i\right), \end{aligned} \tag{1}$$

where

$$A = m + \hat{a}_0 - m\hat{c}_{00} - m\hat{e}_0, \tag{2}$$

$$B_i = -b_i + m\hat{d}_{i0} + \frac{1}{2}\epsilon_{ijk}\left(-m\hat{g}_{jk0} + H_{jk}\right), \tag{3}$$

$$C_i = a_i - m\hat{c}_{0i} - m\hat{c}_{i0} - me_i, \tag{4}$$

$$D_{ij} = -\delta_{ij}\hat{b}_0 + m\delta_{ij}\hat{d}_{00} + md_{ij} + \epsilon_{ijk}\left(\frac{1}{2}(\partial_k\Phi) + m\hat{g}_{0k0} - \hat{H}_{0k}\right), \tag{5}$$

$$E_{ij} = \delta_{ij} - \delta_{ij}\hat{c}_{00} - \delta_{ik}\delta_{jl}(c_{kl} + c_{lk}), \tag{6}$$

$$\begin{aligned} F_{ijk} = {} & \delta_{ij}b_k - \delta_{jk}b_i - m\delta_{jk}\hat{d}_{i0} \\ & + \frac{1}{2}m\epsilon_{klm}\delta_{ij}\hat{g}_{lm0} - \frac{1}{2}\epsilon_{ilm}\delta_{jk}\left(m\hat{g}_{lm0} + H_{lm}\right), \end{aligned} \tag{7}$$

$$G_{ijk} = m\delta_{jk}(\hat{d}_{0i} + \hat{d}_{i0}) + m\epsilon_{ikl}\hat{g}_{0lj} - m\epsilon_{ikl}\hat{g}_{lj0}. \tag{8}$$

In these last expressions m is the particle's mass, p_i are the components of the momentum operator, σ_i are the Pauli matrices, ϵ_{ijk} is the totally antisymmetric tensor with $\epsilon_{123} = 1$, and all the indices run from 1 to 3 with the convention of summing over repeated indices. The SME coefficients for the sector we study are a_μ, b_μ, $c_{\mu\nu}$, $d_{\mu\nu}$, e_μ, f_μ $g_{\mu\nu\rho} = -g_{\nu\mu\rho}$ and $H_{\mu\nu} = -H_{\nu\mu}$, with Greek indices running from 0 to 3. In the case we analyze, the SME coefficients get 'redshifted' by a factor $(1+\Phi)^{-n}$, where n is the number of zero indices in the corresponding coefficient. Thus, to get a compact expression we introduce a caret on top of some of the SME coefficients with the convention that a coefficient with a caret has been redshifted. For example, $\hat{b}_0 \equiv b_0(1+\Phi)^{-1}$ and $\hat{g}_{0i0} \equiv g_{0i0}(1+\Phi)^{-2}$.

The nonrelativistic hamiltonian we have calculated agrees with that of Ref. 5 when the SME coefficients are set to zero and it coincides with the result of Ref. 7 in the limit when $\Phi = 0$. In addition, where there is overlap, it agrees with the nonrelativistic hamiltonian presented in Ref. 3. Even though p_i acts on the SME coefficients when taking the adjoint, it is possible to verify that our hamiltonian is hermitian. Moreover, a_μ and e_μ only appear in the combination $a_\mu - me_\mu$ and the antisymmetric part of $d_{\mu\nu}$ only shows up in $H_{\mu\nu} + m\varepsilon_{\mu\nu}{}^{\rho\sigma}d_{\rho\sigma}/2$ where $\varepsilon_{\mu\nu\rho\sigma}$ is the spacetime volume form. Also, the antisymmetric part of $c_{\mu\nu}$ and f_μ do not enter into our hamiltonian. These facts are consistent with the analysis sketched in Ref. 2 regarding the freedom to redefine the fermionic field.

Note that our hamiltonian is not invariant if we add a constant to Φ. However, it is possible to see that, in the coordinates we work, the point where $\Phi = 0$ has physical meaning: it is where the observer/detector is located. Thus, adding a constant to Φ amounts to relocating the detector at a different height and, in order to verify that the physics is invariant under such transformation, one needs to take into the account the redshift on the detector. In addition, the gravitational effects in the hamiltonian (1) are only relevant for experiments where the particles probe regions with different Φ. Thus, the best candidates to search for the effects of this hamiltonian are interferometry experiments like those of Refs. 8. Also, the nonrelativistic hamiltonian for antiparticles has been calculated and it can be obtained from equation (1) with the replacements given in Ref. 7 that link the particle and antiparticle hamiltonian in the nongravitational case. Furthermore, since each component of the SME coefficients gets redshifted in a different way, our result suggests that by doing experiments at several heights, it should be possible to disentangle the bounds that are usually placed on the linear combinations of these coefficients.

To summarize, we have obtained the nonrelativistic hamiltonian for the minimal matter SME sector in the presence of a uniform newtonian gravitational potential. The spin-dependent terms of this hamiltonian are presented here for the first time, and may lead to new experiments that will allow us to keep testing Lorentz invariance.

Acknowledgments

I wish to thank Alan Kostelecký for many helpful discussions. This work was supported by the Department of Energy under grant DE-FG02-13ER42002 and by the Indiana University Center for Spacetime Symmetries.

References

1. D. Colladay and V.A. Kostelecký, Phys. Rev. D **55**, 6760 (1997); Phys. Rev. D **58**, 116002 (1998).
2. V.A. Kostelecký, Phys. Rev. D **69**, 105009 (2004).
3. V.A. Kostelecký and J.D. Tasson, Phys. Rev. D **83**, 016013 (2011).
4. *Data Tables for Lorentz and CPT Violation,* V.A. Kostelecký and N. Russell, 2013 edition, arXiv:0801.0287v6, Table D28.
5. F.W. Hehl and W.-T. Ni, Phys. Rev. D **42**, 2045 (1990).
6. L.L. Foldy and S.A. Wouthuysen, Phys. Rev. **78**, 29 (1950).
7. V.A. Kostelecký and C.D. Lane, J. Math. Phys. **40**, 6245 (1999).
8. H. Müller, A. Peters, and S. Chu, Nature **463**, 926 (2010); M.A. Hohensee *et al.*, Phys. Rev. Lett. **106**, 151102 (2011); M.A. Hohensee *et al.*, arXiv:1303.2747v2; M.A. Hohensee, these proceedings; H. Müller, these proceedings.

OBTAINING BOUNDS FROM ULTRA-HIGH ENERGY COSMIC RAYS IN ISOTROPIC MODIFIED MAXWELL THEORY

M. SCHRECK

ANKA Storage Ring, Karlsruhe Institute of Technology
76344 Eggenstein-Leopoldshafen, Germany
E-mail: Marco.Schreck@kit.edu

This article reviews the methods used to obtain a two-sided bound on isotropic modified Maxwell theory from experimental data of ultra high-energy cosmic rays in 2008. The bound is updated with results from the HEGRA experiment.

1. Introduction

The advent of quantum mechanics at the beginning of the 20th century marks an important step into a new era of physics. The classical theory was left behind and it became possible to describe physical processes occurring at the atomic length scale of 10^{-10} m. Quantum mechanics was then also applied to atomic nuclei reducing the length scale of its applicability again by five orders of magnitude. In the aftermath, quantum field theory was developed, which for several decades has been used to great success in order to understand physics down to approximately 10^{-18} m.

Perhaps a similar revolution is currently taking place at the beginning of the 21st century. The interest in experiments looking for signs of quantum gravity, which is expected to play a role at the Planck length, i.e., at 10^{-35} m, has been steadily increasing. If Einstein's relativity and quantum physics are assumed to be still valid at this scale, both theories will have to merge into a new theory correctly describing fluctuations of spacetime itself. Currently there is no chance of investigating quantum gravitational effects directly at the Planck energy. However there is a clear signal for such effects that may be visible far below the Planck energy: a violation of Lorentz symmetry.

The latter is motivated by various approaches to a fundamental theory: string theory, loop quantum gravity, noncommutative spacetime, etc.

Since we currently still do not know how such a theory looks like it makes sense to study Lorentz violation within a model-independent framework: the Standard-Model Extension (SME).[1] This is a collection of all field theoretic operators that are symmetric with respect to the gauge group $SU_c(3) \times SU_L(2) \times U_Y(1)$ of the ordinary Standard Model but that violate particle Lorentz invariance. The current goal of experiments is to measure the parameters associated with these operators. A detection of nonvanishing Lorentz-violating coefficients could deliver significant insights on how different particle sectors are indeed affected by quantum gravitational phenomena. So far it has been impossible to detect Lorentz violation. Hence with better and better experiments even stricter constraints on Lorentz-violating parameters can be set.

2. Modified Maxwell theory

In the power-counting renormalizable photon sector of the SME there exist two modifications: Maxwell–Chern–Simons (MCS) theory and modified Maxwell theory. The first is characterized by a dimensionful scale and a background vector field where the second involves a dimensionless fourth-rank tensor background field. The dimensionful parameter of MCS theory was already heavily bounded by astrophysical observations[2] whereas some of the parameters of modified Maxwell theory — especially the isotropic sector — were only weakly bounded in 2008. This was the motivation to consider a modified quantum electrodynamics (QED) in Ref. 3 that results from minimally coupling modified Maxwell theory to a standard Dirac theory of spin-1/2 fermions with charge e and mass M:

$$S_{\text{modQED}} = S_{\text{modMax}} + S_{\text{standDirac}}\,, \tag{1a}$$

$$S_{\text{modMax}} = \int_{\mathbb{R}^4} \mathrm{d}^4x \left\{ -\frac{1}{4} F^{\mu\nu}(x) F_{\mu\nu}(x) - \frac{1}{4} \kappa^{\mu\nu\varrho\sigma} F_{\mu\nu}(x) F_{\varrho\sigma}(x) \right\}, \tag{1b}$$

$$S_{\text{standDirac}} = \int_{\mathbb{R}^4} \mathrm{d}^4x\, \overline{\psi}(x) \Big\{ \gamma^\mu \left[\mathrm{i}\partial_\mu - eA_\mu(x) \right] - M \Big\} \psi(x)\,. \tag{1c}$$

Here $F_{\mu\nu}(x) \equiv \partial_\mu A_\nu(x) - \partial_\nu A_\mu(x)$ is the field strength tensor of the $U(1)$ photon field $A_\mu(x)$ and $\psi(x)$ is the spinor field. The fields are defined in Minkowski spacetime with the metric tensor $(\eta_{\mu\nu}) = \mathrm{diag}(1, -1, -1, -1)$. The second term of Eq. (1b) containing the fourth-rank tensor background field $\kappa^{\mu\nu\varrho\sigma}$ manifestly violates particle Lorentz invariance. This field is a low energy effective description of possible physics at the Planck scale and it defines preferred spacetime directions.

The modified photon action given by Eq. (1b) can be restricted to a sector exhibiting nonbirefringent[a] photon dispersion laws. This is possible via the following ansatz for the background field:

$$\kappa^{\mu\nu\varrho\sigma} = \frac{1}{2}(\eta^{\mu\varrho}\widetilde{\kappa}^{\nu\sigma} - \eta^{\mu\sigma}\widetilde{\kappa}^{\nu\varrho} - \eta^{\nu\varrho}\widetilde{\kappa}^{\mu\sigma} + \eta^{\nu\sigma}\widetilde{\kappa}^{\mu\varrho})\,, \tag{2}$$

where $\widetilde{\kappa}^{\mu\nu}$ is a symmetric and traceless (4×4)-matrix: $\widetilde{\kappa}^{\mu\nu} = \widetilde{\kappa}^{\nu\mu}$, $\widetilde{\kappa}^{\mu}{}_{\mu} = 0$. The isotropic part of modified Maxwell theory is then defined by the choice

$$(\widetilde{\kappa}^{\mu\nu}) \equiv \frac{3}{2}\widetilde{\kappa}_{\text{tr}}\,\text{diag}\left(1, \frac{1}{3}, \frac{1}{3}, \frac{1}{3}\right)\,. \tag{3}$$

From the field equations results a modified photon dispersion law that is isotropic in three-space:

$$\omega(k) = c\sqrt{\frac{1-\widetilde{\kappa}_{\text{tr}}}{1+\widetilde{\kappa}_{\text{tr}}}}\,k\,. \tag{4}$$

Here ω is the photon frequency and k the wave number. Note that c is the maximum attainable velocity of the Dirac particle described by Eq. (1c), which is not affected by Lorentz violation. Dependent on the choice of $\widetilde{\kappa}_{\text{tr}}$ the modified photon velocity does not coincide with the maximum velocity of massive particles any more. This leads to peculiar particle physics processes, which are forbidden in standard QED.

3. Vacuum Cherenkov radiation and photon decay

The modified photon dispersion relation of Eq. (4) allows $\widetilde{\kappa}_{\text{tr}}$ to lie in the interval $(-1,1]$. For $\widetilde{\kappa}_{\text{tr}} \in (0,1]$ the photon velocity is smaller than c and Dirac particles can travel faster than photons. If the latter is the case, a Cherenkov-type process in vacuum takes place leading to energy loss of the Dirac particle by the emission of modified photons $\widetilde{\gamma}$. This process occurs above a certain threshold energy E_{thresh} of the massive particle (e.g., a proton p^+) and its radiated energy rate $\text{d}W/\text{d}t$ far above the threshold is:

$$E_{\text{thresh}}^{\text{p}^+\to\text{p}^+\widetilde{\gamma}} = \frac{M_{\text{p}}c^2}{2}\sqrt{2+\frac{2}{\widetilde{\kappa}_{\text{tr}}}}\,,\qquad \left.\frac{\text{d}W}{\text{d}t}\right|_{E\gg E_{\text{thresh}}^{\text{p}^+\to\text{p}^+\widetilde{\gamma}}}^{\text{p}^+\to\text{p}^+\widetilde{\gamma}} \simeq \frac{7}{12}\alpha\widetilde{\kappa}_{\text{tr}}\frac{E^2}{\hbar}\,. \tag{5}$$

Herein, M_{p} is the proton mass, $\alpha \equiv e^2/(4\pi\varepsilon_0\hbar c)$ the electromagnetic fine structure constant,[b] and $\hbar$ Planck's constant. For vanishing isotropic parameter $\widetilde{\kappa}_{\text{tr}}$ the threshold energy goes to infinity and the radiated energy rate vanishes showing that the process is forbidden in standard QED.

[a] at first order in Lorentz violation.

[b] with the vacuum permittivity ε_0.

Table 1. Ultra-high energy photon and cosmic ray event. For details on the Auger event consult Ref. 3 and references therein. The HEGRA event is obtained from the last bin of Fig. 3 in Ref. 4, which has a significance of 2.7σ. Since the energy is high enough and the source was identified as the Crab nebula, the left-hand bin endpoint with $E \simeq 56\,\mathrm{TeV}$ can be taken as a fiducial photon event. For the energy uncertainty a conservative estimate of 10% is used far above the detector threshold where the uncertainty mainly originates from statistical errors.

Experiment	Observation	Energy E	Energy uncertainty $\Delta E/E$
HEGRA	1997–2002	56 TeV [Fig. 3 of Ref. 4]	10% [p. 12 of Ref. 4]
Auger	ID 737165	212 EeV [see Ref. 3]	25% [see Ref. 3]

For $\widetilde{\kappa}_{\mathrm{tr}} \in (-1, 0)$ the modified photon velocity is larger than the maximum velocity of Dirac particles. In this case a photon may decay preferably into an electron positron pair $\mathrm{e}^+\mathrm{e}^-$. This decay is possible above a certain threshold energy E_{thresh} for the photon and it has the following decay rate Γ:

$$E_{\mathrm{thresh}}^{\widetilde{\gamma}\to\mathrm{e}^+\mathrm{e}^-} = M_{\mathrm{e}}c^2\sqrt{2-\frac{2}{\widetilde{\kappa}_{\mathrm{tr}}}}\,,\quad \Gamma^{\widetilde{\gamma}\to\mathrm{e}^+\mathrm{e}^-}\big|_{E\gg E_{\mathrm{thresh}}^{\widetilde{\gamma}\to\mathrm{e}^+\mathrm{e}^-}} \simeq -\frac{2}{3}\alpha\widetilde{\kappa}_{\mathrm{tr}}E\,, \tag{6}$$

with the electron mass M_{e}. The crucial difference from vacuum Cherenkov radiation is the minus sign appearing together with $\widetilde{\kappa}_{\mathrm{tr}}$. This tells us that the above equations only make sense for negative Lorentz-violating parameters.

4. Updated two-sided bound on the isotropic parameter

If a hadronic primary or a photon is detected on Earth its energy must be smaller than the threshold energy of Eq. (5) and Eq. (6), respectively. Using then the events of Table 1 we obtain the following updated two-sided bound on the isotropic parameter of modified Maxwell theory at the 2σ level:

$$-2\cdot 10^{-16} < \widetilde{\kappa}_{\mathrm{tr}} < 6\cdot 10^{-20}\,. \tag{7}$$

The lower bound has been improved by a factor of 4 in comparison to Ref. 3.

Acknowledgments

It is a pleasure to thank R. Wagner for helpful discussions and for pointing out Ref. 4 during the CPT'13 conference in Bloomington, Indiana.

References

1. D. Colladay and V.A. Kostelecký, Phys. Rev. D **58** 116002 (1998).
2. S.M. Carroll, G.B. Field, and R. Jackiw, Phys. Rev. D **41**, 1231 (1990).
3. F.R. Klinkhamer and M. Schreck, Phys. Rev. D **78**, 085026 (2008).
4. HEGRA Collaboration, F. Aharonian *et al.*, Astrophys. J. **614**, 897 (2004).

MODERN IVES-STILWELL EXPERIMENTS AT STORAGE RINGS: LARGE BOOSTS MEET HIGH PRECISION

G. GWINNER

Department of Physics and Astronomy, University of Manitoba
Winnipeg, R3T 2N2, Canada
E-mail: gwinner@physics.umanitoba.ca

B. BOTERMANN, C. GEPPERT, G. HUBER, S. KARPUK, A. KRIEGER, W. NÖRTERSHÄUSER, and C. NOVOTNY

Johannes Gutenberg-Universität Mainz, 55128 Mainz, Germany

T. KÜHL, R. SÁNCHEZ, and T. STÖHLKER

GSI Helmholtzzentrum für Schwerionenforschung, 64291 Darmstadt, Germany

D. BING, D. SCHWALM, and A. WOLF

Max-Planck Institut für Kernphysik, 69117 Heidelberg, Germany

T.W. HÄNSCH, S. REINHARDT, and G. SAATHOFF

Max-Planck Institut für Quantenoptik, 85748 Garching, Germany

We give a brief overview of time dilation tests using high-resolution laser spectroscopy at heavy-ion storage rings. We reflect on the various methods used to eliminate the first-order Doppler effect and on the pitfalls encountered, and comment on possible extensions at future facilities providing relativistic heavy ion beams at $\gamma \gg 1$.

1. Introduction

Time dilation measurements via precise optical spectroscopy have played a long and distinguished role in testing the predictions of Special Relativity (SR), and have recently also contributed to constraining coefficients in the Lorentz- and CPT-violating Standard-Model Extension (SME).[1,2] The first reported observation of time dilation in SR in 1938, by Ives and Stilwell, was based on measuring the relativistic Doppler effect in fast hydrogen atoms.[3] In the decades after this landmark experiment, other methods tested time dilation with increasing sensitivity, such as lifetime measurements of elementary particles accelerated to relativistic energies, transverse Doppler effect measurements with rotating Mössbauer setups, and comparisons between ground-based and rocket-mounted clocks (for a review and references see, e.g., Ref. 4). Two major breakthroughs allowed optical spectroscopy to re-enter the field, and to dominate it since: tunable, single-mode lasers in the 1970s and heavy-ion storage rings, providing electron-cooled ion beams of exquisite energy definition and transverse emittance, in the 1980s. This has led to the measurement of absolute optical transition frequencies at the 10^{-10} level on near-relativistic beams,[1] possibly the most accurate measurement ever performed with accelerated particle beams.

2. Relativistic Doppler shift measurements

An atom or ion is moving with velocity $\vec{\beta}$ in the laboratory. A transition between two levels in this atom can be driven in its rest frame by radiation of frequency ν_0. A light source in the laboratory frame, according to the relativistic Doppler effect, needs to be tuned to the frequency ν_{lab} such that

$$\nu_0 = \gamma \, (1 - \beta \cos \theta) \, \nu_{\text{lab}}, \tag{1}$$

where β is the relative velocity between the atom and the laboratory frame, and $\gamma = 1/\sqrt{1-\beta^2}$; θ is the angle between $\vec{\beta}$ and the wavevector of the radiation exciting the atom, as measured in the laboratory frame. Testing time dilation, i.e., γ, requires the precise determination of ν_0, ν_{lab}, β, and θ. In reality, it is generally not possible to determine the velocity of an accelerated particle beam with the required precision. Therefore, experiments need to remove the first-order dependence on β. At $\theta = \pi/2$ one could observe a purely transverse Doppler effect, but due to the sensitivity to angular misalignment, this turns out not to be a workable scheme. Geometries with $\theta = (0, \pi)$, while robust against misalignments, appear to suffer from the presence of the full first-order Doppler effect. Nevertheless, atomic spectroscopy provides several well-established techniques to eliminate the first-order Doppler effect, turning a collinear setup into the method of choice.

Commercially available single-mode lasers have much higher monochromaticity ($\Delta\nu/\nu < 10^{-9}$) than accelerated particle beams ($\Delta\beta/\beta \approx 10^{-5}$-$10^{-6}$). In a scenario with two lasers, one (at a fixed frequency) can be used to 'mark' a narrow velocity class within the particle beam's velocity distribution such that a second, frequency-tunable, laser performs spectroscopy only on that specific class of atoms.

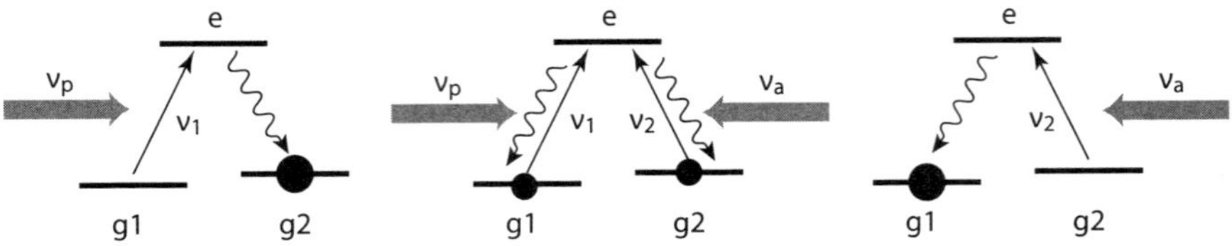

Fig. 1. Λ-spectroscopy: driving a given velocity class only on transition 1 (2) pumps the atoms quickly into the 'dark' state g2 (g1) as shown on the left (right). When both legs are excited, atoms get continually pumped back and forth and strong fluorescence is produced (center).

The first generation of storage ring measurements (1990-94) used Li^+ stored at $\beta = 0.064$ in the TSR storage ring.[5] It used a Doppler-free technique known as Λ-spectroscopy, which is illustrated in Fig. 1. Two lasers with laboratory frequencies ν_a and ν_p counter- and co-propagate with the ions and excite both legs of the Λ at the frequencies ν_1 and ν_2 in the ions' rest frame. Only if both lasers are exciting the *same* velocity class do atoms get pumped back and forth, producing significant fluorescence. Within SR, the resonance conditions $\nu_1 = \gamma(1-\beta)\nu_p$ and $\nu_2 = \gamma(1+\beta)\nu_a$ give the relation $\nu_1\nu_2 = \nu_a\nu_p$. Deviations from relativity can be parametrized in a model-independent way by a velocity-dependent term $\epsilon(\beta^2)$ as

$$\sqrt{\frac{\nu_a\nu_p}{\nu_1\nu_2}} = 1 + \epsilon(\beta^2), \tag{2}$$

which can be expanded, for $\beta \ll 1$, as $\epsilon(\beta^2) \approx \hat{\alpha}\beta^2 + \hat{\alpha}_2\beta^4$. The TSR-1 experiment yielded a new, best time dilation limit of $|\hat{\alpha}| < 7 \times 10^{-7}$.[5] The experimental limitation was an unexpectedly broad lineshape, with a width of 54 MHz, vs. 7.6 MHz expected from the natural linewidth of the transition. The second-generation TSR-2 experiment (2000-06) revealed the problem: the 'marking' of a velocity class by one laser and the 'sampling' of it by the other are generally not concurrent in Λ-spectroscopy and velocity-changing collisions lead to broadening.

The TSR-2 experiment solved this problem by employing a different Doppler-free technique: saturation spectroscopy.[1,6] Its signature feature, the Lamb dip, only occurs when both lasers *simultaneously* interact with the same individual ion. Linewidths approaching the natural linewidth were observed, and a much improved limit was obtained: $|\hat{\alpha}| < 9 \times 10^{-8}$.[1] Ultimately, the method is limited by its reliance on saturation, i.e., the need to scatter many photons. At the intensities required for saturation, the lasers exert a rather strong force on the ions, distorting the velocity distribution significantly. In saturation spectroscopy, we observe a small *drop* in fluorescence (the Lamb dip) on top of a large Doppler-broadened background. In Λ-spectroscopy, we detect a narrow fluorescence *peak*, while the rest of the velocity distribution remains dark.

With the spectroscopy reaching its limitations, higher ion velocities are the obvious avenue for further improvements. At the storage ring ESR, the same experiment can be carried out at $\beta = 0.338$, where the Doppler shifts are extraordinary:[7,8] For $\lambda_{1,2} = 548$ nm, the laser frequencies in the laboratory frame are $\lambda_p = 386$ nm and $\lambda_a = 772$ nm. Saturation spectroscopy turned out not to be viable due to the required intensity for the UV light. Λ-spectroscopy proved robust; while the excess broadening was comparable to the TSR-1 situation (≈ 50 MHz), the much larger β more than compensates and tighter limits result: an upper limit for $|\hat{\alpha}|$ in the lower 10^{-8} range seems in reach.[8] Alternatively, TSR-2 and ESR will be able to set combined limits of $|\hat{\alpha}| < 9 \times 10^{-8}$ and $|\hat{\alpha_2}| < 10^{-6}$.

3. Prospects for future work

Ultimately, it would be intriguing to implement high-resolution spectroscopy at facilities that provide highly relativistic ion beams, like the upcoming High-Energy Storage Ring (HESR) at FAIR. If the rest-frame frequencies $\nu_{1,2}$ are nearly identical and in the visible region, at high boosts ν_p will eventually be beyond the near-UV region. A possible solution would be an *asymmetric* Λ, such that standard lasers can drive it in a low-β measurement (e.g., one in the visible and one in the near/mid-infrared). In the high-boost version, the co-propagating laser will drive the short leg (which will be blue-shifted toward the visible) and the counter-propagating laser will excite the long leg (which will be red-shifted toward the infrared). The challenge will be to identify a suitable transition in a highly charged ion.

References

1. S. Reinhardt *et al.*, Nat. Phys. **3**, 861 (2007).
2. C.D. Lane, Phys. Rev. D **72**, 016005 (2005).
3. H.E. Ives and G.R. Stilwell, J. Opt. Soc. Am. **28**, 215 (1938).
4. G. Gwinner, Mod. Phys. Lett. A **20**, 791 (2005).
5. R. Grieser *et al.*, Appl. Phys. B **59**, 127 (1994).
6. G. Saathoff *et al.*, Phys. Rev. Lett. **91**, 190403 (2003).
7. Ch. Novotny *et al.*, Phys. Rev. A **80**, 022107 (2009).
8. B. Botermann *et al.*, in preparation.

RENORMALIZATION IN QED AND QFT WITH A LORENTZ- AND CPT-VIOLATING BACKGROUND

ILYA L. SHAPIRO

Departamento de Física, ICE, Universidade Federal de Juiz de Fora
Juiz de Fora, Minas Gerais, 36036-900, Brazil
E-mail: shapiro@fisica.ufjf.br

The general features of renormalization and the renormalization group in QED and in general quantum field theories in curved spacetime with additional Lorentz- and CPT-violating background fields are reviewed.

1. Introduction

There is a growing interest in exploring theories with weakly broken Lorentz and CPT symmetries. The importance of such studies should be clear, especially for those who believe that these important symmetries are exact. Indeed, the unique way to confirm this belief is to assume that they can be actually broken and verify the physical consequences (see Ref. 1 for the latest results in this respect). The general parametrization of all possible ways to break Lorentz and CPT symmetries has been formulated by A. Kostelecký *et al.* in a series of papers (see, e.g., Ref. 2) and the focus of attention is now naturally shifting to experimental work with different manifestations of the theories with broken symmetries.

An important aspect of theories with broken symmetries is related to their renormalization structure. The reason is that consistency at the quantum level can impose restrictions on the classical theory and, eventually, restrict the space of parameters for experimental verification. On the other hand, such a study can be helpful in establishing the relation between different parameters related to the breaking of Lorentz and CPT symmetries. In what follows we discuss, in particular, the relation between violations of symmetries in the matter and gravitational sectors of the theory.

2. Renormalization

In the well-studied cases of external fields, such as electromagnetic, metric (see, e.g., Ref. 3,4 for the review and further references) and torsion[5] one can establish the form of possible counterterms by means of covariance and other symmetries from one side and power counting on another side. Naively one may think that these arguments will not work, e.g., for QED with broken Lorentz and CPT symmetries,[2]

$$S = \int d^4x\sqrt{-g}\,\Big\{\,\frac{i}{2}\,\bar{\psi}\Gamma^\mu D_\mu\psi - \frac{i}{2}\,D^\star_\mu\bar{\psi}\Gamma^\mu\psi - \bar{\psi}\,M\,\psi - \frac{1}{4}F_{\mu\nu}\,F^{\mu\nu}$$
$$- \frac{1}{4}\,(k_F)_{\mu\nu\alpha\beta}\,F^{\mu\nu}\,F^{\alpha\beta} + \frac{1}{2}\,(k_{AF})^\rho\,\varepsilon_{\rho\lambda\mu\nu}\,A^\lambda\,F^{\mu\nu}\,\Big\}. \tag{1}$$

In the last formula $\Gamma^\nu = \gamma^\nu + \Gamma^\nu_1$ and $M = m + M_1$, where

$$\Gamma^\nu_1 = c^{\mu\nu}\gamma_\mu + d^{\mu\nu}\gamma_5\gamma_\mu + e^\nu + i\,f^\nu\gamma_5 + \frac{1}{2}\,g^{\lambda\mu\nu}\sigma_{\lambda\mu}, \tag{2}$$

$$M_1 = a_\mu\,\gamma^\mu + b_\mu\,\gamma_5\,\gamma^\mu + i\,m_5\gamma_5 + \frac{1}{2}\,H_{\mu\nu}\,\sigma^{\mu\nu}. \tag{3}$$

All these parameters are experimentally proved to be very weak.[1]

The action (1) is perfectly Lorentz and CPT invariant if we treat all new external parameters as fields which transform according to their Lorentz representations. The symmetries get broken if we fix these fields, for instance, considering them to be constant in a given reference frame. However, the renormalization of the theory can be performed in one given frame, and therefore one can safely treat all the fields a_μ, b_μ, m_5, $H_{\mu\nu}$, $c^{\mu\nu}$, $d^{\mu\nu}$, e^ν, f^ν, and $g^{\lambda\mu\nu}$ as being covariant. For the sake of generality we have included in the action (1) an external metric and, in this way, may use the conventional approach based on covariance.

Consider the renormalization of the theory (1). It is easy to note that the set of external fields includes dimensionless ones in Eq. (2) and dimensional ones in Eq. (3). Furthermore, the renormalization of terms with dimensionless and dimensional terms are problems of different level of complexity.

The form of counterterms constructed from the dimensional fields (3) are strongly restricted by the superficial degree of divergence arguments. At the same time there are no such arguments for the dimensionless fields (2). Without symmetry arguments it is not really possible to take care of renormalization with these terms. Then, the theory with dimensionless fields (2) requires an infinite number of counterterms, hence an infinite number of extra terms in the classical action. However, the situation changes dramatically if we remember that the Lorentz- and CPT-violating terms should be very weak. This means we can safely restrict consideration to the terms linear in Lorentz- and CPT-violating fields and all necessary structures can be controlled by symmetry.

According to the arguments presented above, the action of renormalizable theory in curved space includes: (i) covariant generalization of the standard expression, e.g., as given in Eq. (1) for QED and possible nonminimal terms; (ii) vacuum terms depending on the metric and Lorentz- and CPT-violating parameters, but not on matter fields.

The one-loop calculations have been performed for QED,[6] including in curved space[7] and they are in a perfect agreement with the consideration presented above. For example, this can be observed in the (incomplete) expression for the vacuum divergences of the 1-loop effective action,

$$\Gamma^{(1)}_{\rm div} = -\frac{\mu^{\epsilon}}{(4\pi)^2\,\epsilon}\int d^n x\,\sqrt{-g}\,\Big\{R_{\mu\alpha}\nabla_\rho\nabla_\tau k_F^{\tau\alpha\rho\mu} - \frac{1}{6}R\nabla_\mu\nabla_\nu k_F^{\mu\nu}$$
$$+\frac{1}{3}R_{\mu\rho\alpha\beta}\nabla^\beta\nabla_\tau k_F^{\tau\mu\alpha\rho} - \frac{1}{12}k_F^{\rho\lambda\mu\tau}RR_{\rho\lambda\mu\tau} + \frac{1}{2}k_F^{\alpha\rho\tau\lambda}R^\mu{}_{\rho\tau\lambda}R_{\mu\alpha}\Big\}, \quad (4)$$

where we write $(k_F)^{\mu\lambda\nu}{}_{\lambda} = (k_F)^{\mu\nu}$.

The expression (4) tells us something essential about renormalization of the theories under discussion. In the curvature-dependent low-energy vacuum sector of the theory one could expect, for dimensional reasons, terms of the form[8]

$$S_{\rm full} = \int d^4x\sqrt{-g}\,\left\{\phi R + \phi^{\mu\nu}R_{\mu\nu} + \phi^{\mu\nu\alpha\beta}R_{\mu\nu\alpha\beta}\right\} + S_{\rm HD}, \quad (5)$$

where the term $S_{\rm HD}$ represents possible higher derivative terms. As we can see from Eq. (4), the terms presented in Eq. (5) are indeed necessary if we consider a spacetime dependent parameter $k_F^{\alpha\beta\mu\nu}$. The introduction of the terms (5) is necessary already at the classical level, for otherwise quantum corrections produce such terms with infinite coefficients which cannot be removed by renormalization.

3. Renormalization group

An additional illustration of the relations between different Lorentz- and CPT-violating parameters in the matter and gravity sectors can be obtained within the renormalization group method. The renormalization group equations of the new low-energy sector parameters in the gravitational sector have the form[7]

$$\mu \frac{d\phi}{d\mu} = \frac{1}{96\pi^2} \nabla_\mu \nabla_\nu k_F^{\mu\nu},$$
$$\mu \frac{d\phi^{\mu\nu}}{d\mu} = -\frac{1}{16\pi^2} \nabla_\alpha \nabla_\beta k_F^{\alpha\mu\beta\nu},$$
$$\mu \frac{d\phi^{\mu\nu\alpha\beta}}{d\mu} = -\frac{1}{48\pi^2} \nabla^\beta \nabla_\lambda k_F^{\lambda\mu\alpha\nu}. \tag{6}$$

These equations show that the order of magnitude of the parameters ϕ cannot be too much smaller than that of the variation of the parameter $k_F^{\alpha\mu\beta\nu}$. It may be interesting to perform a systematic analysis of these kinds of relations. In fact, for some particular cases the restrictions coming from renormalization group equations can compete with the direct experimental bounds.

4. Conclusions

We presented a brief and mainly qualitative review of the renormalization properties of quantum field theories in curved spacetime in the presence of Lorentz- and CPT-violating terms. The study of quantum corrections is useful for formulating the theory in a consistent way.

Acknowledgments

This work was partially supported by CNPq, FAPEMIG and ICTP.

References

1. *Data Tables for Lorentz and CPT Violation,* V.A. Kostelecký and N. Russell, 2013 edition, arXiv:0801.0287v6.
2. D. Colladay, and V.A. Kostelecký, Phys. Rev. D **55**, 6760 (1997); Phys. Rev. D **58**, 116002 (1998).
3. I.L. Buchbinder, S.D. Odintsov and I.L. Shapiro, *Effective Action in Quantum Gravity,* IOP Publishing, Bristol, 1992.
4. I.L. Shapiro, Class. Quantum Grav. **25**, 103001 (2008).
5. I.L. Shapiro, Phys. Rep. **357**, 113 (2002).
6. V.A. Kostelecký, C.D. Lane and A.G.M. Pickering, Phys. Rev. D **65**, 056006 (2002); V.C. Zhukovsky, A.E. Lobanov, and E.M. Murchikova, Phys. Rev. D **73**, 065016 (2006).
7. G. de Berredo-Peixoto and I.L. Shapiro, Phys. Lett. **B642**, 153 (2006).
8. V.A. Kostelecký, Phys. Rev. D **69**, 105009 (2004).

CONSTRAINTS ON VIOLATIONS OF LORENTZ SYMMETRY FROM GRAVITY PROBE B

JAMES M. OVERDUIN* and RYAN D. EVERETT

Department of Physics, Astronomy & Geosciences, Towson University
Towson, MD 21252, USA
**E-mail: joverduin@towson.edu*

QUENTIN G. BAILEY

Department of Physics, Embry-Riddle Aeronautical University
Prescott, AZ 86301, USA
E-mail: baileyq@erau.edu

We use the final results from Gravity Probe B to set new upper limits on the gravitational sector of the Standard-Model Extension, including for the first time the coefficient associated with the time-time component of the new field responsible for inducing local Lorentz violation in the theory.

The minimal pure-gravity sector of the Standard-Model Extension (SME) is characterized by nine independent coefficients $\bar{s}^{\mathrm{AB}}$ corresponding to the vacuum expectation values of a new tensor field whose couplings to the traceless part of the Ricci tensor induce spontaneous violations of local Lorentz symmetry.[1] These coefficients are assumed to be constant in the asymptotically flat (Minkowski) limit. Most are constrained either individually or in various combinations by existing experiments and observations,[2] but no limits have yet been placed on the $\bar{s}^{\mathrm{TT}}$ coefficient.

Gravity Probe B (GPB) was a satellite experiment launched in 2004 to measure the geodetic and frame-dragging effects predicted by General Relativity (GR). As shown by Bailey and Kostelecký in 2006,[3] the orientation of a gyroscope in orbit around a spinning central mass like the earth is sensitive to seven of the nine $\bar{s}^{\mathrm{AB}}$ coefficients, *including* $\bar{s}^{\mathrm{TT}}$. Following earlier preliminary work,[4] our goal here is to calculate the resulting constraints using the recently released final results from GPB.[5]

Within GR the geodetic and frame-dragging precession rates of a gyroscope with position $\vec{r}$ and velocity $\vec{v}$ in orbit around a central mass M with

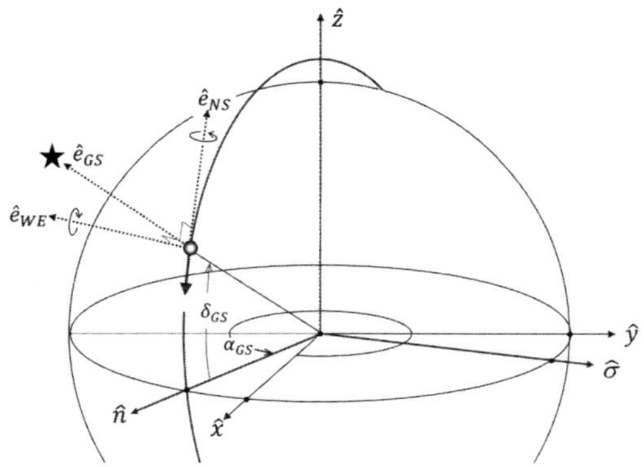

Fig. 1. Experimental results are expressed in GPB coordinates $(\hat{e}_{\rm GS}, \hat{e}_{\rm NS}, \hat{e}_{\rm WE})$. Theoretical SME predictions are derived in the $(\hat{n}, \hat{\sigma}, \hat{z})$ system. Both are ultimately referred to Sun-centered inertial coordinates $(\hat{x}, \hat{y}, \hat{z})$, where $\hat{x}$ points toward the vernal equinox.

moment of inertia I and angular velocity $\vec{\omega}$ are:

$$\vec{\Omega}_{\rm g,GR} = \left(\frac{3}{2}\frac{GM}{c^2 r^3}\right)\vec{r}\times\vec{v} \quad , \quad \vec{\Omega}_{\rm fd,GR} = \frac{GI}{c^2 r^3}\left[\frac{3\vec{r}}{r^2}(\vec{\omega}\cdot\vec{r}) - \vec{\omega}\right] . \tag{1}$$

The combined precession $\vec{\Omega}_{\rm GR} = \vec{\Omega}_{\rm g,GR} + \vec{\Omega}_{\rm fd,GR}$ causes the unit spin vector $\hat{S}$ of the gyroscope to undergo a relativistic drift $\vec{R} \equiv d\hat{S}/dt = \vec{\Omega}_{\rm GR} \times \hat{S}$. Averaging over a circular, polar orbit of radius r_0 around a spherically symmetric central mass, one obtains

$$\vec{R}_{\rm g,GR} = -\frac{3(GM)^{3/2}}{2\,c^2 r_0^{5/2}}\,\hat{e}_{\rm NS} \quad , \quad \vec{R}_{\rm fd,GR} = -\frac{GI\omega\cos\delta_{\rm GS}}{2\,c^2 r_0^3}\,\hat{e}_{\rm WE} , \tag{2}$$

where $\hat{e}_{\rm GS}$ points toward the guide star (located in the orbit plane at right ascension $\alpha_{\rm GS}$ and declination $\delta_{\rm GS}$), $\hat{e}_{\rm WE}$ is an orbit normal pointing along the cross-product of $\hat{e}_{\rm GS}$ and the unit vector $\hat{z}$ (aligned with the earth's rotation axis) and $\hat{e}_{\rm NS}$ is a tangent to the orbit directed along $\hat{e}_{\rm WE} \times \hat{e}_{\rm GS}$ (Fig. 1). The choice of polar orbit orthogonalizes the two effects so that $\vec{R}_{\rm g,GR}$ points entirely along $\hat{e}_{\rm NS}$ and $\vec{R}_{\rm fd,GR}$ points entirely along $\hat{e}_{\rm WE}$.

For GPB with guide star IM Pegasi, r_0 = 7018.0 km, $\delta_{\rm GS} = 16.841°$, $R_{\rm g,GR}$ = 6606.1 mas/yr (including oblateness) and $R_{\rm fd,GR}$ = 39.2 mas/yr where mas=milliarcsecond. The final joint results for all four gyros indicate that $R_{\rm NS,obs} = 6601.8 \pm 18.3$ mas/yr and $R_{\rm WE,obs} = 37.2 \pm 7.2$ mas/yr with 1σ uncertainties.[5] Thus the NS and WE components of relativistic drift rate may deviate from the predictions of GR by at most $\Delta R_{\rm NS} < |R_{\rm g,GR} - R_{\rm NS,obs}|$ = 22.6 mas/yr and $\Delta R_{\rm WE} < |R_{\rm fd,GR} - R_{\rm WE,obs}|$ = 9.2 mas/yr.

Within the SME, Lorentz-violating terms introduce an additional 'anomalous' relativistic drift $\Delta\vec{R}$ whose components along $\hat{n}, \hat{\sigma}$ and $\hat{z}$ are given by Eqs. (158)-(160) of Ref. 3. Here $\hat{n} \equiv \hat{\sigma} \times \hat{z}$ and $\hat{\sigma} = -\hat{e}_{\rm WE}$ is an orbit normal (Fig. 1). These equations may be expressed in the form

$$\Delta\vec{R} = \begin{pmatrix} \frac{1}{2}\,\omega_{\rm GS}(\overline{s}^{\rm YY} - \overline{s}^{\rm XX})\sin 2\alpha_{\rm GS} + \omega_{\rm GS}\overline{s}^{\rm XY}\cos 2\alpha_{\rm GS} \\ \omega_{\rm T}\overline{s}^{\rm TT} + \omega_{\rm NS}(\overline{s}^{\rm XX}\sin^2\alpha_{\rm GS} - \overline{s}^{\rm XY}\sin 2\alpha_{\rm GS} + \overline{s}^{\rm YY}\cos^2\alpha_{\rm GS}) \\ \omega_{\rm WE}(\overline{s}^{\rm YZ}\cos\alpha_{\rm GS} - \overline{s}^{\rm XZ}\sin\alpha_{\rm GS}) \end{pmatrix}, \quad (3)$$

where $\omega_{\rm GS} = \omega_{\rm WE} = \frac{5}{6}(1 - 3I/5Mr_0^2)\,R_{\rm g,GR} = 4603$ mas/yr, $\omega_{\rm T} = \frac{3}{4}(1 - I/3Mr_0^2)\,R_{\rm g,GR} = 4503$ mas/yr, $\omega_{\rm NS} = \frac{1}{12}(1 + 9I/Mr_0^2)\,R_{\rm g,GR} = 1904$ mas/yr and $\alpha_{\rm GS} = 343.26°$. To transform to GPB coordinates, we reflect across the orbit plane and rotate about $\hat{\sigma}$ by $\delta_{\rm GS}$. The resulting drift rates along the GS, NS and WE axes are

$$\Delta\vec{R} = \begin{pmatrix} \omega_{\rm GS}\big[\frac{1}{2}(\overline{s}^{\rm YY} - \overline{s}^{\rm XX})\sin 2\alpha_{\rm GS}\cos\delta_{\rm GS} + \overline{s}^{\rm XY}\cos 2\alpha_{\rm GS}\cos\delta_{\rm GS} \\ \quad - \overline{s}^{\rm XZ}\sin\alpha_{\rm GS}\sin\delta_{\rm GS} + \overline{s}^{\rm YZ}\cos\alpha_{\rm GS}\sin\delta_{\rm GS}\big] \\ -\,\omega_{\rm T}\overline{s}^{\rm TT} - \omega_{\rm NS}(\overline{s}^{\rm XX}\sin^2\alpha_{\rm GS} - \overline{s}^{\rm XY}\sin 2\alpha_{\rm GS} + \overline{s}^{\rm YY}\cos^2\alpha_{\rm GS}) \\ \omega_{\rm WE}\big[\frac{1}{2}(\overline{s}^{\rm XX} - \overline{s}^{\rm YY})\sin 2\alpha_{\rm GS}\sin\delta_{\rm GS} - \overline{s}^{\rm XY}\cos 2\alpha_{\rm GS}\sin\delta_{\rm GS} \\ \quad - \overline{s}^{\rm XZ}\sin\alpha_{\rm GS}\cos\delta_{\rm GS} + \overline{s}^{\rm YZ}\cos\alpha\cos\delta_{\rm GS}\big] \end{pmatrix}. \quad (4)$$

Numerically,

$$\begin{aligned} \Delta R_{\rm GS} &= 1215\overline{s}^{\rm XX} + 3674\overline{s}^{\rm XY} + 384\overline{s}^{\rm XZ} - 1215\overline{s}^{\rm YY} + 1277\overline{s}^{\rm YZ}, \\ \Delta R_{\rm NS} &= -4503\overline{s}^{\rm TT} - 158\overline{s}^{\rm XX} - 1050\overline{s}^{\rm XY} - 1746\overline{s}^{\rm YY}, \\ \Delta R_{\rm WE} &= -368\overline{s}^{\rm XX} - 1112\overline{s}^{\rm XY} + 1269\overline{s}^{\rm XZ} + 368\overline{s}^{\rm YY} + 4219\overline{s}^{\rm YZ}, \end{aligned} \quad (5)$$

where $\Delta R_{\rm NS} < 22.6$ and $\Delta R_{\rm WE} < 9.2$ from GPB (all units in mas/yr). The SME can accommodate precessions greater than those predicted by GR, unlike other extensions of the Standard Model where Einstein's theory is a limiting case.[6] GPB does not constrain the GS component, since the gyro spin axes point along this direction by design. The GS and WE components are linear combinations of $\overline{s}^{\rm XY}, \overline{s}^{\rm XZ}, \overline{s}^{\rm YZ}$ and $(\overline{s}^{\rm XX} - \overline{s}^{\rm YY})$, so they are superseded in any case by existing constraints, which read:[2,7]

$$|\overline{s}^{\rm XY}| < (0.6 \pm 1.5) \times 10^{-9}, \quad (6)$$

$$|\overline{s}^{\rm XZ}| < (2.7 \pm 1.4) \times 10^{-9}, \quad (7)$$

$$|\overline{s}^{\rm YZ}| < (0.6 \pm 1.4) \times 10^{-9}, \quad (8)$$

$$|\overline{s}^{\rm XX} - \overline{s}^{\rm YY}| < (1.2 \pm 1.6) \times 10^{-9}, \quad (9)$$

$$|\overline{s}^{\rm XX} + \overline{s}^{\rm YY} - 2\overline{s}^{\rm ZZ}| < (1.8 \pm 38) \times 10^{-9}. \quad (10)$$

Thus in practice the only new GPB constraint on the SME comes from the NS component of Eq. (5), associated entirely with geodetic precession in standard GR. It reads:

$$|\overline{s}^{\rm TT} + 0.035\overline{s}^{\rm XX} + 0.23\overline{s}^{\rm XY} + 0.39\overline{s}^{\rm YY}| < 5.0 \times 10^{-3}. \qquad (11)$$

To get seven conditions on seven unknowns, we supplement Eqs. (6)-(11) with the requirement that $\overline{s}^{\rm AB}$ be traceless, $|\overline{s}^{\rm TT} - \overline{s}^{\rm XX} - \overline{s}^{\rm YY} - \overline{s}^{\rm ZZ}| = 0$.[3] Inverting, we then find that

$$\overline{s}^{\rm TT} < 4.4 \times 10^{-3}, \quad \overline{s}^{\rm XX}, \overline{s}^{\rm YY}, \overline{s}^{\rm ZZ} < 1.5 \times 10^{-3}.$$

This constitutes the first experimental upper bound on $\overline{s}^{\rm TT}$. (Other tests such as light deflection are also sensitive to this coefficient at similar levels of precision.[8]) It also lifts a degeneracy between other existing limits, allowing us to extract individual upper bounds on $\overline{s}^{\rm XX}$, $\overline{s}^{\rm YY}$ and $\overline{s}^{\rm ZZ}$.

One should also look at the effect of $\overline{s}^{\rm AB}$ on the equation of motion for the gyroscope.[3] This has the effect of rescaling Newton's gravitational constant G, increasing our sensitivity to $\overline{s}^{\rm TT}$ and strengthening our limits by about 5%.[9] If the actual orbit is not perfectly circular, as was the case for GPB (whose gyros remained in essentially perfect free fall around a non-spherically symmetric Earth), then additional $\overline{s}^{\rm AB}$-dependent terms are also introduced in the *leading-order* (GR) expressions for geodetic and frame-dragging precession, Eq. (1). These do not significantly alter the NS or geodetic constraint from GPB, but they do strengthen the WE or frame-dragging constraint so that it may potentially become competitive with existing limits. We will report on these results elsewhere.[9]

References

1. V.A. Kostelecký, Phys. Rev. D69, 105009 (2004).
2. *Data Tables for Lorentz and CPT Violation,* V.A. Kostelecký and N. Russell, 2013 edition, arXiv:0801.0287v6.
3. Q.G. Bailey and V.A. Kostelecký, Phys. Rev. D74, 045001 (2006).
4. J.M. Overduin, in V.A. Kostelecký (ed.), *Fourth Meeting on CPT and Lorentz Symmetry,* Singapore, World Scientific, 2008.
5. C.W.F. Everitt *et al.*, Phys. Rev. Lett. 106, 221101 (2011).
6. J.M. Overduin *et al.*, Gen. Rel. Grav., in press [arxiv:1305.6871].
7. J.B.R. Battat *et al.*, Phys. Rev. Lett. 99, 241103 (2007); H. Müller *et al.*, Phys. Rev. Lett. 100, 031101 (2008); K.-Y. Chung *et al.*, Phys. Rev. D80, 016002 (2009).
8. R. Tso and Q.G. Bailey, Phys. Rev. D84, 085025 (2011).
9. Q.G. Bailey *et al.*, submitted to Class. Quant. Grav. (2013).

THE INFLUENCE OF LORENTZ VIOLATION ON UHE PHOTON DETECTION

G.I. RUBTSOV, P.S. SATUNIN,* and S.M. SIBIRYAKOV

Institute for Nuclear Research of the Russian Academy of Sciences
60th October Anniversary Prospect, 7a, 117312 Moscow, Russia
**E-mail: satunin@ms2.inr.ac.ru*

We show that violation of the Lorentz symmetry in quantum electrodynamics can suppress the rates of the interactions crucial for the formation of photon-induced air showers, such as pair production on nuclei and in the geomagnetic field. As a consequence, the allowed region in the space of Lorentz-violating parameters will be seriously restricted if several photons with energies $\gtrsim 10^{19}$ eV are detected.

1. Introduction

The presence of a photon component in the ultra-high-energy (UHE) cosmic rays is an important open question in astroparticle physics. An answer to it will give information on astrophysical issues such as the composition and origin of the UHE cosmic rays, as well as on the fundamental topic of possible modifications of the spacetime symmetries.[1]

In the standard physical picture, a primary UHE photon reaching the Earth interacts in the atmosphere and produces an extensive air shower of particles with lower energies that can be detected by the ground-based experiments. The characteristics of the shower are sensitive to the altitude, at which the first interaction giving rise to the shower takes place. This, in its turn, is determined by the cross section of the first interaction. At energies $\sim 10^{19}\,$eV the dominant channel of the first interaction is electron-positron pair production on nuclei in the atmosphere, the so-called 'Bethe-Heitler process.'[2] At higher energy ($10^{20}\,$eV and above) the pair production in the geomagnetic field becomes important leading to the formation of a preshower above the atmosphere. The photon showers initiated by these processes can be identified by the cosmic ray detectors using various observables.

The cross section of the first interaction, and consequently the characteristics of the shower, may be affected by new physics. In this paper we study the influence of possible violation of the Lorentz invariance (or Lorentz violation (LV) for short) on the aforementioned processes, and the bounds on LV parameters that can be established if the UHE photons are detected.

2. The model

To make the quantitative predictions we study the model of LV quantum electrodynamics with the following lagrangian:

$$\mathcal{L} = \bar{\psi}\left(i\gamma^\mu D_\mu - m\right)\psi - \frac{1}{4}F_{\mu\nu}F^{\mu\nu} + i\varkappa\bar{\psi}\gamma^i D_i\psi + \frac{ig}{M^2}D_j\bar{\psi}\gamma^i D_i D_j\psi + \frac{\xi}{4M^2}F_{kj}\partial_i^2 F^{kj}, \qquad (1)$$

where $\varkappa$, g and ξ are dimensionless parameters, the covariant derivative D_μ is defined in the standard way, $D_\mu\psi = (\partial_\mu + ieA_\mu)\psi$, and M is the Planck mass. Summation over repeated indices with Minkowski metrics is understood. This lagrangian contains all inequivalent operators of dimension up to 6 that are rotationally invariant in the preferred frame, gauge invariant, and CPT- and P-even. The requirements on the theory and their motivation are discussed in detail in Ref. 3.

From the lagrangian (1) one obtains the dispersion relations for photons and electrons/positrons,

$$E_\gamma^2 = k^2 + \frac{\xi k^4}{M^2}, \qquad E_e^2 = m^2 + p^2\left(1 + 2\varkappa\right) + \frac{2gp^4}{M^2}. \qquad (2)$$

Also the Feynman rules for the model (1) are modified compared to the Lorentz invariant case.[3] For processes with an electron-positron pair in the final state one introduces the combination

$$\omega_{\rm LV}(x) = -\varkappa k - \frac{gk^3}{4M^2}(1+3x^2) + \frac{\xi k^3}{2M^2}\,, \qquad (3)$$

which characterizes the energy transfer in the process. Here $x \in [-1.1]$ denotes the asymmetry between the momenta of the produced electron and positron (see Ref. 3 for the precise definition). If $\omega_{\rm LV}(x)$ is larger than $2m^2/k$ for some x, vacuum photon decay becomes kinematically allowed. UHE photons decay very quickly into electron-positron pairs and do not reach the Earth. Below we restrict to the values of $\omega_{\rm LV}$ outside this range.

3. Pair production on a nucleus

Let us consider the Bethe-Heitler process. The standard result for the cross section reads[2] $\sigma_{\rm BH} = 28Z^2\alpha^3/9m^2 \log\left(183/Z^{1/3}\right)$, where m is the electron mass, α is the fine structure constant and Z is the nucleus charge. Lorentz violation significantly suppresses the cross section for negative $\omega_{\rm LV}(x) < -m^2/k$. For example, in the case $1 \ll k|\omega_{\rm LV}|/m^2 \ll \alpha^{-4}Z^{-4/3}$ the modified cross section takes the form[a3]

$$\sigma_{\rm BH}^{\rm LV} \simeq \frac{8Z^2\alpha^3}{3k|\omega_{\rm LV}(1)|} \log\frac{1}{\alpha Z^{1/3}} \cdot \log\frac{k|\omega_{\rm LV}(1)|}{m^2}. \qquad (4)$$

This expression is suppressed roughly by a factor $m^2/k\,|\omega_{\rm LV}(1)|$ with respect to the LI result.

A future UHE photon detection by cosmic ray experiments would imply that the cross section of the first interaction is not too much suppressed compared to the standard expectation: otherwise the photon would go through the atmosphere without developing a shower. Conservatively, we will assume that the cross section does not differ by more that an order of magnitude. This gives the bound $|\omega_{\rm LV}| < 10m^2/k$ for negative $\omega_{\rm LV}$. Similar (actually, even stronger) bound for positive $\omega_{\rm LV}$ follows from the absence of the vacuum photon decay. Barring accidental cancellations we find that a prospective detection of photons with energies $k \sim 10^{19}$ eV will allow to constrain the LV parameters at the level

$$|\varkappa| \lesssim 10^{-25}\ ; \qquad |g|, |\xi| \lesssim 10^{-7}\ . \qquad (5)$$

It is worth stressing that these bounds will be insensitive to any astrophysical assumptions as to the origin and propagation of cosmic rays through the interstellar medium.

4. Photon decay in magnetic field

Next we turn to the photon decay in geomagnetic field. We use[4] the method of 'worldline instantons,' proposed by Affleck, Alvarez and Manton[5] to study the Schwinger effect. The photon decay in a weak magnetic field is represented as a tunneling process whose width is given by a statistical sum of a certain auxiliary quantum-mechanical system. This statistical sum can be evaluated in the saddle-point approximation and yields the following

[a]Unlike the standard case, the cross section is saturated at the maximal asymmetry between the momenta of the pair, $x = \pm 1$.

exponential behavior for the photon decay width,

$$\Gamma \propto \exp\left[-\frac{8m^3}{3keH\sin\varphi}\left(1-\frac{k\cdot\omega_{\mathrm{LV}}(0)}{2m^2}\right)^{3/2}\right]. \tag{6}$$

Here H denotes the magnetic field, φ is an angle between the photon momentum and the magnetic field. In the Lorentz invariant case, $\omega_{\mathrm{LV}}(0)=0$, this answer coincides with the standard one.[6] We see that the process is exponentially suppressed till the photon energy reaches $k\sim 8m^3/3eH\sin\varphi$, which for the Earth magnetic field and $\sin\varphi\sim 1$ gives $k\sim 10^{20}\,\mathrm{eV}$. In the presence of LV with $\omega_{\mathrm{LV}}(0)<0$ the exponent is enhanced and the effective threshold energy is shifted upwards. Requiring that the shift is not too large again implies $|\omega_{\mathrm{LV}}|\lesssim m^2/k$. We conclude that a detection of UHE photons at $k\sim 10^{20}\,\mathrm{eV}$ with the preshower signature will yield the constraints,[b]

$$|\varkappa|<10^{-27}\;;\qquad |g|,|\xi|<10^{-11}\,. \tag{7}$$

5. Conclusion

We have shown that a future UHE photon detection can establish very strong two-sided bounds on the parameters describing deviations from LI in quantum electrodynamics. These bounds will be insensitive to any astrophysical assumptions about the origin of the UHE photons and their propagation towards the Earth.

Acknowledgments

This work has been supported in part by the Grants of the Russian Ministry of Education and Science 8412 and 14.B37.21.0457, the Grants of the President of Russian Federation NS-5590.2012.2, MK-1170.2013 and by the RFBR grants 11-02-01528, 12-02-01203, 12-02-91323.

References

1. M. Galaverni and G. Sigl, Phys. Rev. D **78**, 063003 (2008).
2. H. Bethe and W. Heitler, Proc. Roy. Soc. Lond. A **146**, 83 (1934).
3. G. Rubtsov, P. Satunin and S. Sibiryakov, Phys. Rev. D **86**, 085012 (2012).
4. P. Satunin, Phys. Rev. D **87**, 105015 (2013).
5. I.K. Affleck, O. Alvarez and N.S. Manton, Nucl. Phys. B **197**, 509 (1982).
6. N.P. Klepikov, Zh. Exp. Theor. Phys. **26**, 19 (1954).

[b] The bound on the positive ω_{LV} again follows from the absence of vacuum photon decay.

LIMITS ON LORENTZ VIOLATION IN NEUTRAL-KAON DECAY

K.K. VOS,* H.W. WILSCHUT and R.G.E. TIMMERMANS

Kernfysisch Versneller Instituut, University of Groningen
9747 AA Groningen, The Netherlands
**E-mail: K.K.Vos@rug.nl*

The KLOE collaboration recently reported bounds on the directional dependence of the lifetime of the short-lived neutral kaon K_S with respect to the cosmic microwave background dipole anisotropy. We interpret their results in a general framework developed to probe Lorentz violation in the weak interaction. In this approach a Lorentz-violating tensor $\chi_{\mu\nu}$ is added to the standard propagator of the W boson. We derive the K_S decay rate in a naive tree-level model and calculate the asymmetry for the lifetime. By using the KLOE data the real vector part of $\chi_{\mu\nu}$ is found to be smaller than 10^{-2}. We briefly discuss the theoretical challenges concerning nonleptonic decays.

1. Testing Lorentz violation in the weak interaction

Recently Lorentz violation in the weak interaction has been studied in allowed[1,2] and forbidden[3] β-decay. A general theoretical framework was developed for these tests,[1] in which the standard W-boson propagator is modified by a general Lorentz-violating tensor $\chi_{\mu\nu}$. In the low-energy limit the modified propagator is

$$\langle W^{\mu+}(q)W^{\nu-}(-q)\rangle = -i\,\frac{g^{\mu\nu}+\chi^{\mu\nu}}{M_W^2}\,. \tag{1}$$

To constrain $\chi_{\mu\nu}$ we derive the K_S decay rate with Lorentz violation and use data from the KLOE collaboration,[4] which searched for the directional dependence of the lifetime of K_S mesons.

2. KLOE lifetime asymmetry

The KLOE collaboration measured the lifetime asymmetry of the neutral K_S meson, by transforming the K_S momenta event-by-event to galactic coordinates $\{\ell, b\}$, where ℓ is the galactic longitude and b is the galactic latitude. The lifetime asymmetry is defined as

$$\mathcal{A} = \frac{\tau^+ - \tau^-}{\tau^+ + \tau^-}, \tag{2}$$

where $\tau^{+(-)}$ is the lifetime parallel (antiparallel) to a specific direction in space. Three directions were studied: CMB0, the direction of the dipole anisotropy in the Cosmic Microwave Background (CMB), and two perpendicular directions, CMB1 and CMB2. Only events inside a cone of 30° opening angle were used. The measured asymmetries are given in Table 1, where the errors are mainly statistical.

Table 1. Observed K_S lifetime asymmetry.

$\{\ell, b\}$	$\mathcal{A} \times 10^3$	Ref.
CMB0 $= \{264°, 48°\}$	-0.2 ± 1.0	4
CMB0 $= \{264°, 48°\}$	-0.13 ± 0.4	5
CMB1 $= \{174°, 0°\}$	0.2 ± 1.0	4
CMB2 $= \{264°, -42°\}$	0.0 ± 0.9	4

3. $\Delta I = 1/2$ rule

The description of nonleptonic processes is somewhat more involved than leptonic and semileptonic processes, due to the importance of gluon loops. These QCD corrections can be of the same order as tree-level contributions. Within the Standard Model, nonleptonic $\Delta S = 1$ decays are usually described by an effective lagrangian,[6] which contains different operators and their coefficients. An important consequence of the gluon loops is that they generate a so-called penguin diagram, depicted in Fig. 1, which can effectively be described by an operator that also contains right-handed quarks. For K_S decays there are two isospin I final states, $I = 0$ and $I = 2$. Experimentally it is found that the former, a $\Delta I = 1/2$ transition, is enhanced compared to the latter, a $\Delta I = 3/2$ transition. This enhancement is an order of magnitude larger than expected from theoretical calculations with an effective lagrangian, and is known as the $\Delta I = 1/2$ rule. The penguin diagram generates a $\Delta I = 1/2$ operator $\mathcal{O}_5$ that explains part of the enhancement due to the coupling to right-handed quarks, but the enhancement is not sufficient to explain the $\Delta I = 1/2$ rule.[7]

The derivation of the complete Lorentz violating lagrangian is more involved, due to mixing between operators, and lies beyond the scope of

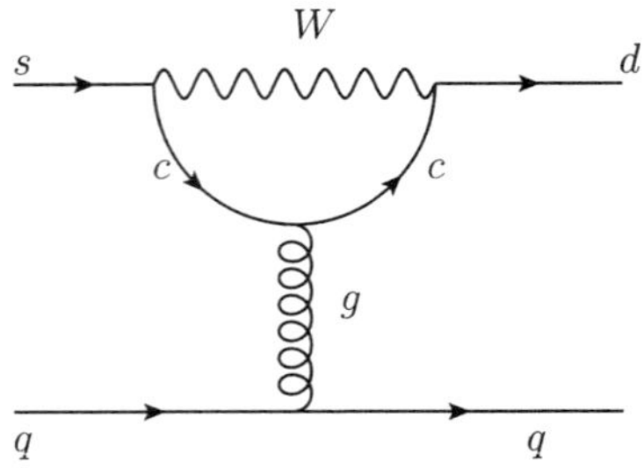

Fig. 1. The penguin diagram for an s to d quark transition.

this work. The penguin diagram does not mix and we are able to derive the Lorentz violating operator associated with this diagram by integrating out the W boson, following Ref. 8. We find,[9]

$$\mathcal{O}_5^{LV+SM} = -\frac{1}{2}\bar{d}_L t^a \left[-2g_{\mu\nu} + \chi_{\mu\nu} + \chi_{\nu\mu} + \tilde{\chi}_{\mu\nu}\right] \gamma^\nu s_L \left(\bar{q}_R t^a \gamma^\mu q_R\right) + \text{h.c.}, \tag{3}$$

where the $g_{\mu\nu}$ part is the Standard Model part and $\tilde{\chi}_{\mu\nu} \equiv i\epsilon^{\alpha\beta\mu\nu}\chi_{\alpha\beta}$. Further discussion and implications of $\mathcal{O}_5$ are given in Ref. 9. The outstanding problem of the $\Delta I = 1/2$ rule makes our calculation with Lorentz violation challenging. We therefore first explore the possibilities of the KLOE data with a naive tree-level model.

4. Tree-level model

To explore the possible bounds from the K_S decay we first use a tree-level model, in which we assume that the dominant process for K_S decay is the tree-level decay. We thus treat this process naively as being semileptonic. For the tree-level amplitude with the modified W-boson propagator we find

$$\mathcal{M} \sim \left\langle \pi^+ | \bar{u}_L \gamma^\mu d_L | 0 \right\rangle \left(g_{\mu\nu} + \chi^*_{\mu\nu}\right) \left\langle \pi^- | \bar{s}_L \gamma^\nu u_L | K^0 \right\rangle . \tag{4}$$

After integrating, we find for the asymmetry,

$$\mathcal{A}_{\vec{n}} = -\frac{\frac{4}{3} + \frac{2}{3}\frac{m_\pi^2}{m_K^2}}{(1-\beta_K^2)\left(1 - \frac{m_\pi^2}{m_K^2}\right)} (\chi^r_{i0} + \chi^r_{0i})\beta^i_K = -0.343(\chi^r_{i0} + \chi^r_{0i})\,\hat{\beta}^i_K, \tag{5}$$

where $2\chi^r_{\mu\nu} = \chi_{\mu\nu} + \chi^*_{\mu\nu}$ and $\hat{\beta}_K$ is the normalized kaon velocity, where for the last part we used $\beta = 0.2$. The asymmetry thus gets a γ^2 enhancement and is sensitive to the symmetric real vector part of $\chi_{\mu\nu}$.

For our final results we transform $\chi_{\mu\nu}$ to the Sun-centered reference frame[10] and find at 95% confidence level,[9]

$$\begin{aligned} |X^r_{10} + X^r_{01}| &< 3 \times 10^{-3}, \\ |X^r_{20} + X^r_{02}| &< 6 \times 10^{-3}, \\ |X^r_{30} + X^r_{03}| &< 6 \times 10^{-3}. \end{aligned} \tag{6}$$

5. Conclusion and outlook

We investigated Lorentz violation in neutral K_S decays by modifying the W-boson propagator. This nonleptonic decay is theoretically more challenging than semileptonic or leptonic decays. Using a tree-level model we are able to put bounds on the real vector part of $\chi_{\mu\nu}$ of the order of 10^{-2}. These bounds complement bounds from allowed[2] and forbidden[3] β-decay, but are several orders of magnitude less strict than bounds in other sectors.[10] In order to significantly improve these bounds a possibility would be to use the γ^2 enhancement that occurs in decay asymmetries. From the theoretical point of view we argued that nonleptonic decays are much more involved due to QCD corrections, making experiments with leptonic or semileptonic decays preferable.

Acknowledgments

This research was supported by the Dutch Stichting voor Fundamenteel Onderzoek der Materie (FOM) under Programmes 104 and 114 and project 08PR2636.

References

1. J.P. Noordmans, H.W. Wilschut, and R.G.E. Timmermans, Phys. Rev. C **87**, 055502 (2013).
2. H.W. Wilschut *et al.*, arXiv:1303.6419; S.E. Müller *et al.*, in preparation.
3. J.P. Noordmans *et al.*, arXiv:1308.5829.
4. F. Ambrosino *et al.*, Eur. Phys. J. C **71**, 1604 (2011).
5. A. De Angelis *et al.*, Nuov. Cim. C **034N3**, 323 (2011).
6. M. Shifman *et al.*, Nucl. Phys. B **120**, 316 (1977).
7. J.F. Donoghue, Phys. Rev. D **30**, 1499 (1984).
8. M.B. Wise and E. Witten, Phys. Rev. D **20**, 1216 (1979).
9. K.K. Vos *et al.*, in preparation.
10. *Data Tables for Lorentz and CPT Violation,* V.A. Kostelecký and N. Russell, 2013 edition, arXiv:0801.0287v6.

SEARCH FOR LORENTZ VIOLATION IN TOP QUARK PAIR PRODUCTION AND DECAY

D.W. WHITTINGTON

Department of Physics, Indiana University
727 E. Third St.
Bloomington, IN 47405, USA
E-mail: dwwhitti@indiana.edu

We present a summary of the first search for Lorentz violation in the top quark sector. This study examined the $t\bar{t}$ production and decay process using data collected with the D0 detector at the Fermilab Tevatron Collider. In this process, violation of Lorentz invariance would manifest as a periodic variation in the rate at which $t\bar{t}$ events occur. This variation was quantified using the Standard-Model Extension framework, and the first limits were set on some of the coefficients $(c_Q)_{\mu\nu 33}$ and $(c_U)_{\mu\nu 33}$ parametrizing Lorentz violation in the top quark sector.

1. Violation of Lorentz invariance in the top quark sector

The top quark sector of the Standard Model (SM) represents one of the frontiers of particle physics, remaining accessible only through experiments at high energy particle collider facilities. Members of the D0 collaboration recently carried out the first investigation of Lorentz violation (LV) in the top quark sector in $t\bar{t}$ events using data collected at the Fermilab Tevatron Collider.[1] While violation of this symmetry has been studied in many particle species through a variety of processes,[2] until this result LV in the top sector had remained unexplored.

The Standard-Model Extension[3] (SME) introduces Lorentz- and CPT-violating terms into the SM Lagrange density. Here we consider only those terms affecting the top quark fields. The relevant SME operators couple the third generation left-handed doublet (containing t_L and b_L) with the right-handed singlet field t_R. Two of the associated coefficients, $(a_Q)_{\mu 33}$ and $(a_U)_{\mu 33}$, are odd under CPT transformations and do not contribute observable effects to the $t\bar{t}$ production and decay process at leading order. Elements of the CPT-even coefficients $(c_Q)_{\mu\nu 33}$ and $(c_U)_{\mu\nu 33}$ lead to modifications of the SM $t\bar{t}$ cross section which can introduce sidereal time dependence as a consequence of Lorentz violation.

At leading order in the SME coefficients, the matrix element for the production and decay of a $t\bar{t}$ pair involves only $c_{\mu\nu}$-type coefficients. The $q\bar{q} \to t\bar{t}$ process is the primary mode[4] (approximately 85%) in $p\bar{p}$ collisions at the Tevatron operating energy. The relevant expression, including SME modifications from the relevant $c_{\mu\nu}$-type coefficients, was evaluated by M. Berger and A. Kostelecký using the narrow-width approximation.[5] Their result for the matrix element describing this process is

$$\begin{aligned} |\mathcal{M}|^2 = {} & PF\bar{F} + (c_L + c_R)_{\mu\nu}(\delta P)^{\mu\nu}F\bar{F} \\ & + (c_L)_{\mu\nu}(P(\delta F)^{\mu\nu}\bar{F} + PF(\delta\bar{F})^{\mu\nu}). \end{aligned} \tag{1}$$

Here, we adopt the notation $(c_L)_{\mu\nu} = (c_Q)_{\mu\nu 33}$ and $(c_R)_{\mu\nu} = (c_U)_{\mu\nu 33}$. The terms P, F and $\bar{F}$ describe the SM production and decay processes. δP describes the contribution from SME terms in the process $q\bar{q} \to t\bar{t}$, while δF and $\delta\bar{F}$ result from SME terms in the t and $\bar{t}$ decays, respectively.

2. LV signature at collider detectors

By comparing the matrix element of Eq. (1) to the SM component alone through the ratio $|\mathcal{M}|^2/PF\bar{F}$, the time-dependent effects on the $t\bar{t}$ cross section take the form

$$\sigma(t) \approx \sigma_{\text{ave}}[1 + f_{\text{SME}}(t)], \tag{2}$$

where σ_{ave} is the observed (time averaged) cross section for the $t\bar{t}$ production and decay process, and

$$f_{\text{SME}}(t) = [(c_L)_{\mu\nu} + (c_R)_{\mu\nu}]R^{\mu}_{\alpha}(t)R^{\nu}_{\beta}(t)A_P^{\alpha\beta} + (c_L)_{\mu\nu}R^{\mu}_{\alpha}(t)R^{\nu}_{\beta}(t)A_F^{\alpha\beta}. \tag{3}$$

The matrices of c-type coefficients are defined by convention in the canonical Sun-centered reference frame. We adopt the index ordering convention $\mu, \nu = \{T, X, Y, Z\}$ to refer to coordinates in the Sun-centered frame and $\alpha, \beta = \{t, x, y, z\}$ to refer to coordinates in the apparatus frame. The terms $A_P^{\alpha\beta}$ and $A_F^{\alpha\beta}$ in Eq. (3) depend on the kinematics of the particles involved in the production and decay of each $t\bar{t}$ event, respectively. These terms are evaluated using $t\bar{t}$ Monte Carlo events which pass detector acceptance, trigger, event reconstruction, and analysis selections (modeled by a full simulation of the detector). The characteristic sidereal time dependence is introduced through the rotation matrix $R^{\mu}_{\alpha}(t)$ which transforms $A_P^{\alpha\beta}$ and $A_F^{\alpha\beta}$ from the detector frame to the Sun-centered frame.

We estimate one possible component of each SME coefficient at a time, imposing the requirements that each matrix is symmetric and traceless (choosing, e.g., $(c_L)_{YY} = -(c_L)_{XX}$ to satisfy the latter condition). We find that the TT and ZZ coefficients couple to expressions that only affect the total cross section; we did not attempt to extract these coefficients. Each of the Ti coefficients combines with the small off-diagonal elements of $A_P^{\alpha\beta}$ and $A_F^{\alpha\beta}$, resulting in poor sensitivity. However, the XX and XY coefficients couple to expressions that depend on sidereal time while the XZ and YZ terms couple to expression that vary twice with the sidereal period; we find sufficient sensitivity to explore the effects of these SME coefficients.

3. Limits on SME coefficients from the D0 experiment

To search for this unique sidereal variation in the $t\bar{t}$ event rate, we examined data collected by the D0 detector[6] at the Fermilab Tevatron Collider, corresponding to 5.3 fb^{-1} of integrated luminosity collected between August 2002 and June 2009. In the process $t\bar{t} \to W^+ b W^- \bar{b}$, we consider events where one W boson decays hadronically (to $q\bar{q}'$) and the other decays to a charged lepton and associated neutrino. Both the presence of b quarks and the single charged lepton provide a powerful discriminants against non-$t\bar{t}$ events. We employed the same event selection as described in the measurement of the $t\bar{t}$ cross section in this channel performed by the D0 collaboration.[7]

We expect the background rate (principally W+jets events) to be proportional only to the luminosity. To search for a variation in the $t\bar{t}$ event rate with sidereal time, we sum the contributions to each of twelve time bins (corresponding to two sidereal hours each) for all events:

$$N_i \approx N_{\text{tot}} \frac{\mathcal{L}_i}{\mathcal{L}_{\text{int}}} [1 + f_S f_{\text{SME}}(t)], \tag{4}$$

where N_{tot} is the total number of signal ($t\bar{t}$) and background (non-$t\bar{t}$) events corresponding to the total integrated luminosity $\mathcal{L}_{\text{int}}$, $\mathcal{L}_i$ is the integrated luminosity over the appropriate time bin, and f_S is the average fraction of signal events in the data (estimated from the cross section measurement mentioned previously). By choosing events containing more than three jets, we find $f_S(e\text{+>3-jets}) = 0.78 \pm 0.12$ and $f_S(\mu\text{+>3-jets}) = 0.76 \pm 0.11$.

To simplify fitting $f_{\text{SME}}(t)$, we define a variable R for each bin:

$$R_i \equiv \frac{1}{f_S} \left(\frac{N_i/N_{\text{tot}}}{\mathcal{L}_i/\mathcal{L}_{\text{int}}} - 1 \right). \tag{5}$$

This expression can be compared directly to $f_{\text{SME}}(t)$. In the absence of any significant sidereal time dependence, all the R_i values should be consistent with zero.

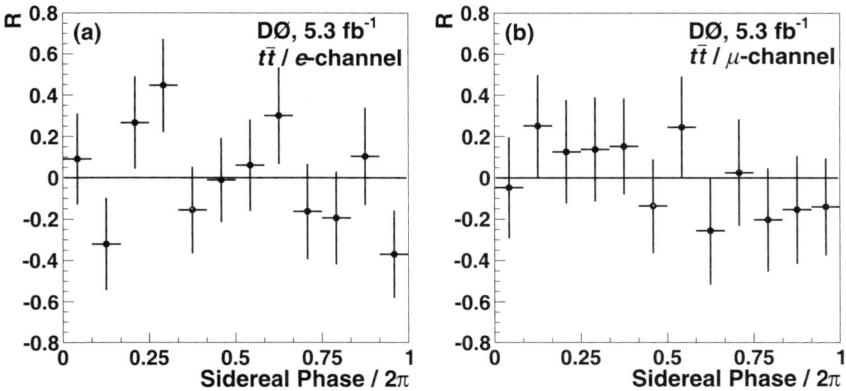

Fig. 1. The dependence of R, as defined in Eq. (5), on sidereal phase for[1] (a) e+>3-jets $t\bar{t}$ candidates, and (b) μ+>3-jets $t\bar{t}$ candidates.

We extracted values for each of the SME coefficients accessible through this method by performing simultaneous fits to these distributions. All measured coefficients were found to be consistent with zero within uncertainties, and we defined the observed limits as the extracted values ± 2 standard deviations. Upper limits on the magnitudes of all measured coefficients ranged between 0.14 and 0.34.[1]

References

1. V.M. Abazov *et al.* (The D0 Collaboration), Phys. Rev. Lett. **108**, 261603 (2012).
2. V.A. Kostelecký and N. Russell, *Data Tables for Lorentz and CPT Violation,* 2013 edition, arXiv:0801.0287v6.
3. D. Colladay and V.A. Kostelecký, Phys. Rev. D **58**, 116002 (1998).
4. R. Kehoe, M. Narain, and A. Kumar, Int. J. Mod. Phys. A **23**, 353 (2008).
5. M.S. Berger and V.A. Kostelecký, *Notes on Lorentz and CPT Violation in Top-Quark Production and Decay,* IUHET 530.
6. V.M. Abazov *et al.* (D0 Collaboration), Nucl. Instrum. Methods Phys. Res. A **565**, 463 (2006); M. Abolins, *et al.* Nucl. Instrum. Meth. A **584**, 75 (2008); R. Angstadt *et al.* (D0 Collaboration), Nucl. Instrum. Meth. A **622**, 298 (2010).
7. V.M. Abazov *et al.* (D0 Collaboration), Phys. Ref. D **84** 012008 (2011).

THE KÄLLÉN-LEHMANN REPRESENTATION IN LORENTZ-VIOLATING FIELD THEORY

R. POTTING

CENTRA and Physics Department
Universidade do Algarve, FCT
8005-139 Faro, Portugal
E-mail: rpotting@ualg.pt

We derive a Källén-Lehmann representation for the propagator in two models, one consisting purely of scalars, the other also involving fermions, that couple to a set of constant background coupling coefficients transforming as a symmetric observer Lorentz two-tensor. In particular, we establish the form of the one-particle poles, which are crucial to define consistently the external states in S-matrix amplitudes.

1. Motivation

Consider a general time-ordered correlation function for a scalar field ϕ, $\langle 0|T\phi(x_1)...\phi(x_n)|0\rangle$, represented graphically in Fig. 1. To get the physical

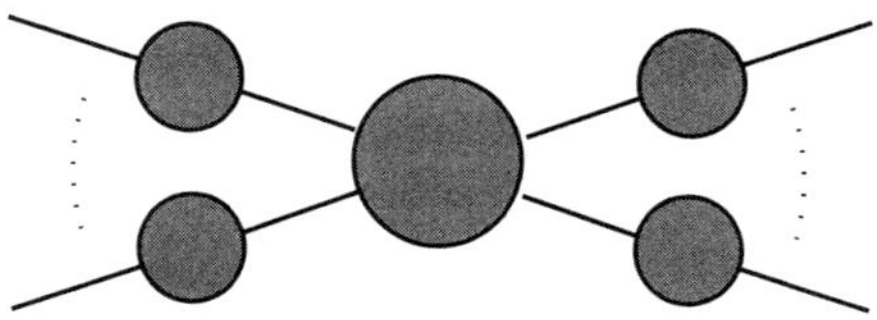

Fig. 1. General n-point amplitude; the external propagators are dressed.

S-matrix elements we take the dressed external propagators with momenta on the physical mass shell, and consider the amputated diagram with the external legs removed (the Lehmann-Symanzik-Zimmermann reduction formula). The only information we need for this is the one-particle pole of the dressed propagator, defined (for scalar field theory) by (i) the dispersion relation, which, if Lorentz invariance is applicable, dictates $p^2 - m^2 = 0$ for some m, and (ii) its residue, the wavefunction renormalization constant Z.

Källén and Lehmann showed in the 1950s that in interacting quantum field theory the propagator can always be written as a sum of free propagators:[1]

$$D_F(x-y) = \langle 0|T\phi(x)\phi(y)|0\rangle$$
$$= \int_0^\infty ds \int \frac{d^4p}{(2\pi)^4} \frac{i\,\rho(s)}{p^2 - s + i\epsilon}\, e^{-ip\cdot(x-y)}, \qquad (1)$$

where ρ denotes a real, non-negative, distribution, the spectral density:

$$\rho(s)/(2\pi)^3 \equiv \sum_\alpha \delta^{(4)}(p-p_\alpha)|\langle 0|\phi(0)|\alpha\rangle|^2. \qquad (2)$$

Here α is summed over a complete set of physical states (not necessarily one-particle). If the spectrum includes a stable one-particle state, ρ can be written into a one-particle and multiparticle continuum contrbutions:

$$\rho(s) = \rho^{\text{1-part}}(s) + \sigma(s), \qquad \rho^{\text{1-part}}(s) = Z\delta(s-m^2), \qquad (3)$$

where Z denotes the wave-function renormalization constant.

The questions we will try to answer in this talk are: (i) can the Källén-Lehmann representation be generalized to Lorentz-violating field theories; and (ii) if so, what form does it take?

2. Lorentz-violating scalar field theory

Consider the Lorentz-violating scalar-field action[2]

$$S_2 = \int d^4x \frac{1}{2}\Big((\partial_\mu\phi_1)^2 + c^{\mu\nu}\partial_\mu\phi_1\partial_\nu\phi_1 + (\partial_\mu\phi_2)^2$$
$$- m_1^2\phi_1^2 - m_2^2\phi_2^2 - V(\phi_1,\phi_2)\Big), \qquad (4)$$

with $c^{\mu\nu}$ a set of constant background coupling coefficients transforming as a symmetric traceless rank-two observer Lorentz tensor. It should be regarded as a property of the vacuum. Note that the vacuum continues to be translation invariant, an important requirement in the construction of the Källén-Lehmann representation.

In trying to generalize the Källén-Lehmann representation for the model defined by Eq. (4), we have to take into account that the spectral density cannot be expected anymore to depend solely on $s \equiv p^2$, as the vacuum is no longer Lorentz invariant, as it depends on the tensor quantity $c^{\mu\nu}$. However, we can expect observer invariance to continue to hold. It then follows that ρ should depend on any observer scalars we can build from $c^{\mu\nu}$ and p_μ. These are p^2 as well as the quantities $(c^n)^p_p$, where

$$p\cdot c^n \cdot p = p^\mu \eta_{\mu\alpha_1} c^{\alpha_1\beta_1} \eta_{\beta_1\alpha_2} c^{\alpha_2\beta_2} \ldots \eta_{\beta_n\nu} p^\nu \equiv (c^n)^p_p. \qquad (5)$$

In four dimensions we can in fact restrict ourselves to the values $n \leq 3$, as $(c^n)_p^p$ with $n \geq 4$ can be functionally expressed in terms of the $(c^n)_p^p$ with $n \leq 3$ (note that $p^2 \equiv (c^0)_p^p$).

Taking into account the above it then follows that:[2]

$$D_F(x-y) = \int ds \int \frac{d^4p}{(2\pi)^4} \frac{i\,\rho\big[s;(c^i)_p^p\big]}{p^2 - s + i\epsilon}\, e^{-ip\cdot(x-y)}. \tag{6}$$

We can split ρ into contributions of a stable one-particle state (if present) and a continuum of multi-particle states:

$$\rho\big[s;(c^i)_p^p\big] = \rho^{\text{1-part}}\big[s;(c^i)_p^p\big] + \sigma\big[s;(c^i)_p^p\big]\,. \tag{7}$$

The one-particle state should satisfy an equation of motion, that is, having support on a mass shell. It then follows that

$$\rho^{\text{1-part}}\big[s;(c^i)_p^p\big] = f\big[(c^i)_p^p\big]\,\delta\big(s - g\big[(c^i)_p^p\big]\big)\,. \tag{8}$$

The functions f and g can be expanded in a Taylor series:

$$f\big[(c^i)_p^p\big] = Z\left(1 + \sum_{1\leq i_1 \leq \ldots i_r \leq 3} f_{i_1\ldots i_r} \prod_{j=1}^{r} (c^{i_j})_p^p\right), \tag{9}$$

$$g\big[(c^i)_p^p\big] = m^2 - \sum_{1\leq i_1 \leq \ldots i_r \leq 3} g_{i_1\ldots i_r} \prod_{j=1}^{r} (c^{i_j})_p^p, \tag{10}$$

generalizing the Lorentz-invariant case $f\big[(c^i)_p^p\big] = Z$ and $g\big[(c^i)_p^p\big] = m^2$. The propagator now takes the Källén-Lehmann-like form

$$D_F(x-y) = i \int \frac{d^4p}{(2\pi)^4}\, e^{-ip\cdot(x-y)} \times \left(\frac{f\big[(c^i)_p^p\big]}{p^2 - g\big[(c^i)_p^p\big] + i\epsilon} + \int_{M^2}^{\infty} ds\, \frac{\sigma\big[s;(c^i)_p^p\big]}{p^2 - s + i\epsilon} \right). \tag{11}$$

To first order in $c^{\mu\nu}$ one finds the one-particle equation of motion:

$$\left(\Box + m^2 + g_1 c_\partial^\partial + g_2 (c^2)_\partial^\partial + g_{11}\big(c_\partial^\partial\big)^2 + \ldots\right)\phi_{1,in}(x) = 0\,. \tag{12}$$

Here $c_\partial^\partial \equiv c^{\mu\nu}\partial_\mu\partial_\nu$. The coefficients g_i can be computed perturbatively. Thus, equation of motion satisfied by external states includes higher-dimensional terms not present in the original Lagrangian!

An unexpected consequence of Eq. (11) is that the wave-function renormalization constant Z becomes a momentum-dependent quantity:

$$Z \to Z(1 + f_1\, c_p^p + \ldots). \tag{13}$$

The coefficients f_i can be computed perturbatively.[3]

By evaluating the expectation value of the field commutator $[\phi(x), \phi(y)]$ sum rules can be derived, generalizing known results in the Lorentz-symmetric case.[2]

As a second example, consider the following action with a Dirac fermion and a scalar field:[2]

$$S_3 = \int d^4x \Big(\bar{\psi}(i\gamma^\mu \partial_\mu + ic^{\mu\nu}\gamma_\mu\partial_\nu - m)\psi + \frac{1}{2}(\partial_\mu\phi)^2 - \frac{1}{2}(m')^2\phi^2 + \lambda\phi\bar{\psi}\psi \Big). \tag{14}$$

Indicating the time-ordered product $\langle 0|T\psi_\alpha(x)\bar{\psi}_\beta(y)|0\rangle$ by $S_F(x-y)_{\alpha\beta}$ one can show by a careful analysis that

$$S_F(x-y) = \int ds \int \frac{d^4p}{(2\pi)^4} \frac{i.e^{-ip\cdot(x-y)}}{p^2 - s + i\epsilon} \Big(\rho_0 + \sum_{k=0}^{3} (c^k)^p_\gamma \, \rho_{k+1} \Big). \tag{15}$$

where the spectral density functions ρ_k depend on s and $(c^i)^p_p$ as above.

In case the spectrum includes a stable one-particle state, the density functions can be split into a one-particle pole satisfying a mass-shell condition and a multi-particle continuum, analogous to Eq. (7). The one-particle contributions can then be written in the form (8) and expanded as in Eqs. (9) and (10). The upshot is that the (free) fermion equation of motion includes Lorentz-violating terms not present in the lagrangian (14):

$$\Big(\gamma^\mu\partial_\mu - \bar{m}\big[(c^i)^\partial_\partial\big] + \bar{x}_1\big[(c^i)^\partial_\partial\big]\, c^\partial_\gamma + \bar{x}_2\big[(c^i)^\partial_\partial\big]\, (c^2)^\partial_\gamma + \dots \Big)\psi(x) = 0, \tag{16}$$

where the quantities $\bar{m}$, $\bar{x}_i$ include radiative corrections depending on the scalar quantities $(c^n)^\partial_\partial \equiv \partial \cdot c^n \cdot \partial$ (note that, without radiative corrections, only $\bar{m} = m$ and $\bar{x}_1 = 1$ are nonzero). Moreover, one finds that the wave-function renormalization function takes again the momentum-dependent form (13).

Acknowledgments

Financial support by the Portuguese Fundação para a Ciência e a Tecnologia is gratefully acknowledged.

References

1. G. Källén, Helv. Phys. Acta **25**, 417 (1952); H. Lehmann, Nuov. Cim. **11**, 342 (1954).
2. R. Potting, Phys. Rev. D **85**, 045033 (2012).
3. M. Cambiaso, R. Lehnert, and R. Potting, to appear; M. Cambiaso, these proceedings.

TIME–VARYING NUCLEAR DECAY PARAMETERS AND DARK MATTER

J. NISTOR,* E. FISCHBACH,*¶ J.T. GRUENWALD,* D. JAVORSEK,†
J.H. JENKINS,‡ and R.H. LEE§
*Purdue University, Physics, West Lafayette, IN 47906, USA
†Air Force Test Center, Edwards Air Force Base, CA 93524, USA
‡Texas A&M University, Nuclear Engineering, College Station, TX 77843, USA
§United States Air Force Academy, Physics, Colorado Springs, CO 80920, USA
¶E-mail: ephraim@purdue.edu

Recently published data suggest a possible solar influence on some nuclear decay rates, including evidence for an annual variation attributed to the varying Earth-Sun distance. Here, we consider the possibility that the annual signal seen by the DAMA collaboration, and interpreted by them as evidence for dark matter, may in fact be due to the radioactive contaminant ^{40}K, which is known to be present in their detector. We also consider the possibility that part of the DAMA signal may arise from relic big-bang neutrinos.

1. Introduction

Nuclear decay measurements reported by independent groups in 1986[1] and in 1998[2] exhibited an anomalous oscillatory behavior. Once known systematic causes were ruled out, the results suggested that the decay constants λ in some radioactive isotopes might themselves be time dependent. If this were the case, then the usual decay law would assume the form $dN(t)/dt = -\lambda(t)N(t)$, where λ and $T_{1/2} = \ln 2/\lambda$ would be explicitly time dependent. Interestingly, a significant number of the decay-rate variations observed to date can be attributed in one way or another to the Sun: e.g., an annual signal presumed to arise from the annual variation of the Earth-Sun distance,[3] R, a (10–15) yr^{-1} variation associated with the Sun's rotation,[4,5] a Rieger periodicity,[6] and evidence for an association between solar storms and the decay rate of ^{54}Mn.[7] A summary of existing data suggesting a time dependence of $\lambda(t)$ is presented in Table 2 of Ref. 8.

Although some of the decay-rate variations in that table may be due in part to seasonal variations in the sensitivities of various detectors (i.e., arising from variations in temperature, pressure, and humidity), it seems unlikely that such variations could account for the fact that experiments at different locations, using a variety of isotopes and detector types, report similar effects. Moreover, seasonal effects would not be expected to account for periodicities in the (10–15) yr^{-1}range, nor for short-term changes in decay rates coincident with solar storms. In what follows we will thus assume that the observed annual periodic dependence of $\lambda(t)$ for some nuclear decays arises from the Sun through some as-yet unknown mechanism.

2. Dark matter

There are numerous reasons to believe in the existence of dark matter (DM), such as the observation of flat galactic rotation curves. A recent review of the literature can be found in Ref. 9. Among the many ongoing DM searches, the DAMA collaboration has presented the strongest evidence to date for the presence of a DM signal.[10] However, when their results are parametrized in terms of the DM particle mass and its cross section for scattering off nuclei, the region allowed by DAMA in the resulting exclusion plot is ruled out by other experiments.[a] This has raised the question of whether the observed DAMA signal could have an alternate explanation.

Since DAMA utilizes a large NaI(Tl) detector, which is known to have potassium contamination, the possibility has been raised that the DAMA signal may be attributable to the decay of naturally occurring ^{40}K which has a half-life of 1.26×10^9 yr. To understand the relevance of ^{40}K, we note that the DAMA signal is an annually varying count rate in their detector which may be consistent with the fact that the Earth's speed relative to the hypothesized DM halo reaches a maximum around June 2, when the Earth's velocity adds maximally to that of the Sun, and a minimum on December 2 when the Earth's velocity points the opposite direction. The resulting annual variation in the DM flux (known as the annual modulation signature) could thus explain the annual signal reported by DAMA.

In contrast, the ^{40}K hypothesis attributes the annual signal seen by DAMA in their NaI(Tl) crystals to X-ray photons emitted in the detector itself from ^{40}K decays. There are 3 dominant decay modes of ^{40}K, of which two are most relevant. The first is an electron capture (EC) mode to the 1461 keV excited state of ^{40}Ar, followed by the electromagnetic decay to the ground state of ^{40}Ar. The signal for this mode, which has a branching

[a]Some models allowing for inelastic WIMP-nucleon scattering and spin dependence may still account for the DAMA signal.

ratio of 10.55%, is the 1461 keV gamma accompanied by a ~3.2 keV X-ray photon arising from the electromagnetic cascade which results from filling the K-shell hole left by the captured electron. The second mode of interest is the EC decay of ^{40}K directly to the ground state, again with the emission of a 3.2 keV X-ray photon but without an accompanying 1461 keV de-excitation gamma.[11] This mode, which is estimated to have a small branching ratio (~0.2%), has not yet been seen directly but may be the most interesting (see below). As it turns out, the annually varying DAMA signal is detected for X-ray photons in the 2–4 keV range, with a peak intensity near 3 keV, which is just the energy expected from the cascade following EC in ^{40}K. Although this may be coincidental, it may also be a clue to why the DAMA signal is not seen in other DM experiments.

Another clue comes from the count rate observed by DAMA. Given the concentration of potassium that DAMA acknowledges to be present in their NaI(Tl) detectors, we estimate that ^{40}K would produce an event rate in the range of ~1000 counts per day (cpd) from EC decay to the excited ^{40}Ar state, whereas for DAMA/LIBRA the total count rate in the 2–4 keV region is ~530 cpd. Thus ^{40}K could correctly account for both the energy range in which DAMA see their signal and their observed count rate.

The preceding picture is suggestive, but far from complete, for the following reasons. (1) First, it presumes that ^{40}K exhibits an annual modulation in the EC decay mode, similar to other isotopes, for which there is no evidence at present. Fortunately, an experiment under way by Lang and collaborators should settle this question in the next year or so. (2) Secondly, the phase of the DAMA signal (i.e., the calendar date where their count rate is maximum) is approximately mid-May, whereas the isotopes exhibiting annual variations which have been studied to date have phases that are typically in the January-February or July-August time frames. This could be a problem for the ^{40}K hypothesis or, as we discuss below, a hint at other new physics. (3) Another problem is the claim by DAMA that they see no annual modulation in the signal for those events in which both a 1461 keV gamma and a 3.2 keV X-ray (the expected signal for the EC decay mode) are detected.[12] Since the data to back up this claim have not been made available by the DAMA collaboration, their claim cannot be further evaluated at this time. Irrespective of this, it is worth noting that the modulation in the low-energy events accompanied by ^{40}K decay is *not* necessarily correlated with a modulation in the 1461 keV event rate. Those events which are associated with the EC decay directly to the ground state of ^{40}Ar (on the order of ~20 cpd) are not associated with the emission of higher energy photons — and therefore represent an *irreducible background* in the energy region in which DAMA claims a signal. The activity from this decay may be large enough to account for DAMA's observed annual modulation in count rate (amplitude of ~10 cpd).

3. Discussion

Although the ^{40}K hypothesis as a mechanism to explain the DAMA data remains open, there may be a hint in the phase of the DAMA data of even more new physics. Consider a system sensitive to several influences with a common frequency ω, but different amplitudes and phases, described by a function $f(t)$:

$$f(t) = f_{\max}\left[1 + \sum_i \delta_i \cos(\omega t - \phi_i)\right] \equiv f_{\max}\left[1 + \alpha\cos(\omega t - \beta)\right]. \quad (1)$$

In Eq. (1), $\tan\beta = (\sum \delta_i \sin\phi_i)/(\sum \delta_i \cos\phi_i) \equiv B/A$, and $\alpha = A/\cos\beta = B/\sin\beta$, and $\delta_i \ll 1$ is assumed. It is thus possible that the 'anomalous' DAMA phase is the result of detecting several annual signals. These could include a solar influence (phase Jan-Feb or July-Aug), dark matter (May-Jun), and big bang relic neutrinos (~Dec 10). The possibility that DAMA could also be detecting relic neutrinos follows from the similarity of the fluxes of solar and relic neutrinos, as noted in Refs. 7 and 13, notwithstanding their differences in energy.

References

1. D.E. Alburger, G. Harbottle, and E.F. Norton, Earth Planet. Sci. Lett. **78**, 168 (1986).
2. H. Siegert, H. Schrader, and U. Schötzig, Appl. Radiat. Isot. **49**, 1397 (1998).
3. J.H. Jenkins, E. Fischbach, J.B. Buncher, J.T. Gruenwald, D.E. Krause, and J.J. Mattes, Astropart. Phys. **32**, 42 (2009).
4. P.A. Sturrock *et al.*, Astropart. Phys. **34**, 121 (2010).
5. P.A. Sturrock *et al.*, Solar Phys. **267**, 251 (2010).
6. P.A. Sturrock, E. Fischbach, and J.H. Jenkins, Solar Phys. **272**, 1 (2011).
7. J.H. Jenkins, E. Fischbach, Astropart. Phys. **31**, 407 (2009).
8. D. O'Keefe *et al.*, Astrophys. Space Sci. **344**, 297 (2013).
9. *Particle Dark Matter, observations, models and searches*, G. Bertone, ed., Cambridge University Press, Cambridge, 2010.
10. R. Bernabei *et al.*, Eur. Phys. J. C **67**, 39 (2010).
11. J. Pradler, B. Singh, and I. Yavin, Phys. Lett. B**270**, 399 (2013).
12. R. Bernabei *et al.*, arXiv:0912.0660.
13. A.G. Parkhomov, arXiv:1010.1591.

LORENTZ VIOLATION AND TOPOLOGICAL DEFECTS

MICHAEL D. SEIFERT

Department of Physics, Williams College
Williamstown, MA 01267, USA
E-mail: ms10@williams.edu

If Lorentz symmetry is broken, it must have occurred dynamically, via a vector or tensor field whose potential energy forces it to take on a nonzero background expectation value 'in vacuum.' If the set of minima of this potential (the *vacuum manifold*) has a nontrivial topology, then there can arise *topological defects*: stable solutions in which the field approaches different potential minima as we go to infinity in different directions. I discuss the current status of research into these topological defects in the context of Lorentz symmetry breaking, including recent results concerning the birefringent light bending of monopole solutions, and the search for models supporting cosmic-string and domain-wall defects.

It has been known for some time now[1] that if Lorentz symmetry is to be broken and spacetime is to be well described by riemannian geometry, then the breaking of Lorentz symmetry must be spontaneous. The lagrangian for a model in which this occurs will have the general form

$$S = \int d^4x \sqrt{-g} \left[\frac{R}{16\pi G} - (\nabla \mathcal{T})(\nabla \mathcal{T}) - V(\mathcal{T}) \right], \tag{1}$$

where R is the Ricci scalar and $\mathcal{T}$ is a tensor field (whose indices have been suppressed for the sake of generality). If the potential term $V(\mathcal{T})$ is constructed in such a way that

$$V(\bar{\mathcal{T}}) = \frac{\delta V(\bar{\mathcal{T}})}{\delta g^{ab}} = \frac{\delta V(\bar{\mathcal{T}})}{\delta \mathcal{T}} = 0 \tag{2}$$

for some nonzero tensor $\bar{\mathcal{T}}$, then a solution of the equations of motion is $g_{ab} = \eta_{ab}$ and $\mathcal{T} = \bar{\mathcal{T}}$ everywhere in spacetime. In other words, we have Minkowski spacetime, but with an additional geometric structure (due to the constant but nonzero tensor field $\mathcal{T}$) upon which we can 'hang' Lorentz-violating effects.

Since the lagrangian (1) is Lorentz and diffeomorphism invariant, the background value $\bar{\mathcal{T}}$ of the tensor field in such a theory will not be unique. Rather, there will be some set of possible tensor values that will satisfy the conditions (2). This set of tensor values forms a manifold in field space, called the *vacuum manifold.* Its dimension and shape will depend on the precise form of the potential. As an example, suppose that $\mathcal{T}$ is a vector field A^a, and that $V(A^a) = \lambda(A^a A_a - \mathcal{C})^2/4$. The vacuum manifold will then be a three-dimensional hyperboloid in the four-dimensional field space, consisting of all vectors of a given norm $A^a A_a = \mathcal{C}$. If $\mathcal{C} > 0$, then all these vectors will be spacelike and the hyperboloid will be homeomorphic to $S^2 \times \mathbb{R}$; if $\mathcal{C} < 0$, then the hyperboloid will consist of two disconnected hypersurfaces, each homeomorphic to $\mathbb{R}^3$ (i.e., the past-directed and future-directed vectors.)

If the topology of the vacuum manifold is nontrivial, it is possible that static *topological defect* solutions can arise.[2] These solutions have the property that the field takes on different vacuum values at different 'points at infinity,' but that the field configuration at infinity cannot be smoothly extended to all of space without the field leaving the vacuum manifold. There results a configuration which is (at least) metastable, but whose stress-energy tensor is nonzero. The type of solution that arises will depend on the topology of the vacuum manifold; if there is a noncontractible S^n in the vacuum manifold (with $n \leq 2$), localized topological defect solutions are possible. A noncontractible S^0 can give rise to *domain walls* separating two regions with different vacuum field values; a noncontractible S^1 can lead to filament-like *cosmic strings*; and a noncontractible S^2 can support point-like *monopoles.* (In more precise mathematical terms, these cases correspond to the vacuum manifold having nonvanishing homotopy groups π_0, π_1, or π_2, respectively.)

The question then arises of whether the vacuum manifold for a Lorentz-violating tensor field $\mathcal{T}$ could have the topology necessary to support a topological defect. This question was answered in the affirmative in my previous work.[3] If we assume that the vacuum manifold is of the form $\mathcal{T}_{abc\ldots}\mathcal{T}^{abc\ldots} = \mathcal{C}$, only three types of irreducible tensor fields have the right topology to support topological defects: vectors, antisymmetric two-tensors, and symmetric trace-free two-tensors.

Of these types of fields, topological defects have been found for one of them: an antisymmetric tensor field B_{ab} with a flat spacetime lagrangian of the form

$$\mathcal{L} = -\frac{1}{12}F_{abc}F^{abc} - \frac{\lambda}{2}(B^{ab}B_{ab} - b^2)^2, \tag{3}$$

where $F_{abc} \equiv 3\partial_{[a}B_{bc]}$. A spherically symmetric static monopole solution exists in this theory, and its properties have been studied.[4] In this solution, $B_{\theta\phi}$ interpolates smoothly from zero at $r = 0$ to the vacuum manifold as $r \to \infty$; all other components of B_{ab} vanish. We can also 'turn on' gravity by including the Ricci scalar in the above action. The spacetime curvature which results due to the stress-energy tensor of B_{ab} would then be detectible due to its influence on light rays. It can be shown that the effects of gravitational redshift would be within a couple of orders of magnitude of $\epsilon \equiv 4\pi G b^2$, and that angular deflection $\delta\phi$ of light rays is constant to leading order in the impact parameter.[4]

More recently,[5] Kamuela Lau and I have been investigating the behavior of light rays travelling near a monopole when there is a direct coupling between the Maxwell field and the monopole field B_{ab}:

$$\mathcal{L} \supset -\frac{1}{4}F_{ab}F^{ab} - (k_F)^{abcd}F_{ab}F_{cd}, \tag{4}$$

where F_{ab} is the Maxwell field strength,

$$(k_F)^{abcd} = \xi\left(B^{ab}B^{cd} - \frac{1}{12}(\eta^{ac}\eta^{bd} - \eta^{ad}\eta^{bc})B^{ef}B_{ef} - B^{[ab}B^{cd]}\right), \tag{5}$$

and ξ is a coupling constant. The modified Maxwell's equations that arise are equivalent to Maxwell's equations in an inhomogeneous birefringent medium.[6] We can then use a geometric-optics approximation to study the paths of light rays. The result is that a distant source will be split into two images in the presence of a monopole solution, each image being entirely polarized. One polarization will travel through the monopole in a straight line; the other will be bent, and so will appear at a different location on the sky. The angle between these two images will generically be of order ξb^2, where b is the vacuum magnitude of the field B_{ab}.

Another student, Brandon Ling, has been conducting a more general survey of the possible types of vacuum manifolds and topological defects that might be found. Specifically, he has examined the topology of the vacuum manifolds for an antisymmetric tensor field B_{ab} with a quartic potential of the form

$$V(B_{ab}) = \alpha X^2 + \beta XY + \gamma Y^2 + \delta X + \lambda Y, \tag{6}$$

where $X = B_{ab}B^{ab}$ and $Y = B_{ab}B_{cd}\epsilon^{abcd}$. (Note that the potential in (3) corresponds to the case $\beta = \gamma = \lambda = 0$, up to an additive constant.) He has found that the only noncontractible spheres in these vacuum manifolds are two-spheres; no noncontractible circles can exist, and the vacuum manifolds are always connected. This implies that antisymmetric tensors cannot give rise to cosmic strings or domain walls. Remarkably, it can be shown that in most cases the vacuum manifold for the potential (6) is homeomorphic to TS^2, the tangent bundle on the two-sphere. We are now exploring which parameter values in this allow for monopole solutions, and what the properties of these solutions will be.

Acknowledgments

I would like to thank Williams College for their financial support in attending this conference.

References

1. V.A. Kostelecký, Phys. Rev. D **69**, 105009 (2004).
2. A. Vilenkin and E.P.S. Shellard, *Cosmic Strings and Other Topological Defects*, Cambridge University Press, Cambridge, 1994.
3. M. Seifert, Phys. Rev. D **82**, 125015 (2010).
4. M. Seifert, Phys. Rev. Lett. **105**, 201601 (2010); X. Li, P. Xi, and Q. Zhang, Phys. Rev. D **85**, 085030 (2012).
5. K. Lau and M. Seifert, in preparation.
6. V.A. Kostelecký and M. Mewes, Phys. Rev. D **66**, 056005 (2002).

TESTING LORENTZ INVARIANCE WITH A DOUBLE-PASS OPTICAL RING CAVITY

Y. MICHIMURA,*§ N. MATSUMOTO,* N. OHMAE,†
W. KOKUYAMA,‡ Y. ASO,* M. ANDO,* and K. TSUBONO*

**Department of Physics, University of Tokyo, Bunkyo, Tokyo 113-0033, Japan*
†Department of Applied Physics, University of Tokyo, Bunkyo, Tokyo 113-8656, Japan
‡National Metrology Institute of Japan, National Institute of Advanced Industrial Science and Technology (AIST), Tsukuba, Ibaraki 305-8563, Japan
§E-mail: michimura@granite.phys.s.u-tokyo.ac.jp

We have developed an apparatus to test Lorentz invariance in the photon sector by measuring the resonant frequency difference between two counterpropagating directions of an asymmetric optical ring cavity using a double-pass configuration. No significant evidence for the violation was found at the level of $\delta c/c \lesssim 10^{-14}$. Details of our apparatus and recent results are presented.

1. Introduction

Resonant cavities have been frequently used to test Lorentz invariance in photons. These cavity experiments can be classified into two types, depending on the parity of the cavity used. Even-parity experiments have put limits on even-parity light speed anisotropy to the $\delta c/c \lesssim 10^{-17}$ level.[1,2]

Our experiment is one of the few odd-parity experiments. We look for a nonzero resonant frequency difference between two counterpropagating directions of an asymmetric optical ring cavity by making use of a double-pass configuration. We have put limits on odd-parity anisotropy to the $\delta c/c \lesssim 10^{-14}$ level,[3] which is an order of magnitude improvement over previous results from cavity experiments,[2,4] and is comparable with a result from a Compton scattering experiment.[5]

2. Experiment

A simple way to express the odd-parity Lorentz violation is to express the speed of light with $c = 1 + \alpha \cos\theta$, where θ is the direction of the light propagation with respect to the preferred frame and α is an anisotropy

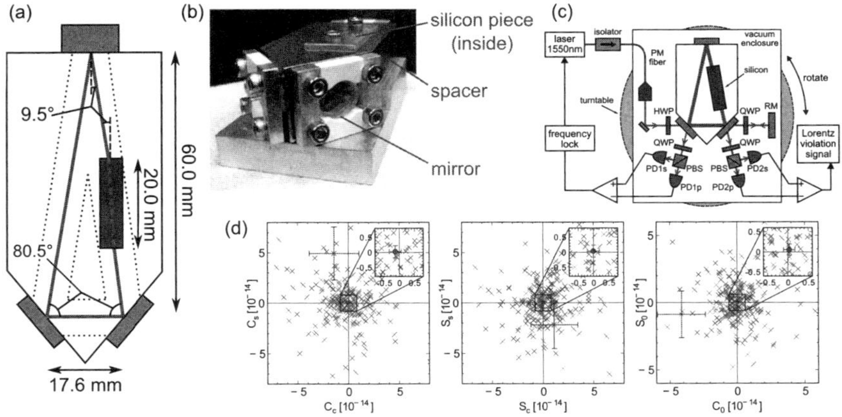

Fig. 1. (a) Dimensions and (b) picture of the optical ring cavity. (c) Schematic of the experimental setup. (d) Sidereal modulation amplitudes determined from 1 day data. The mean values and standard errors (shown as a dot in each inset) of all 232 points are $C_{\rm c} = -1.3 \pm 2.0$, $C_{\rm s} = 0.3 \pm 2.0$, $S_{\rm c} = -0.1 \pm 1.9$, $S_{\rm s} = 0.4 \pm 2.0$, $C_0 = 0.3 \pm 1.4$, and $S_0 = -0.6 \pm 1.4$ (all values $\times 10^{-15}$).

parameter that is zero when there is no violation. From this expression, the resonant frequency shift ($\delta\nu$) of a cavity due to the anisotropy is calculated as

$$\frac{\delta\nu}{\nu} = \alpha \oint n(\theta) \cos\theta r {\rm d}\theta, \tag{1}$$

where $n(\theta)$ is the refractive index through the path, and the integral is taken over the round-trip path of the cavity. $\delta\nu$ will be nonzero if $n(\theta)$ changes asymmetrically. Also, since the signs of $\delta\nu$ will be opposite between the round-trip integral taken clockwise and counterclockwise, measuring the resonant frequency difference between two counterpropagating directions gives us a null measurement of the anisotropy parameter α. This differential measurement is highly insensitive to environmental disturbances because the effects of cavity length fluctuations are common to both resonances.

Our ring cavity, shown in Fig. 1(a) and (b), is a triangular cavity constructed from three half-inch mirrors. These mirrors are mechanically fixed on a spacer made of Super Invar. The spacer has through holes for the optical path. The radius of the holes are 4 mm and the beam radius inside the cavity is 260 um at maximum. The spacer also has a hole for placing a silicon piece along one side of the triangle. This silicon piece is rectangular, and its size is $5 \times 10 \times 20$ mm. Also, this silicon piece is antireflection coated ($R < 0.5\%$/surface), and the incident beam to this piece is slightly angled

($\theta_{\text{in}} = 9.5°$) in order to avoid the cross coupling between the counterpropagating beams. Silicon has high transmittance and a large refractive index (measured value $n = 3.69$) at wavelength $\lambda = 1550$ nm. The round-trip length of our cavity is 14 cm and the finesse is about 120 for p-polarized light, with the silicon piece inside the cavity.

We use a double-pass configuration for comparing the resonant frequencies between the counterpropagating directions[6] [see Fig. 1(c)]. We use a single-frequency laser source with a wavelength of 1550 nm (Koheras AdjustiK C15). Measured relative frequency noise and relative intensity noise without any external stabilization servo was 2×10^{-10} $/\sqrt{\text{Hz}}$ and 1×10^{-3} $/\sqrt{\text{Hz}}$ at 0.1 Hz respectively. No external intensity stabilization was employed. The laser beam is fed into the ring cavity in the counterclockwise direction via a polarization maintaining fiber. A collimator (Thorlabs PAF-X-5-C) was used to align and mode-match the incident beam to the ring cavity. The incident beam power to the ring cavity is about 1 mW.

The frequency of the laser beam is stabilized to the counterclockwise resonance using a piezoelectric actuator attached on the laser cavity. We used the Hänsh-Coulillaud method[7] to obtain the error signal for the laser frequency servo. By taking the differential of two PD outputs to obtain an error signal, we reduce the effect of laser intensity fluctuation.

The transmitted light of the counterclockwise beam is then reflected back into the cavity in the clockwise direction by a reflection mirror (RM). We obtain the second error signal, which is proportional to the resonant frequency difference, and in this signal we search for the Lorentz violation. Measured relative frequency noise of the second error signal was 4×10^{-13} $/\sqrt{\text{Hz}}$ at 0.1 Hz. To obtain the second error signal, we again used the Hänsch-Couillaud method. We did not use Pound-Drever-Hall method because we did not want to introduce spatial mode distortion of the beam from phase modulation.

This double-pass configuration enables a null measurement of the resonant frequency difference with a fairly simple setup. Another way to measure the resonant frequency difference is to inject the laser beam from both directions, stabilize the frequency of each beam to each resonance, and compare the frequency of counterpropagating beams.[4] However, this configuration needs additional actuation (e.g., acousto-optic modulator) and servo, which could introduce additional noise. Also, stabilizing the frequencies of two counterpropagating beams of a ring cavity has the possibility of lock-in behavior,[8] which is a common effect in ring laser gyroscope.

All the optics are placed in a $30 \times 30 \times 17$ cm vacuum enclosure ($\sim$1 kPa) to realize a stable operation. This enclosure is fixed on a turntable together with the laser source, and rotated with a direct drive servo motor (Nikki Denso NMR-CAUIA2A-151A). Positive and reverse rotations of 420° are repeated alternately in order to avoid twist of the electrical cables. We used thin cables (0.1 mm dia.) for reducing vibrations introduced through the cables. The rotational speed is $\omega_{\rm rot} = 30°/\text{sec}$ ($f_{\rm rot} = 0.083$ Hz), and S-curve acceleration and deceleration was used when flipping the sign of the rotations in order to avoid sudden rotational speed change. For data analysis, we only used an interval of 360° in the middle of each rotation where the rotational speed is constant. Measured rotational speed fluctuation was less than 1 mrad/sec/$\sqrt{\text{Hz}}$ at 0.1 Hz.

3. Recent results and outlook

Since August 2012, we are continuing the Lorentz violation search. Figure 1(d) shows recent results from the data taken for 232 days between August 2012 to April 2013. Six sidereal modulation amplitudes defined in Ref. 3 are plotted. Standard errors of each averaged amplitude are $\sim 2 \times 10^{-15}$, which is factor of $\sim$ 1.4 improvement over our previous results.[3] No deviation from zero by more than 1σ was found. This means that there is no significant evidence for anisotropy in the speed of light in a sidereal frame at the level of $\delta c/c \lesssim 10^{-14}$.

We are expecting to achieve a year-long run in August 2013. By using the year-long data, we are planning to put first limits on odd-parity higher-order coefficients in SME.[9]

Acknowledgments

We thank Matthew Mewes for useful discussions. This work was supported by Grant-in-Aid for JSPS Fellows number 25·10386.

References

1. Ch. Eisele *et al.*, Phys. Rev. Lett. **103**, 090401 (2009).
2. S. Herrmann *et al.*, Phys. Rev. D **80**, 105011 (2009).
3. Y. Michimura *et al.*, Phys. Rev. Lett. **110**, 200401 (2013).
4. F.N. Baynes *et al.*, Phys. Rev. Lett. **108**, 260801 (2012).
5. J.-P. Bocquet *et al.*, Phys. Rev. Lett. **104**, 241601 (2010).
6. B.J. Cusack *et al.*, Class. Quantum Grav. **19**, 1819 (2002).
7. T.W. Hänsch and B. Couillaud, Opt. Commun. **35**, 441 (1980).
8. F. Zarinetchi and S. Ezekiel, Opt. Lett. **11**, 401 (1986).
9. M. Mewes, Phys. Rev. D **85**, 116012 (2012).

LASER TIME-TRANSFER AND SPACE-TIME REFERENCE IN ORBIT

P. BERCEAU and L. HOLLBERG*

Department of Physics and Hansen Experimental Physics Laboratory
Stanford University, Stanford, CA 94305-4085
**E-mail: leoh@stanford.edu*

A high performance Space-Time Reference in orbit could be realized using a stable atomic clock in a precisely defined orbit and linking that to high accuracy atomic clocks on the ground using a laser based time-transfer link. This would enhance performance of existing systems and provide unique capabilities in navigation, precise timing, earth sciences, geodesy and the same approach could provide a platform for testing fundamental physics in space. Precise laser time- and frequency-transfer from the ground to an orbiting satellite would make it possible to improve upon the current state of the art in timing (about 1 to 30 ns achieved with GPS) by roughly a factor of 1000 to the 1 ps level.

1. Motivation

There have been tremendous advances in the performance of atomic frequency standards (clocks) over the past 40 years, and, for compelling reasons, there are growing efforts to put more advanced atomic clocks into space. Prominent examples are the PHARAO cold-cesium atomic clock that is part of the European ACES mission[1] scheduled to fly on the International Space Station about 2016, the compact Hg^+ ion standard of JPL designed for space applications, and other promising systems under development for the future (the ESA Space Optical Clock,[2] the DARPA Slow Beam Optical Clock[3]). Advanced laser systems have already vastly improved the performance of atomic clocks and optical frequency synthesis and division. Lasers can do the same for time transfer to space.

2. System concept

The basic system concept is to take advantage of the very high stability and high accuracy available from cold-atom atomic clocks, as well as the precise timing and optical frequency division provided by femotosecond lasers, and

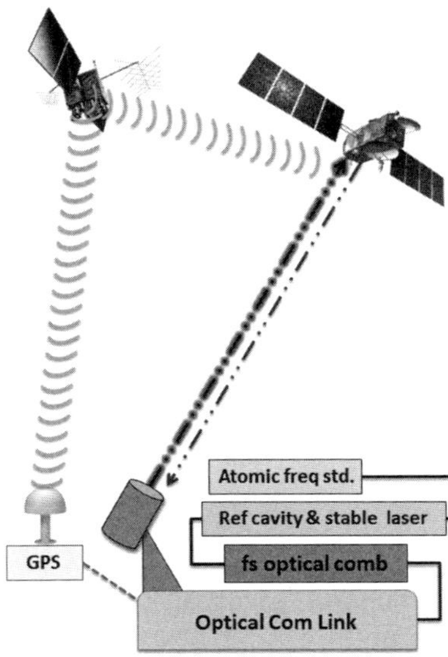

Fig. 1. Diagram of a high performance optical time/frequency transfer link between the ground and an STR satellite. Also shown is a GPS satellite in a higher orbit. GPS signals detected on the STR satellite provide information about the satellite position and orbit, while the STR measures the GPS clocks relative to the ground clock with high precision and from above the troposphere and ionosphere.

then leverage instrumentation and technologies developed for other applications in space to achieve a robust high precision time transfer system. The other technologies required include: laser communication (LaserCom) links between ground and space,[4] ultrafast electronics and photodetectors, and robust fiber-optic telecom technologies. LaserCom links to the satellite can transport the high performance atomic time from state-of-the-art ground clocks to the satellite. Integrating these systems and technologies would allow us to create a very high performance inertial Space-Time Reference (STR) satellite with unprecedented performance, that would serve as a stable and accurate reference for coordinate position and time. A block diagram of the major system components is shown in Fig. 1.

The precise time and frequency reference on the ground comes from a high performance atomic frequency standard (clock) as found at a national standards laboratory. A stable CW laser (or microwave source) is locked to the narrow atomic transition; the stable CW laser, in turn, stabilizes the self-referenced fs-optical frequency comb. The stabilized fs-optical comb is a train of optical pulses that are phase-coherent with the optical comb and the stable CW laser. The train of ultra-short optical pulses serves as the heart of the optical time transfer system. The precise timing information available from the comb and optical pulses are then transferred to a two-way LaserCom optical link to the satellite. The individual sub-systems, depicted by boxes in Fig. 1, have been developed in national standards laboratories and universities and have demonstrated outstanding performance for different specific applications. Many critical scientific questions and technical challenges must be addressed, including: optimal encoding of precise timing and time (epoch) information on the optical pulses, pointing and satellite tracking, signal acquisition, mitigation of atmospheric turbulence and group delay effects and dispersion, determination of time (epoch) relative to GNSS time and coordinate systems and orbit determination.

3. Goals expected and outcome

Our current studies are investigating how to optimally transfer precise atomic time (epoch), time-interval, and frequency from high accuracy atomic clocks on the ground to satellites in space (or alternatively between satellites in space). We aim to achieve the following: (i) $\leq$1 ps timing precision between high accuracy atomic frequency standards on the ground and the stable clock in orbit, (ii) the atomic clock on the STR should support timing stability at the 1 ps level for time intervals between updates from ground clocks, (iii) the precision laser time-transfer and with an orbit above the atmosphere and most of the ionosphere the STR satellite will provide higher precision measurements of GNSS clocks in orbit, (iv) sub-mm orbit determination from a combination of the two-way LaserCom system and signals from a multitude of GNSS satellites, and perhaps augmented by other systems (two-way microwave link, VLBI), and (v) high precision time- and frequency-transfer, enabling comparison of ground clocks at the 10^{-17} fractional frequency. With the right combination of existing clocks, lasers and related technologies it is now feasible to improve the timing precision in space and time distribution around the world by roughly a factor of a thousand, from the few nanosecond from GNSS systems to the picosecond level. In the laboratory we have been evaluating system configurations, architectures and components that can give high performance in timing, time and frequency and that are robust and of sufficiently low power that it is reasonable to expect that these could be used in space in the near future. Our optical timing and time-transfer system is based on a self-referenced fs mode-locked Erbium fiber laser (Menlo Systems) that can be locked to either an RF or optical frequency reference. Currently we simulate the time transfer link using a 4 km round-trip optical fiber between buildings on campus. Even without fiber link stabilization we can achieve timing noise densities of much less than 1 ps on the optical pulses and timing noise densities of several fs/$\sqrt{\text{Hz}}$ on the optical carrier.

Dedicated optical links on the ground have already demonstrated timing precision at the femtosecond level over fiber and short free-space distances, so the stability of the laser systems and ground clocks would not be the limiting factor.[5] Achieving ps timing precision between ground and space would be a major advance for satellite systems and could significantly improve the performance of existing space navigation systems.

This project is motivated by trying to bring the capabilities of advanced atomic clocks to space. The widespread practical utility of precise time and space reference naturally bring potentially synergistic connections to existing and growing GNSS and applications, including proposed geodetic and geoscience missions where precise orbit determination and relative position measurements are critical for validating terrestrial reference frames, understanding earth plate motions, water/ice movement, and geodetic and earth science missions. This type of approach could serve as an exceptional platform for testing aspects of spacetime dependence of fundamental physics including General Relativity and searches for new physics beyond the Standard Model. For some fundamental physics experiments and earth science applications it could be advantageous to operate the STR satellite as a drag-free satellite using an inertial test mass reference to further reduce external perturbations to the orbit.[6]

Acknowledgments

We gratefully acknowledge support from the NASA Fundamental Physics and DARPA-QUASAR programs.

References

1. P. Delva *et al.*, arXiv:1206.6239.
2. S. Schiller *et al.*, arXiv:1206.3765.
3. Tom Loftus, SBOC, AOSense, private communication.
4. H. Hemmati, *Earth Laser Communications*, CRC Press, Florida, 2009.
5. R. Bluhm *et al.*, Phys. Rev. Lett. **88**, 090801 (2002).
6. J. Lipa, S. Buchman, S. Sharaff, and B. Byer, private communication.

EQUIVALENCE PRINCIPLE IN COSMOLOGY

SERGEI KOPEIKIN

Department of Physics and Astronomy
University of Missouri, Columbia, MO 65211, USA
E-mail: kopeikins@missouri.edu

We analyse the Einstein equivalence principle (EEP) for a Hubble observer in Friedmann-Lemaître-Robertson-Walker (FLRW) spacetime. We show that the affine structure of the light cone in the FLRW spacetime should be treated locally in terms of the optical metric $g_{\alpha\beta}$ which is not reduced to the Minkowski metric $\mathtt{f}_{\alpha\beta}$ due to the nonuniform parametrization of the local equations of light propagation with the proper time of the observer's clock. The physical consequence of this difference is that the Doppler shift of radio waves measured locally is affected by the Hubble expansion.

1. Introduction

Experimental gravity in the solar system is done under the assumption that spacetime is asymptotically flat.[1] In contrast, cosmology makes use of the FLRW metric[2]

$$ds^2 = -dt^2 + a^2\delta_{ij}dy^i dy^j \ , \tag{1}$$

where t is the cosmological time, y^i are global coordinates, Latin indices $i, j, k, ..$ take values $1, 2, 3$, and the scale factor $a = a(t)$ is found from Einstein's equations. We use a geometric system of units in which $G = c = 1$. The metric (1) can be reduced to a conformally flat metric by introducing a conformal time $\eta = \eta(t)$ defined by $dt/a(t) = d\eta$. It brings Eq. (1) to

$$ds^2 = a^2\mathtt{f}_{\alpha\beta}dy^\alpha dy^\beta \ , \tag{2}$$

where $y^\alpha = (y^0, y^i) = (\eta, y^i)$ and $\mathtt{f}_{\alpha\beta} = \mathrm{diag}(-1, 1, 1, 1)$ is the Minkowski metric. The cosmological time t is physical and can be measured by the Hubble observers with the help of clocks while the conformal time η is merely a convenient coordinate parameter.[2] Typically, the cosmological metric (2) is applied to describe the properties of spacetime on cosmological distances. We shall assume that metric (2) describes the background geometry of spacetime on all scales.

2. Einstein's equivalence principle

The Einstein equivalence principle (EEP) states that at each point on a manifold with an arbitrary gravitational field, it is possible to chose local inertial coordinates such that, within a sufficiently small region of the point in question, the laws of Nature take the same form as in nonaccelerated cartesian coordinates.[3] According to EEP each Hubble observer carries out local inertial coordinates (LIC), $x^\alpha = (x^0, x^i) = (\tau, x^i)$ such that the physical metric $g_{\alpha\beta} = a^2(\eta)\mathtt{f}_{\alpha\beta}$, given by Eq. (2), is reduced to the Minkowski metric, $\mathtt{f}_{\alpha\beta}$, and the affine connection is nil on the worldline of the observer. EEP also asserts that free fall of electrically neutral test particles and photons is geodesic and their equations of motion in LIC are given locally by the same equation

$$\frac{d^2x^\alpha}{d\tau^2} = 0 \;, \tag{3}$$

where τ is the coordinate time of the LIC, with all tidal terms neglected. EEP does not hold in conformal spacetime as shown in Eq. (6) below.

3. Local inertial coordinates

Let us construct LIC in the vicinity of the worldline of a Hubble observer who is located at the origin of LIC, $x^i = 0$. The observer carries out an ideal clock that measures the parameter on the observer's worldline which is identified with the cosmic time t in Eq. (1). EEP suggests that the physical spacetime interval (2) written down in LIC, reads

$$ds^2 = \mathtt{f}_{\alpha\beta}dx^\alpha dx^\beta \;, \tag{4}$$

where we neglected all tidal terms. In the linearised Hubble approximation the metric (4) can be obtained from Eq. (2) with the help of the special conformal transformation

$$y^\alpha = \Omega^{-1}(x)\left(x^\alpha - b^\alpha x^2\right) \;, \tag{5}$$

where $b^\alpha = H/2u^\alpha$, $u^\alpha = (1, 0, 0, 0)$ is four-velocity of observer in local frame, $\Omega(x) = 1 - b_\alpha x^\alpha + b^2x^2$, $b^2 = \mathtt{f}_{\alpha\beta}b^\alpha b^\beta$, and $H = \dot{a}/a$ is the Hubble constant. All operations of rising and lowering indices are completed with the Minkowski metric. It can be checked that with sufficient accuracy $\Omega = a(\tau) + O(H^2)$. The reader should notice that transformation (5) yields $a(\tau)\eta = x^0 + H(x^0)^2/2$ on the worldline of the observer, defined by $x^i = 0$. However, $a(\tau)\eta = x^0$ for light geodesics, because $x^2 = 0$ on the light cone. Comparison of these two equations reveals that when $x^0 = \tau$ on a timelike geodesic, the parameter $x^0 = \tau + H\tau^2/2$ on a light geodesic, and it does not coincide with the proper time τ of the observer.

4. Optical metric

The above consideration suggests that the equation of light geodesic parametrized with the proper time τ of the observer reads

$$\frac{d^2x^\alpha}{d\tau^2} = H\left(\frac{dx^\alpha}{d\tau} - u^\alpha\right) . \tag{6}$$

The Christoffel symbols entering this equation are not all nil. They can be deduced from the optical metric

$$ds^2 = -a^2(\tau)d\tau^2 + \delta_{ij}dx^i dx^j . \tag{7}$$

This metric is degenerated in the sense that it is applied only to the events located on the light cone hypersurface connected by the condition $ds = 0$. Solution of this equation for a radial propagation of light from/to the origin of the coordinates is

$$r = r_0 \pm \tau + \frac{H}{2}\tau^2 , \tag{8}$$

where the $\pm$ signs correspond to outgoing/incoming light rays, $\tau \geq 0$, and r_0 corresponds to the position of the light-ray particle at the instant of time $\tau = 0$. The coordinate speed of light for the outgoing ray, $v = dr/d\tau = 1 + H\tau$, exceeds the fundamental value $c = 1$ for $\tau > 0$. There is no violation of special relativity here because this effect is nonlocal. The local value of the speed of light measured at time of emission, $\tau = 0$, is equal to $c = 1$.

5. Doppler effect

A monochromatic electromagnetic wave propagates on a light cone hypersurface of a constant phase $\varphi = \varphi(x^\alpha)$. The wave vector is $k_\alpha = \partial_\alpha\varphi$, and frequency of the wave measured by an observer moving with 4-velocity, u^α, is[1] $\omega = -k_\alpha u^\alpha$. We denote the point of emission P_1, the point of observation P_2, and the emitted and observed frequencies as ω_1 and ω_2, respectively. Their ratio $\omega_2/\omega_1 = (k_\alpha u^\alpha)_{P_2} / (k_\alpha u^\alpha)_{P_1}$ quantifies the Doppler effect. Locally, the four-velocity of static observers is $u^\alpha = (1, 0, 0, 0)$. Hence,

$$\frac{\omega_2}{\omega_1} = \frac{k_0(\tau_2)}{k_0(\tau_1)} , \tag{9}$$

where τ_1 and τ_2 are the instants of time of emission and observation of light respectively. The parallel transport of k_0 along the light geodesic is computed with the optical metric (7). The equation of the parallel transport of k_α in the local coordinates is a lightlike geodesic that reads

$$\frac{dk_\alpha}{d\lambda} = -\frac{1}{2}\frac{\partial g^{\mu\nu}}{\partial x^\alpha} k_\mu k_\nu \,, \tag{10}$$

where λ is the affine parameter along the light ray, and the wave vector of light is $k^\alpha = dx^\alpha/d\lambda$. The solution of this equation is $k_0(\tau)/a(\tau) =$ const. It yields the Doppler shift of frequency of electromagnetic wave in the expanding Universe as measured in the local inertial frame,

$$\frac{\omega_2}{\omega_1} = \frac{a(\tau_2)}{a(\tau_1)} = 1 + H(\tau_2 - \tau_1) \,. \tag{11}$$

This equation tells us that the cosmological Doppler shift in the local coordinates is *blue* ($\omega_2 > \omega_1$) because $\tau_2 > \tau_1$ and $a(\tau_2) > a(\tau_1)$ due to the Hubble expansion. This may be the reason for the anomalous Doppler shift measured in Pioneer spacecraft mission.[4]

It is tempting to apply the optical metric formalism to microwave cavity resonator. One might expect the same Doppler drift of the resonator's frequency as given in Eq. (11). At the same time, the frequency of an atomic clock is not affected by the Hubble expansion.[4] This type of experiment has been conducted by Storz *et al.*[5] They did not find any relative drift between the two frequencies. The reason is that microwave in the cavity is a standing wave that is locked to the boundary of the cavity. As the size of the cavity is not affected by the Hubble expansion locally,[4] the frequency of the standing wave remains constant.

Acknowledgments

I thank A. Kostelecký for an invitation to give a talk at CPT'13, and the Physics Department at Indiana University, Bloomington for hospitality. I am grateful to S. Schiller for useful discussions and references.

References

1. S. Kopeikin, M. Efroimsky and G. Kaplan, *Relativistic Celestial Mechanics of the Solar System*, Wiley-VCH, Weinheim, 2011.
2. S. Weinberg, *Cosmology*, Oxford University Press, Oxford, 2008.
3. C.M. Will, *Theory and Experiment in Gravitational Physics*, Cambridge University Press, Cambridge, 1993.
4. S. Kopeikin, Phys. Rev. D **86**, 064004 (2012).
5. R. Storz, C. Braxmaier, K. Jäck, O. Pradl, and S. Schiller, Opt. Lett. **23**, 1031 (1998).

UNITARITY, GHOSTS, AND NONMINIMAL TERMS IN LORENTZ-VIOLATING QED

CARLOS M. REYES

Departamento de Ciencias Básicas, Universidad del Bío Bío
Chillan, Casilla 447, Chile
E-mail: creyes@ubiobio.cl

The unitarity of a Lorentz-invariance violating QED model with higher-order Myers and Pospelov photons coupled to standard fermions is studied. As expected, we find ghost states associated to the higher-order terms that may lead to the loss of unitarity. An explicit calculation to check perturbative unitarity in the process of electron-positron scattering is performed and it is found to be possible to be preserved.

1. Introduction

In recent years, higher-order operators have become the object of intense study in the search for possible effects of Lorentz invariance violation.[1–3] These Planck-mass suppressed higher-order operators allow us to describe new physics beyond those obtainable from renormalizable operators, that is, operators with mass dimension four or less.[4,5] For example, the higher-order effective theory may involve additional degrees of freedom associated to ultra-high energies which do not converge perturbatively to the normal ones when taking the limit of the dimensionless parameters in the effective terms to zero. Lee and Wick studied these exotic modes in the context of negative metric theories[6] and in spite of the ghost states that appear, they showed that unitarity can be preserved by demanding all stable particles to be positive norm states.[6,7] Here we check perturbative unitarity in a QED consisting of higher-order Myers-Pospelov photons[1] and standard fermions.

2. The Myers and Pospelov model

The Myers-Pospelov Lagrange density for photons is given by

$$\mathcal{L} = -\frac{1}{4}F^{\mu\nu}F_{\mu\nu} - \frac{\xi}{2M_P}n_\mu \epsilon^{\mu\nu\lambda\sigma}A_\nu (n\cdot\partial)^2 F_{\lambda\sigma}, \tag{1}$$

where n is a four-vector defining a preferred reference frame, M_P is the Planck mass and ξ is a dimensionless parameter.

We can always select a real basis of four-vectors $e_\mu^{(a)}$ to be orthonormal and to satisfy the properties described in Ref. 8. In analogy with the left and right handed polarizations of usual electrodynamics we can switch to a basis of complex four-vectors ε_μ^λ and define the orthogonal projectors $P_{\mu\nu}^\lambda$ as

$$\varepsilon_\mu^\lambda = \frac{1}{\sqrt{2}}(e_\mu^{(1)} + i\lambda\, e_\mu^{(2)}), \qquad P_{\mu\nu}^\lambda = -\varepsilon_\mu^\lambda \varepsilon_\nu^{*\lambda}, \tag{2}$$

where $\lambda = \pm$. To derive the dispersion relation we can expand the gauge field in term of this complex basis and replace it in the equations of motion to arrive at

$$(k^2)^2 - 4g^2(n\cdot k)^4\left((n\cdot k)^2 - n^2k^2\right) = 0, \tag{3}$$

in agreement with the work in Ref. 9.

3. Unitarity

Here we check perturbative unitarity in the process of electron-positron scattering $e^+e^- \to e^+e^-$. For this we use the optical theorem which relates the imaginary part of the forward scattering amplitude $\mathcal{M}_{ii}$ with the total cross section as

$$2\,\mathrm{Im}\mathcal{M}_{ii} = \sum_m \int d\Pi_m \left|\mathcal{M}_{im}\right|^2, \tag{4}$$

where the sum runs over all intermediate physical states.

Considering the QED extension, the amplitudes that contribute to the S-matrix are the direct amplitude

$$\mathcal{M}^{\mathrm{dir}} = (-ie)^2 \int d^4k\, \delta^4(p_1 - p_1' - k)\widehat{U}^\mu U^\nu G_{\mu\nu}(k), \tag{5}$$

and the exchange amplitude

$$\mathcal{M}^{\mathrm{ex}} = (-ie)^2 \int d^4k\, \delta^4(p_1 + p_2 - k)\widehat{V}^\mu V^\nu G_{\mu\nu}(k), \tag{6}$$

where $\widehat{U}^\mu = N_{p_2}N_{p_2'}\bar{v}(p_2)\gamma^\mu v(p_2')$, $U^\nu = N_{p_1'}N_{p_1}\bar{u}(p_1')\gamma^\nu u(p_1)$ and $\widehat{V}^\mu = N_{p_1'}N_{p_2'}\bar{u}(p_1')\gamma^\mu v(p_2')$, $V^\nu = N_{p_2}N_{p_1}\bar{v}(p_2)\gamma^\nu u(p_1)$ and where $N_p = \sqrt{m/E_p}$ is the usual fermionic normalization constant.

Let us start with the left hand side of the unitarity condition (4). A similar calculation has been given in the minimal sector of the Standard-Model Extension, see Ref. 10. To simplify we will consider the lightlike case where we have a ghost state with frequencies ω_0^λ and two photons with frequencies $\omega_{1,2}^\lambda$, see Ref. 8, and the propagator

$$G_{\mu\nu}(k) = -\sum_\lambda \frac{P^\lambda_{\mu\nu}(k)}{k^2 + 2g\lambda(n\cdot k)^3 + i\epsilon}, \tag{7}$$

where we have included the $i\epsilon$ prescription.

We are interested in the imaginary part of the forward scattering amplitude, therefore let us set $p_1' \to p_1$ and $p_2' \to p_2$. Moreover, we can see that the direct process does not contribute since the virtual photon can never be on shell for non-zero external momenta, hence $\mathrm{Im}[\mathcal{M}^{\mathrm{dir}}] = 0$. Let us find the contribution of the exchange process and substitute the propagator (7) in Eq. (6),

$$\mathcal{M}^{\mathrm{ex}} = e^2 \int \frac{d^4\vec{k}}{(2\pi)^4}\delta^4(p_1 + p_2 - k)V^\mu V^{*\nu} \sum_\lambda \frac{P^\lambda_{\mu\nu}(k)}{k^2 + 2g\lambda(n\cdot k)^3 + i\epsilon}. \tag{8}$$

Because only the poles can contribute to the imaginary part and due to energy conservation encoded in $\delta^4(p_1 + p_2 - k)$, we have that only the positive poles of the virtual photon have a chance to contribute. We can discard the ghost contribution since its energy $|\omega_0^\lambda| \sim 1/2g$ lies beyond the region of validity of the effective theory. That is, the external fermions will always satisfy the condition $p_{01} + p_{02} < |\omega_0^\lambda|$. Hence, we have

$$\begin{aligned}
&2\mathrm{Im}[\mathcal{M}^{\mathrm{ex}}] \\
&= -e^2 \int dk^0 \int \frac{d^3\vec{k}}{(2\pi)^3}\delta^4(p_1 + p_2 - k)V^\mu V^{*\nu} \sum_\lambda \frac{P^\lambda_{\mu\nu}\delta(k_0 - \omega_1^\lambda)}{2g\lambda(k_0 - \omega_0^\lambda)(k_0 - \omega_2^\lambda)}, \\
&= e^2 \int \frac{d^3k}{(2\pi)^3}\delta^4(p_1 + p_2 - k)V^\mu V^{*\nu} \sum_\lambda \frac{\varepsilon^\lambda_\mu \varepsilon^{*\lambda}_\nu}{2g\lambda(\omega_1^\lambda - \omega_0^\lambda)(\omega_1^\lambda - \omega_2^\lambda)}, \\
&= \int \frac{d^3k}{(2\pi)^3}\delta^2(p_1 + p_2 - k)\sum_\lambda |\mathcal{M}_\lambda|^2,
\end{aligned} \tag{9}$$

where we have used the notation $\mathcal{M}_\lambda = (-ie)N_{k,\lambda}V^\mu\varepsilon^\lambda_\mu$ for the physical process $\mathcal{M}_{\mathrm{phys}}(e^+e^- \to \gamma)$ and we have introduced the normalization constant $N_{k,\lambda} = 1/\sqrt{2g\lambda(\omega_1^\lambda - \omega_0^\lambda)(\omega_1^\lambda - \omega_2^\lambda)}$. Finally we have

$$2\mathrm{Im}[\mathcal{M}] = \int \frac{d^3k}{(2\pi)^3}\delta^2(p_1 + p_2 - k)\,|\mathcal{M}_{\mathrm{phys}}|^2, \tag{10}$$

and therefore the unitarity condition is satisfied in this scattering process.

4. Conclusions

With an explicit calculation we have verified that the unitarity condition in the process of electron-positron scattering at tree level order is satisfied.

A next step is to verify unitarity to order e^2, which will require analyzing more diagrams. Some of them contain loops where the ghosts can appear off-shell, thus introducing an extra difficulty. Checking the unitarity condition to this order will give us robust support to make physical predictions in the theory.

Acknowledgments

This work was supported by the Dirección de Investigación de la Universidad del Bío-Bío grant number 123809 3/R and FAPEI.

References

1. R.C. Myers and M. Pospelov, Phys. Rev. Lett. **90** (2003) 211601; P.A. Bolokhov and M. Pospelov, Phys. Rev. **D 77**, 025022 (2008).
2. V.A. Kostelecky and M. Mewes, Phys. Rev. D **80**, 015020 (2009); V.A. Kostelecky and M. Mewes, Phys. Rev. D **85**, 096005 (2012).
3. T. Mariz, Phys. Rev. D **83**, 045018 (2011); T. Mariz, J.R. Nascimento and A.Y. Petrov, Phys. Rev. D **85**, 125003 (2012); R. Casana, M.M. Ferreira, E.O. Silva, E. Passos and F.E.P.d. Santos, Phys. Rev. D **87**, 047701 (2013); C.M. Reyes, L.F. Urrutia, and J.D. Vergara, Phys. Rev. D **78**, 125011 (2008).
4. D. Colladay and V.A. Kostelecky, Phys. Rev. D **55**, 6760 (1997); D. Colladay and V.A. Kostelecky, Phys. Rev. D **58**, 116002 (1998).
5. *Data Tables for Lorentz and CPT Violation,* V.A. Kostelecký and N. Russell, 2013 edition, arXiv:0801.0287v6.
6. T.D. Lee and G.C. Wick, Nucl. Phys. B **9**, 209 (1969); T.D. Lee, G.C. Wick, Phys. Rev. D **2**, 1033-1048 (1970).
7. R.E. Cutkosky, P.V. Landshoff, D.I. Olive, and J.C. Polkinghorne, Nucl. Phys. B **12**, 281-300 (1969).
8. C.M. Reyes, Phys. Rev. D **87**, 125028 (2013).
9. C.M. Reyes, Phys. Rev. D **82**, 125036 (2010).
10. M. Schreck, Phys. Rev. D **86**, 065038 (2012); F.R. Klinkhamer and M. Schreck, Nucl. Phys. B **848**, 90 (2011).

LABORATORY SEARCH FOR SPIN-DEPENDENT SHORT-RANGE FORCE FROM AXION-LIKE PARTICLES USING OPTICALLY POLARIZED ^{3}HE GAS

P.-H. CHU

Triangle Universities Nuclear Laboratory and Department of Physics
Duke University, Durham, NC 27708, USA
E-mail: pc102@duke.edu

The possible existence of short-range forces between unpolarized and polarized spin-$\frac{1}{2}$ particles has attracted the attention of physicists for decades. These forces are predicted in various theories and provide a possible new source for parity (P) and time reversal (T) symmetry violation. We use an ensemble of polarized ^{3}He gas in a cell with a 250 μm thickness glass window to search for a force from pseudoscalar boson exchange over sub-millimeter ranges. This interaction would produce a nuclear magnetic resonance frequency shift as an unpolarized mass is moved near and far from the polarized ensemble. Recently, we reported a new upper bound with a factor of 10-30 improvement on the product of the scalar coupling to the fermions in the unpolarized mass and the pseudoscalar coupling of the polarized neutron in the ^{3}He nucleus for force ranges from 10^{-4} to 10^{-2} m, which corresponds to a mass range of 2×10^{-3} to 2×10^{-5} eV for the pseudoscalar boson. This represents the most sensitive published search that sets a direct limit in this important axion window. Currently, a new experiment is being designed and constructed to improve the sensitivity by another factor of 10-100 in the similar region.

1. Introduction

The potential existence of spin-dependent short-range forces (SDSRF) related to unpolarized and polarized particles can provide a possible new source for P and T violation. Moody and Wilczek[1] proposed that a new macroscopic force with an interaction range from cm to μm can couple to scalar and pseudoscalar vertices through the exchange of spin-0 bosons. The scalar coupling is spin-independent and only depends on the fermion density. The pseudoscalar coupling is entirely spin-dependent. The resulting SDSRF has a Yukawa-type interaction potential from one boson exchange of the form

$$V(r) = \frac{g_s g_p \hbar^2}{8\pi m_p}(\hat{\sigma}\cdot\hat{r})(\frac{1}{r\lambda}+\frac{1}{r^2})\exp(-r/\lambda), \tag{1}$$

where $\hat{r}$ is the unit vector from the unpolarized particle to the polarized particle, $\hat{\sigma}$ is the spin of the polarized particle, m_p is the polarized particle mass, $g_s g_p$ is the product of couplings at the scalar and pseudoscalar vertices, and λ is the force range. Such forces may be induced by pseudoscalar bosons like the axion,[2] axion-like-particles[3] or a very light spin-1 boson,[4] which are candidates for cold dark matter.[5]

2. Experiment

A potential of the form $\hat{\sigma}\cdot\hat{r}$ can introduce a shift in the precession frequency of a polarized particle in the presence of an unpolarized particle, similar to that of a magnetic dipole moment in an external magnetic field $\propto \vec{\mu}\cdot\vec{B}$. The first experiment,[6] which used the idea with polarized ^{3}He, achieved a precession frequency sensitivity of 5×10^{-3} Hz and restricted the coupling strength close to the upper limit at the time[7,8] in the force range from 0.01 to 1 cm. A high-pressure (7 amg) ^{3}He cell is applied having an optical pumping chamber and a target chamber connected by a glass tube. The ^{3}He gas is polarized using spin-exchange optical pumping[9] in the spherical pumping chamber of radius 4.3 cm. The polarized ^{3}He atoms diffuse into the lower 40-cm long cylindrical chamber, which possesses two hemispherical glass windows at both ends with a thickness of 250 μm. The thin windows are used to minimize the distance between the unpolarized mass and polarized ^{3}He. An unpolarized mass block is placed at the end of a cylindrical cell containing polarized ^{3}He gas along the axis of the cylinder. The mass block can be moved close to (mass-in) and away from (mass-out) the ^{3}He cell. No mass is placed at the other end of the cylindrical cell. The precession frequency at this end is treated as a magnetometer. After the precession frequency is calibrated by using the magnetometer, the effect of the short-range force can be determined by comparing the precession frequencies between the mass-in and the mass-out conditions. Systematic uncertainties from the magnetic properties of the mass and polarized ^{3}He nuclei are reduced by flipping both the magnetic field and polarization directions.

In the following experiment,[10] a number of improvements were applied. The magnetic field uniformity is improved by tuning the position of the window of the ^{3}He cell. Two pickup coils are put together at the same

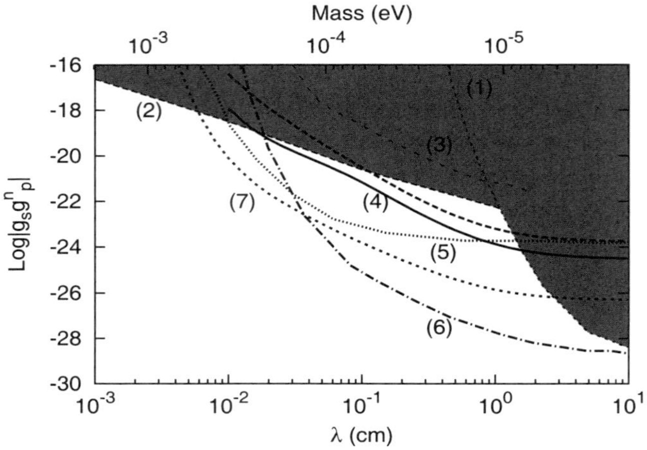

Fig. 1. Constraints on the coupling strength $g_s g_p^n$ as a function of the force range λ and the equivalent mass of the axion-like particles. The dark gray area is the region excluded by previous works. (1) is from A. N. Youdin *et al.*[7] (2) is from A. Petukhov *et al.*[8] (3) is the constraint of the phase I.[6] The dashed (solid) curve of (4) is the constraint of the salt water (ceramic) sample within one standard deviation in the phase II.[10] (5) is from M. Bulatowic *et al.*[12] (6) is from K. Tullney *et al.*[13] (7) is the projection of the phase III with assumed sensitivity 10^{-6} Hz.

end of the cell in order to have a better correlation. One pickup coil is mounted below the window to measure the precession frequency shift of the polarized ^{3}He nuclei due to SDSRF from the unpolarized mass. Another pickup coil is positioned to be insensitive to SDSRF; its signal is used to monitor the holding-field drift. A 24-kHz rf pulse is applied to tip the spins by a small angle with negligible polarization loss. The precessing polarized ^{3}He nuclei induce electromotive forces in the pickup coils which are digitized and recorded. Precession frequency is determined by applying a Fourier transform to a signal $s(t)$ in the time domain and obtaining the signal in the frequency domain $S(f)$. The Fourier transform is numerically calculated using Richardson extrapolation. The reference frequency f is then varied with a 10^{-6} Hz step to locate the maximum of $S(f)$, which is the precession frequency.[11] Two samples are used as the unpolarized masses: a Macor ceramic mass block, and a liquid mixture of 1.02% $MnCl_2$ in pure water. These samples are chosen for their different nucleon densities, low magnetic impurities and magnetic susceptibilities, and minimal influence on the nuclear magnetic resonance measurement procedure. The paramagnetic salt of $MnCl_2$ is added in order to compensate for the diamagnetism of the water. The magnetic susceptibility of this mixture is measured to be less than 5% of that of pure water. As shown in Fig. 1, this experiment reached the sensitivity of the precession frequency of 2×10^{-5} Hz and constrained the coupling strength to a new upper limit at the time in the force range from 0.01 to 1 cm.

3. Future plans

Two groups[12,13] have reported new upper limits since our result[10] was published. Currently, we explore new techniques searching for SDSRF in the region that cannot be detected by other methods. Several improvements will be applied, including a more uniform magnetic field generated by a solenoid with μ-metal shielding, a shorter cell of length 8 cm with thinner windows (150 μm), and denser unpolarized masses, like BGO, copper, etc. Systematic uncertainties will be suppressed by placing two masses on both ends of the cell. In Fig. 1, a projected sensitivity based on the new proposed method shows that a significant improvement compared to the current result[10] is achievable.

Acknowledgments

We acknowledge support from Duke University, the Indiana University Center for Spacetime Symmetries, the IU STARS program, the Department of Energy under grant number DE-FG02-03ER41231, and the National Science Foundation under grant number PHY-1068712.

References

1. J.E. Moody and Frank Wilczek, Phys. Rev. D **30**, 130 (1984).
2. R.D. Peccei and H.R. Quinn, Phys. Rev. Lett. **38**, 1440 (1977).
3. J. Jaeckel and A. Ringwald, Annu. Rev. Nucl. Part. Sci. **60**, 405 (2010).
4. P. Fayet, Class. Quant. Grav. **13**, A19 (1996).
5. E.W. Kolb and M.S. Turner, *The Early Universe,* Addison-Wesley, Redwood, CA, 1990.
6. W. Zheng *et al.*, Phys. Rev. D **85**, 031505 (2012).
7. A.N. Youdin *et al.*, Phys. Rev. Lett. **77**, 2170 (1996).
8. A. Petukhov *et al.*, Phys. Rev. Lett. **105**, 3 (2010).
9. T.G. Walker and W. Happer, Rev. Mod. Phy. **69**, 629 (1997).
10. P.-H. Chu *et al.*, Phys. Rev. D **87**, 011105 (2013).
11. H. Yan *et al.*, arXiv:1305.3676.
12. M. Bulatowicz *et al.*, arXiv:1301.5224.
13. K. Tullney *et al.*, arXiv:1303.6612.

LOCAL LORENTZ-SYMMETRY BREAKING AND GRAVITY

Q.G. BAILEY

Physics Department, Embry-Riddle Aeronautical University
3700 Willow Creek Road, Prescott, AZ 86301, USA
E-mail: baileyq@erau.edu

The lagrangian-based Standard-Model Extension framework offers a broad description of possible gravitational effects from local Lorentz violation. In this talk, I review the status of the theoretical and phenomenological work in this area. The extension of previous results in linearized gravity to the nonlinear regime is discussed.

1. Introduction

Although the Standard Model of particle physics and General Relativity (GR) provide a successful description of all observable physics, it is widely believed that a unified description exists that contains both theories as limiting cases. This theory remains largely unknown so far and direct experimental clues are sparse.

Signals coming from an underlying theory that are potentially detectable in sensitive experiments include minuscule violations of local Lorentz symmetry.[1] The Standard-Model Extension (SME) is a comprehensive effective field theory framework that describes observable signals of Lorentz violation.[2] Much theoretical and experimental work on the SME has involved the Minkowski-spacetime limit.[3] Lorentz violation in the gravity sector has been explored more recently[4,5] and analyses have been performed.[6] In this talk, we discuss the SME framework incorporating gravity, with emphasis on spontaneous Lorentz-symmetry breaking.

2. SME theory

With zero torsion, the SME Lorentz-violating couplings linear in the curvature tensor are given by

$$\mathcal{L} = \tfrac{1}{2\kappa}\left(-uR + s_{\mu\nu}(R_T)^{\mu\nu} + t_{\mu\nu\kappa\lambda}C^{\mu\nu\kappa\lambda}\right) + \mathcal{L}'. \qquad (1)$$

Twenty independent 'coefficient fields' are contained in u, $s_{\mu\nu}$, and $t_{\mu\nu\kappa\lambda}$, which couple to the Ricci scalar, traceless Ricci tensor, and the Weyl curvature tensor, respectively. Dynamical terms for these fields are contained in $\mathcal{L}'$. Under the assumption of spontaneous Lorentz-symmetry breaking, the coefficient fields acquire vacuum expectation values $\overline{u}$, $\overline{s}_{\mu\nu}$, and $\overline{t}_{\mu\nu\kappa\lambda}$.

Previous work focused on the limit of linearized gravity assuming the metric can be expanded around a Minkowski background, $g_{\mu\nu} = \eta_{\mu\nu} + h_{\mu\nu}$, where $h_{\mu\nu}$ are the metric fluctuations. It is possible to make several assumptions on the dynamics of the coefficient fields in order to extract an effective linearized equation for $h_{\mu\nu}$ that depends only on the Minkowski metric $\eta_{\mu\nu}$ and the vacuum values of the coefficients $\overline{u}$, $\overline{s}_{\mu\nu}$, and $\overline{t}_{\mu\nu\kappa\lambda}$.[4,7] The linearized ($L$) field equations can then be written in the compact form

$$(G_L)_{\mu\nu} = \kappa(T_M)_{\mu\nu} - \overline{s}^{\alpha\beta}(\mathcal{G}_L)_{\mu\alpha\beta\nu}. \tag{2}$$

The coefficient $\overline{u}$ can be removed from the equations at this level as an unobservable scaling, while the contribution from the $\overline{t}_{\mu\nu\kappa\lambda}$ coefficients vanishes by a tensor identity. We use in Eq. (2) the linearized double dual of the Einstein tensor $(\mathcal{G}_L)_{\mu\alpha\beta\nu}$.

The field equation (2) can be shown to satisfy the conservation laws associated with local Lorentz symmetry and diffeomorphism symmetry, as expected for spontaneous symmetry breaking. At this level, the equations must satisfy the Bianchi identities $\partial^\mu(G_L)_{\mu\nu} = 0$ and be symmetric in the indices. The phenomenology associated with the solutions to Eq. (2), including the post-newtonian expansion to $PNO(3)$, is discussed elsewhere,[4] and experimental limits have been placed on many of the $\overline{s}_{\mu\nu}$ coefficients.[6]

It would be of interest to determine the metric component g_{00} to $PNO(4)$, which conventionally contains the first terms that exhibit the nonlinearity of GR. To date, the analysis producing Eq. (2) has not been extended to second order in $h_{\mu\nu}$. In fact, a completely 'decoupled' equation to second order, involving only the vacuum values of the coefficients, and not also their dynamical fluctuations, may not be obtainable without solving the complete system of equations for the metric fluctuations and the coefficient fields. The process of obtaining general results for the metric $h_{\mu\nu}$ at higher order is therefore likely to be highly model dependent and large in scope.

3. Models of spontaneous Lorentz-symmetry breaking

Alternatively, one may study specific models of spontaneous Lorentz-symmetry breaking and try to generalize the results. Several types of models exist in the literature that have a connection to the formalism described above. One class of models involves an antisymmetric tensor field.[8] When nonminimal couplings to gravity are included, these models can produce effective $\overline{s}_{\mu\nu}$ coefficients and match the form Eq. (2).

The so-called 'bumblebee models' involve a dynamical vector field B_μ that acquires a vacuum expectation value b_μ via a potential term in the lagrangian.[9] Consider the bumblebee model lagrangian

$$\mathcal{L}_B = \frac{1}{2\kappa}(R + \xi B^\mu B^\nu G_{\mu\nu}) - \frac{1}{4}B^{\mu\nu}B_{\mu\nu} - V + \mathcal{L}_{\rm M}, \tag{3}$$

where the field strength is $B_{\mu\nu} = \partial_\mu B_\nu - \partial_\nu B_\mu$. The gravitational field equations can be written in the form

$$G_{\mu\nu} = \kappa(T_M)_{\mu\nu} + \kappa(B_\mu{}^\alpha B_{\nu\alpha} - g_{\mu\nu}B^{\alpha\beta}B_{\alpha\beta} - Vg_{\mu\nu} + 2V'B_\mu B_\nu) + \kappa\xi(T_\xi)_{\mu\nu}. \tag{4}$$

The potential energy V is a function of the scalar $X = B^\mu B_\mu - x$, where x is a constant real number, thus $V = V(X)$ and $V' = dV/dX$. The terms proportional to ξ are those generated by the nonminimal couplings.

The field equations for the vector field are given by

$$D^\mu B_{\mu\nu} = 2V'B_\nu - \frac{\xi}{\kappa}B^\mu G_{\mu\nu}. \tag{5}$$

The covariant divergence of the left-hand side vanishes identically ($D^\mu D^\nu B_{\mu\nu} = 0$), which implies a constraint on the right-hand side:

$$D^\nu \left(2V'B_\nu - \frac{\xi}{\kappa}B^\mu G_{\mu\nu}\right) = 0. \tag{6}$$

Upon expanding around the vacuum values for the metric and vector field, $B_\mu = b_\mu + \mathcal{E}_\mu$, at linear order in $h_{\mu\nu}$ and $\mathcal{E}_\mu$, the constraint Eq. (6) becomes $b^\mu\partial_\mu(2V') = 0$, for which the obvious boundary condition choice is $(V')_L = 0$ for both the spacelike and timelike vacuum values. The linearized limit of this model can be shown to match the form of Eq. (2).[4,7]

Solving the Eqs. (4) and (5) beyond the linearized limit introduces a number of complexities. In particular, the constraint Eq. (6) becomes $b^\mu\partial_\mu(2V') = (\xi/\kappa)G^{\mu\nu}D_\mu B_\nu$, for which the right-hand side has terms that do not vanish at second order in the fluctuations. If the derivative of the potential V' were set to zero (vanishing massive mode condition) then the right-hand side would also have to be zero, which is clearly the case for vanishing nonminimal coupling $\xi = 0$. However, in order to explore the solutions for nonzero ξ, it appears that we cannot consistently make the choice $V' = 0$.

In the post-newtonian limit, Eq. (6) becomes

$$b^\mu \partial_\mu (2V') = \xi \rho b_j \partial_j U + O(\xi^2), \tag{7}$$

where ρ is the mass density and U is the newtonian potential. Absent any information about specific boundary conditions, we can construct the general solution to Eq. (7) for the spacelike b_μ case by integrating over the coordinate $z = \vec{x} \cdot \hat{b}$ along the vacuum value direction.

The massive mode combination V' contributes as an effective source on the right-hand side of the Einstein equations:

$$G_{\mu\nu} \supset \kappa 2V' b_\mu b_\nu. \tag{8}$$

This source has an intriguing distribution in space for an isolated matter source ρ. Outside the matter source where $\rho = 0$, the right-hand side of (7) vanishes and the massive mode combination must be independent of one direction (in agreement with other results[10]). On the other hand, there is a finite contribution to the integral of Eq. (7), so this remaining source function does not generally vanish as $z \to \infty$. The equation for the portion of the metric g_{00} that has V' as its source is effectively Poisson's equation. Detailed solutions including this term, the complete $PNO(4)$ metric, and the relevant phenomenology will be presented elsewhere.

References

1. R. Lehnert, Hyperfine Int. **193**, 275 (2009).
2. D. Colladay and V.A. Kostelecký, Phys. Rev. D **55**, 6760 (1997); Phys. Rev. D **58**, 116002 (1998); V.A. Kostelecký, Phys. Rev. D **69**, 105009 (2004).
3. *Data Tables for Lorentz and CPT Violation,* V.A. Kostelecký and N. Russell, 2013 edition, arXiv:0801.0287v6.
4. Q.G. Bailey and V.A. Kostelecký, Phys. Rev. D **74**, 045001 (2006).
5. V.A. Kostelecký and J.D. Tasson, Phys. Rev. Lett. **102**, 010402 (2009); Phys. Rev. D **83**, 016013 (2011).
6. J.B.R. Battat *et al.*, Phys. Rev. Lett. **99**, 241103 (2007); H. Müller *et al.*, Phys. Rev. Lett. **100**, 031101 (2008); J. Overduin *et al.*, these proceedings.
7. M.D. Seifert, Phys. Rev. D **79**, 124012 (2009).
8. B. Altschul *et al.*, Phys. Rev. D **81**, 065028 (2010).
9. V.A. Kostelecký and S. Samuel, Phys. Rev. D **40**, 1886 (1989).
10. R. Bluhm *et al.*, Phys. Rev. D **77**, 065020 (2008).

GENERATION OF AXION COUPLINGS VIA LORENTZ VIOLATION

A.F. FERRARI
Federal University of ABC (UFABC)
Santo André, SP, Brazil
E-mail: alysson.ferrari@ufabc.edu.br

Axion-like particles (ALP) are light and weakly interacting pseudoscalars, much studied for solving some technical problems in the Standard Model, and for being viable candidates for dark matter. Direct experimental searches for ALP usually try to detect the interaction with the photon. In a recent work, we pointed out the generation of this coupling as an effect of quantum corrections, originating from an underlying Lorentz violating background. This mechanism depends on the existence of a fermion field which couples to the pseudoscalar and to the photon with specific Lorentz violating interactions. We show that as a consequence of this couplings, the photon effective action contains an ALP-photon interaction that turns out to be Lorentz invariant, thus mimicking the standard coupling studied in current ALP experiments. This consideration shows that violations of spacetime symmetries, much studied as possible consequences of physics in the very high energy scales, might infiltrate in other contexts in unsuspected ways. We thus hope to point out a new connection between theories involving Lorentz violating backgrounds and experiments.

The Standard-Model Extension (SME) of Kostelecký and Colladay[1] provides a rich environment for testing the validity of relativistic symmetry in different phenomena, as can be seen from other contributions to this volume. In essence, in the SME one incorporates in the Standard Model lagrangian all possible covariant, renormalizable and gauge invariant terms that involve constant background tensors. One well-studied example is the Carroll-Field-Jackiw (CFJ) term $k_\mu \epsilon^{\mu\nu\rho\sigma} A_\nu \partial_\rho A_\sigma$.[2] Dynamical generation of the CFJ term has been shown to exist if we assume the usual minimal coupling of the photon to a charged fermion field, which itself presents a Lorentz breaking coupling of the form $b_\mu \overline{\psi} \gamma_5 \gamma^\mu \psi$. t One nontrivial feature of such a calculation is that the constant of proportionality between k_μ and

b_μ depends on the value of a Feynman integral which is finite, but regularization dependent (see, for example, Refs. 3,4 and references therein).

On the other hand, a very active area of experimental work in the last decades is the search for light pseudoscalars, or axion-like particles (ALP).[5–7] These searches typically focus on the detection of the interaction

$$G_{a\gamma\gamma} a F_{\mu\nu} \tilde{F}^{\mu\nu} \sim G_{a\gamma\gamma} a \vec{E} \cdot \vec{B} \,, \tag{1}$$

where $F_{\mu\nu}$ is the electromagnetic field strength, $\tilde{F}^{\mu\nu} = \varepsilon^{\mu\nu\alpha\beta} F_{\alpha\beta}$ its dual, and $a\,(x)$ is the pseudoscalar field. Integrating by parts Eq. (1), one obtains

$$-\frac{1}{2} G_{a\gamma\gamma} \left(\partial_\mu a\right) A_\nu \tilde{F}^{\mu\nu} \,. \tag{2}$$

Assuming then $a\,(x)$ to be a slowly varying field, such that $(\partial_\mu a) = K_\mu$, K_μ being a constant vector, one ends up essentially with a CFJ term. This is how Lorentz violation (LV) has typically been considered in models containing ALP.

In Ref. 8, we pointed out another surprising way LV can contribute to ALP phenomenology, which is the generation of the coupling (1) — which itself is Lorentz invariant — via quantum corrections from an underlying LV background. This finding is relevant in pointing out that possible violation of spacetime symmetries might infiltrate in unsuspected ways in the context where it is not typically taken into account, direct searches for ALP being the case discussed here.

The simplest lagrangian which implements this mechanism is

$$\mathcal{L} = \bar{\psi} \left[i\gamma^\mu \partial_\mu - m - \gamma^\mu (e A_\mu + g F_{\mu\nu} b^\nu) - \gamma_5 \gamma^\mu b_\mu \phi \right] \psi \,, \tag{3}$$

in which the electromagnetic field interacts minimally with a fermionic field and also nonminimally via the Lorentz breaking interaction $g F_{\mu\nu} b^\nu \overline{\psi} \gamma^\mu \psi$. Within this model, b_μ is the constant vector responsible for the Lorentz symmetry breaking.

The one-loop correction to the effective action of the gauge field can be calculated in the standard way. Explicit calculation shows that this correction is finite (yet it involves the result of an integral which is regularization dependent) and contains the term

$$S_{\text{eff}}[b, A] \supset C\, eg\, \phi \epsilon^{\rho\mu\nu\lambda} F_{\rho\mu} b_\lambda b^\kappa F_{\kappa\nu} \,. \tag{4}$$

Then, by noticing that

$$\epsilon^{\rho\mu\nu\lambda} F_{\rho\mu} b_\lambda b^\kappa F_{\kappa\nu} = 2b^2 \vec{E} \cdot \vec{B} \,, \tag{5}$$

we conclude that

$$S_{\text{eff}}[b, A] \supset 2C\, egb^2\, \phi \vec{E} \cdot \vec{B}\,. \tag{6}$$

More details of this calculation can be found in Ref. 8.

It is surprising to notice that even if the model starts with explicit Lorentz violation, the generated interaction depends only on the scalar $b^2 = b^\mu b_\mu$. That means the underlying LV background that we considered generated a *Lorentz invariant* interaction, which mimics precisely the effects of the ALP interactions studied in the Lorentz preserving context.

In summary, the connection between Lorentz violation and the phenomenology of axion-like particles can have very interesting consequences both for the searches of possible violations of standard spacetime symmetries, and for the search of the elusive light pseudoscalars that could solve many theoretical puzzles in our current understanding of the Universe. The fact that an underlying LV could also generate Lorentz invariant interactions could open new possibilities for the incorporation of LV effects in physics.

Acknowledgments

The author acknowledges financial support from the Brazilian agency CNPq.

References

1. D. Colladay and V.A. Kostelecký, Phys. Rev. D **58**, 116002 (1998).
2. S.M. Carroll, G.B. Field, and R. Jackiw, Phys. Rev. D **41**, 1231 (1990).
3. R. Jackiw and V.A. Kostelecký, Phys. Rev. Lett. **82**, 3572 (1999).
4. F.A. Brito, J.R. Nascimento, E. Passos, and A.Yu. Petrov, Phys. Lett. B **664**, 112 (2008).
5. P. Sikivie, Phys. Rev. Lett. **51**, 1415 (1983); **52** 695 (1984).
6. J. Jaeckel and A. Ringwald, Ann. Rev. Nucl. Part. Sci. **60**, 405 (2010).
7. A. Ringwald, Phys. Dark Univ. **1**, 116 (2012).
8. L.H.C. Borges, A.G. Dias, A.F. Ferrari, J.R. Nascimento, and A.Y. Petrov, arXiv:1304.5484.

FINITE ONE-LOOP RADIATIVE CORRECTIONS IN THE LORENTZ- AND CPT-VIOLATING QED EXTENSION

D.H.T. FRANCO and A.H. GOMES*

Universidade Federal de Viçosa, Viçosa, 36570-000, Minas Gerais, Brazil
**E-mail: andre.herkenhoff@ufv.br*

We report on our current progress in the study of finite one-loop radiative corrections in the minimal Lorentz- and CPT-violating QED extension.

1. Introduction

Quantum gravity effects coming from Planck scale energies may appear as small violations of fundamental laws in the limit of low energies. One possibility is the breaking of both Lorentz and CPT symmetries. A comprehensive investigation of this case is made viable, for instance, in the framework known as the Standard Model Extension,[1] the most general effective model comprising particle Lorentz and CPT violation but coordinate invariant. The minimal extension respects the Standard Model $SU(3)\otimes SU(2)\otimes U(1)$ gauge symmetry and is power-counting renormalizable. In this proceedings contribution we report our progress in the study of finite one-loop radiative corrections in the minimal Lorentz- and CPT-violating QED extension.

2. The QED extension and setup for one-loop evaluations

The Lorentz- and CPT-violating QED extension for one Dirac fermion can be described by the Lagrange density[1]

$$\begin{aligned}\mathcal{L} &= i\overline{\psi}(\gamma^\mu + \Gamma_1^\mu)D_\mu\psi - \overline{\psi}(m + M_1)\psi - \frac{1}{4}F^{\mu\nu}F_{\mu\nu} \\ &\quad -\frac{1}{4}(k_F)_{\mu\nu\rho\sigma}F^{\mu\nu}F^{\rho\sigma} + \frac{1}{2}(k_{AF})^\mu\epsilon_{\mu\nu\rho\sigma}A^\nu F^{\rho\sigma}, \qquad (1)\end{aligned}$$

where Lorentz violation (LV) is parameterized by coefficients in the form of constant background fields — see the last two terms of Eq. (1) and the definitions

$$\Gamma_1^\mu \equiv c^{\lambda\mu}\gamma_\lambda + d^{\lambda\mu}\gamma_5\gamma_\lambda + e^\mu + if^\mu\gamma_5 + \frac{1}{2}g^{\kappa\lambda\mu}\sigma_{\kappa\lambda}, \tag{2}$$

$$M_1 \equiv a^\mu\gamma_\mu + b^\mu\gamma_5\gamma_\mu + \frac{1}{2}H^{\mu\nu}\sigma_{\mu\nu}. \tag{3}$$

We also add to (1) a photon mass term $\frac{1}{2}\lambda^2 A^2$ to control infrared divergences and a gauge-fixing piece $-\frac{1}{2\alpha}(\partial_\mu A^\mu)^2$.

We follow the guidelines presented in Ref. 2 to deal with one-loop radiative corrections: evaluations are performed in the so-called concordant frames to avoid issues with the perturbative expansion; nonlinear LV is disregarded because it could lead to effects comparable to multiloop ones, which we do not consider; and LV pieces of the Lagrange density (1) are treated as new interaction vertices, effectively entering as propagator or vertex insertions, as allowed by the experimental smallness of LV.

3. Results and discussion

As a first step, we considered only the b_μ and $(k_{AF})_\mu$ coefficients. Results for the photon self-energy can be found in Ref. 1, so our goal was first the vertex correction diagram, Fig. 1(a). The electron self-energy was used so far only to check the related gauge identity (more details below), and its evaluation will be our next goal. The LV contribution with one b_μ insertion comes from two diagrams, Figs. 1(b) and 1(c), with the new Feynman rule

$$= -ib_\mu\gamma_5\gamma^\mu, \tag{4}$$

and the contribution from $(k_{AF})_\mu$ comes from the one in Fig. 1(d), where

μ  ν

$$= 2(k_{AF})^\alpha\varepsilon_{\alpha\mu\beta\nu}k^\beta. \tag{5}$$

These insertions lower the superficial degree of divergence of vertex-correction loop integrals, making them actually finite, avoiding complications related to γ_5.

Aiming for generality, we computed both cases with fermions and photon off shell, i.e., $P^2 \neq m^2 \neq p^2$ and $q^2 \neq 0$, and we made no use of the Dirac equation. Since Gordon identities are unavailable to make further simplifications — they are valid only on shell — the final result itself is not very illuminating because of its size and structure, but it is worth mentioning that because we do not use $P^2 = m^2 = p^2$ our expressions contain not only pieces symmetric under $P \leftrightarrow p$ but also pieces that are antisymmetric. On the other hand, symmetry under $P \leftrightarrow p$ is recovered when the fermions go

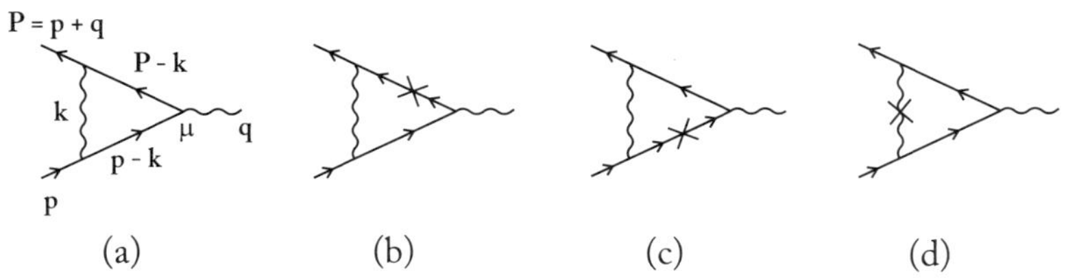

Fig. 1. (a) Conventional QED vertex correction and (b)–(d) Lorentz-violating insertions.

on shell, and the final expressions are greatly simplified. For instance, the LV correction due to $(k_{AF})_\mu$ has the structural form

$$\Gamma^\mu_{k_{AF}} \sim \Big\{ \not{k}_{AF}\gamma_5(P+p)^\mu,\ \gamma_5\sigma^{\mu\nu}(P+p)_\nu k_{AF}\cdot(P+p),\\ \gamma^\mu\gamma_5 k_{AF}\cdot(P+p),\ \gamma_5\sigma^{\kappa\lambda}(k_{AF})_\kappa(P+p)_\lambda(P+p)^\mu \Big\}, \qquad (6)$$

and the result from the b^μ insertion has the same structure but with the additional form $\gamma_5\sigma^{\mu\nu}b_\nu$, which vanishes for the $(k_{AF})_\mu$ contribution.

As a consistency check, we verified the Ward-Takashi identity, $q_\mu\Gamma^\mu = \Sigma(p+q)-\Sigma(p)$, for both cases up to order q^2 with arbitrary IR regulator λ, but so far with fermions on shell — the off shell case is still to be done. The Ward-Takashi identity demands the evaluation of LV contributions to the electron self energy, which are divergent. Nevertheless, only the finite piece of this integral is momentum dependent and contributes to the right-hand side of the Ward-Takashi identity. It is interesting to notice that, different from the conventional QED case, $q_\mu\Gamma^\mu \neq 0$ even when the Dirac equation is used for the external leg spinors: even in this situation the momentum shift in the electron self energy still contributes to the identity.

The next step is to establish the finite contributions coming also from the divergent piece of the electron self energy, which may require a careful choice of subtraction points. Applications in physical processes will also be envisaged, as well as the analysis for other LV coefficients.

Acknowledgments

Financial support was provided by CNPq (Brazil).

References

1. D. Colladay, and V.A. Kostelecký, Phys. Rev. D **58**, 116002 (1998).
2. V.A. Kostelecký, C.D. Lane, and A.G.M. Pickering, Phys. Rev. D **65**, 056006 (2002).

MACROSCOPIC OBJECTS, INTRINSIC SPIN, AND LORENTZ VIOLATION

DAVID W. ATKINSON,* MCCOY BECKER, and JAY D. TASSON

Department of Physics and Astronomy, Carleton College
Northfield, MN 55057, USA
**E-mail: atkinsod@carleton.edu*

The framework of the Standard-Model Extension (SME) provides a relativistic quantum field theory for the study of Lorentz violation. The classical, nonrelativistic equations of motion can be extracted as a limit that is useful in various scenarios. In this work, we consider the effects of certain SME coefficients for Lorentz violation on the motion of macroscopic objects having net intrinsic spin in the classical, nonrelativistic limit.

1. Introduction

The most successful description of physics today is provided by General Relativity and the Standard Model; however, the combined description is unacceptable as one approaches the Planck scale. The Standard-Model Extension (SME) seeks to provide relevant experimental guidance in addressing this issue by introducing a framework of all possible Lorentz-violating terms.[1] The quest to experimentally detect or constrain the Lorentz violating terms in the SME requires the ability to detect the impact of these terms at scales reasonable for experiment.

One way to search for relativity violations is to study the effects of these couplings on macroscopic objects at low energies. The goal of this work is to search for the impact of relativity violating terms in the SME on the acceleration of macroscopic objects with intrinsic spin in the nonrelativistic limit. This extends work investigating torques on such objects,[2] work considering spin-independent Lorentz-violating accelerations,[3] and work on the classical relativistic theory.[4]

The initial theoretical tool for our investigation is the Foldy-Wouthuysen transformation.[5] The motivation for this transformation is to extract the nonrelativistic hamiltonian from its relativistic counterpart. The Foldy-

Wouthuysen transformation is designed to reduce the off-diagonal portions of the Dirac hamiltonian in a systematic fashion, so that the nonrelativistic (top 2×2 block) component of the hamiltonian can be taken on its own to some degree of accuracy.

2. Free hamiltonian and generic force

In considering the nonrelativistic free particle hamiltonian, we consider results proportional to two powers of momentum.[6] These take the form of

$$H = \frac{p^2}{2m} + ... - \frac{1}{m} C_{jk} p_j p_k + ..., \tag{1}$$

where

$$C_{jk} = c_{jk} + \frac{1}{2} c_{00} \delta_{jk} - \Bigg\{ \Bigg[(d_{0j} + d_{j0}) - \frac{1}{2m} \Bigg(b_j + m d_{j0} + \frac{1}{2} m \epsilon_{jmn} g_{mn0} + \frac{1}{2} \epsilon_{jmn} H_{mn} \Bigg) \Bigg] \delta_{kl} + \frac{1}{2m} \Bigg(b_l + \frac{1}{2} m \epsilon_{lmn} g_{mn0} \Bigg) \delta_{jk} - \epsilon_{jlm} (g_{m0k} + g_{mk0}) \Bigg\} \sigma^l \tag{2}$$

to first order in Lorentz violation. Using Hamilton's equations of motion, we can repackage the Lorentz violating piece into an effective mass term m_{jk}, given by

$$m_{jk} = m \delta_{jk} + 2m C_{(jk)}, \tag{3}$$

where $C_{(jk)}$ denotes symmetrization on the indices jk; $C_{(jk)} = \frac{1}{2}(C_{jk} + C_{kj})$. We produce a modified form of Newton's second law

$$F_j = m_{jk} a_k, \tag{4}$$

where the mass is now a two-tensor dependent on spin orientation with respect to the coefficients for Lorentz violation.

Figure 1 illustrates two examples of motion resulting from a nonzero b_j in Eq. (2). In the left-hand image, the particle has spin perpendicular to the direction of $\vec{b}$. The net acceleration of this object is

$$\vec{a} = \frac{\vec{F}}{m} - \frac{\vec{b}(\vec{\sigma} \cdot \vec{F})}{2m^2} - \frac{\vec{\sigma}(\vec{b} \cdot \vec{F})}{2m^2}, \tag{5}$$

with directionality affected by the orientation of $\vec{F}$ with respect to $\vec{b}$ and $\vec{\sigma}$. The right-hand image shows the particle in the case of a force applied perpendicular to the direction of $\vec{b}$ and $\vec{\sigma}$. The net acceleration of the object is given by

$$\vec{a} = \left(\frac{1}{m} + \frac{\vec{b} \cdot \vec{\sigma}}{m^2} \right) \vec{F}. \tag{6}$$

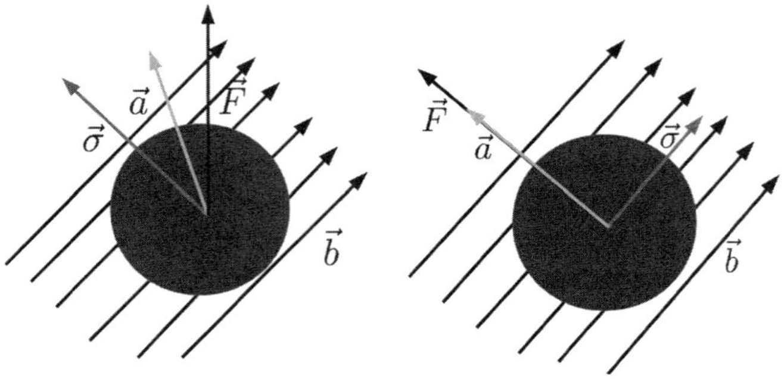

Fig. 1. Two different cases of a free particle with some intrinsic spin, $\vec{\sigma}$, in the presence of an arbitrary force $\vec{F}$ and the Lorentz violating field $\vec{b}$.

Note that in this case, the acceleration is in the usual direction, but the magnitude is altered by $\vec{b} \cdot \vec{\sigma}$.

3. Interactions

In general, the force appearing in Eq. (4) may contain additional Lorentz violation and should be handled by considering interactions in the hamiltonian. As an example, we consider electromagnetism with b_μ as the only nonzero coefficient for Lorentz violation. We find a hamiltonian of the form

$$H = \frac{1}{2m}\left(\vec{p} - q\vec{A}\right)^2 - \frac{q}{2m}B_j\sigma^j + qA^0 - b_j\sigma^j + \frac{1}{m}(b_0 p_j\sigma^j - qb_0A_j\sigma^j), \quad (7)$$

including terms to order in 1/m in the Foldy-Wouthuysen expansion. To this order we find no additional b_μ contributions associated with the interaction, and the force appropriate for insertion into Eq. (4) takes the usual form:

$$F_j = qE_j + q\epsilon_{jkl}v_kB_l. \quad (8)$$

A less trivial example is provided by the case of gravitational interactions.[7]

References

1. D. Colladay and V.A. Kostelecký, Phys. Rev. D **58**, 116002 (1998).
2. R. Bluhm and V.A. Kostelecký, Phys. Rev. D **84** 1381 (2000).
3. V.A. Kostelecký and J.D. Tasson, Phys. Rev. D **83** 016013 (2011); T. Bertschinger *et al.*, in preparation; T. Bertschinger and J.D. Tasson, these proceedings.
4. V.A. Kostelecký and N. Russell, Phys. Lett. B **693**, 443 (2010).
5. L.L. Foldy and S.A. Wouthuysen, Phys. Rev. **78**, 29 (1950).
6. V.A. Kostelecký and C. Lane, J. Math. Phys. **40**, 6245 (1999).
7. D.W. Atkinson *et al.*, in preparation; Y. Bonder, these proceedings.

APPLICATION OF LASER FREQUENCY COMBS AND NITROGEN VACANCY DIAMOND MAGNETOMETERS TO SEARCHES FOR NEW PHYSICS

D.F. PHILLIPS and R.L. WALSWORTH

Harvard-Smithsonian Center for Astrophysics, Cambridge, MA 02138, USA
E-mail: dphil@cfa.harvard.edu, rwalsworth@cfa.harvard.edu

Searches for new physics often benefit from improved technologies. Here we discuss possible applications of two emerging technologies to searches for physics beyond the Standard Model. First, laser frequency combs enable broad spectral coverage and coherent conversion between optical and RF signals. We are investigating tests of the nonminimal Standard-Model Extension using frequency combs coupled to broadband optical cavities. Second, nitrogen vacancy centers in diamond enable precision nanoscale magnetometry with applications from imaging to quantum science. We are investigating their use in searches for short-range spin-spin couplings.

1. Laser frequency combs

Laser frequency combs[1] provide high stability optical signals referenced to atomic clocks and provide large optical bandwidths of individual narrow frequency teeth. The absolute optical frequencies of the comb lines are determined by $f = f_{\text{ceo}} + m \times f_{\text{rep}}$, where f_{rep} is the laser's repetition rate, f_{ceo} is the offset frequency and m is an integer labeling the comb teeth. We have recently developed broadband optical filter cavities coupled to frequency combs for use as wavelength calibrators for astrophysical spectrographs.[2] The cavities are capable of covering > 100 nm of optical bandwidth at visible wavelengths that are resonant with the laser comb teeth (Fig. 1), effectively increasing f_{rep}. The optical cavities are composed of complementary-chirped mirror[3] pairs: two mirrors with opposite group delay dispersion (GDD) capable of forming a cavity with minimal GDD (measured phase error < 13 mrad) over the optical bandwidth.

Optical filter cavities coupled to laser frequency combs may enable study of Lorentz violation beyond the minimal Standard-Model Extension (mSME) by searching for time variation in offsets between optical cavity

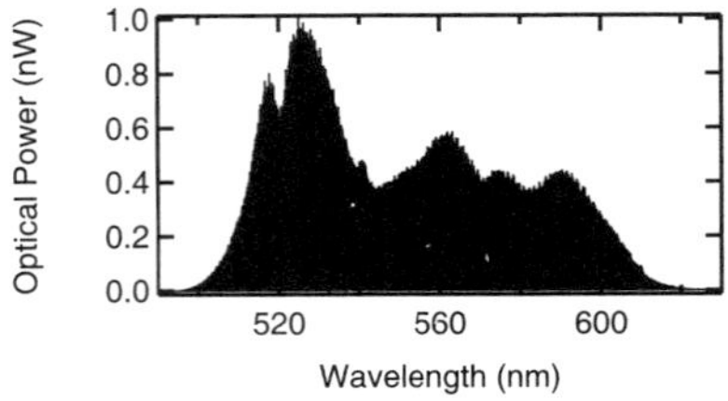

Fig. 1. Optical spectrum of visible wavelength laser frequency comb after propagating through a broadband optical filter cavity.

resonances and comb teeth. While Lorentz violation in the photon sector has been tightly constrained in the mSME,[4] the SME has recently been extended to include higher order, nonrenormalizable terms,[5,6] which may produce different effects than the minimal Lorentz violating parameters.

Nonbirefringent Lorentz violating terms beyond the mSME are not constrained by typical astrophysical measurements[6] and thus must be constrained by laboratory tests. Such measurements have recently been performed at microwave frequencies.[7] These nonbirefringent terms are parameterized by a series of coefficients $(c_F^{(d)})^{(0E)}_{njm}$ where d is the dimension of the underlying operator (and must be even) and n, j, and m are an angular momentum expansion (with $0 < n \leq d - 2$). The shifts induced in the resonant frequency of an optical cavity by these higher order, nonbirefringent terms scale as $c_F^{(d)} \sim \omega^{d-4}$. One may then separate constraints on the $c_F^{(d)}$ terms of different dimension d by measuring cavity resonances across a range of frequencies. Additionally, by measuring 10^5 spectral lines with an optical cavity tuned to the comb repetition rate, one may realize improved sensitivity. We are currently investigating the feasibility of such experiments using laser frequency combs and optical cavities we have developed for astrophysical applications.

2. Nitrogen vacancy diamond magnetometers

Nitrogen vacancy (NV) centers in diamond (see Fig. 2) are a powerful tool for precision spin measurements[8] with applications ranging from quantum information to biosensing.[9] NV-diamond magnetometers have recently obtained sensitivities in the pT/Hz$^{1/2}$ regime.[10] We are investigating application of NV centers as detectors for short range (nanoscale), anomalous spin-spin couplings. Such precision measurements will require a nearby spin source to couple to the NV electronic and/or nuclear spins. Possibilities include substitutional nitrogen defects near NV centers within the diamond

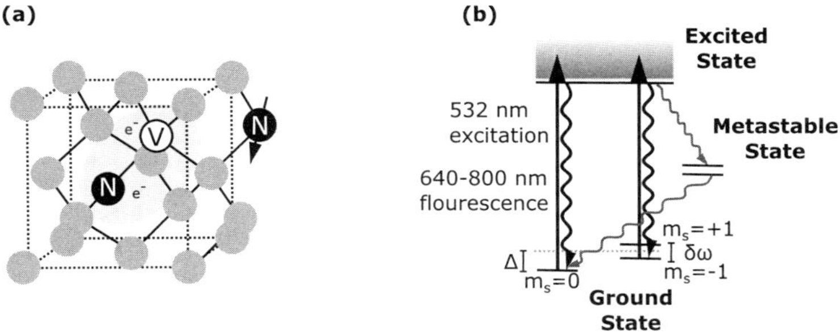

Fig. 2. (a) Nitrogen vacancy (NV) center in diamond showing a substitutional N atom which may be used as a nanoscale spin source. (b) Energy level structure of an NV center. The NV spin may be optically polarized and read out via excitation with green light and detection of red fluorescence, allowing precision nanoscale magnetometry.

lattice (Fig. 2(a)), which can be created via implantation and annealing of molecular nitrogen ions in the creation of NVs.[11] The nuclear and/or electronic spins of the nitrogen defects can be controlled via RF techniques independently from the electronic spin of the NV, which can be addressed through a combination of optical and microwave fields. Alternatively, NV centers with good properties for magnetometry can be made reliably within 2-10 nm of the surface of a diamond. Additional spin sources can then be deposited on the surface and controlled using RF while the NV centers sense the effects of the spins. Possible comagnetometry techniques include the use of NV nuclear spins as well as their electronic spin. Such techniques may allow the controlled study of spin forces on the nm scale.

References

1. S.T. Cundiff and J. Ye, Rev. Mod. Phys. **75**, 325 (2003).
2. C.-H. Li *et al.*, Nature **452**, 610 (2008).
3. L.-J. Chen *et al.*, in M. Chergui *et al.*, eds., *Ultrafast Phenomena XVII,* Oxford University Press, New York, 2011.
4. *Data Tables for Lorentz and CPT Violation,* V.A. Kostelecký and N. Russell, 2013 edition, arXiv:0801.0287v6.
5. V.A. Kostelecký and M. Mewes, Phys. Rev. D **80**, 015020 (2009).
6. M. Mewes, Phys. Rev. D **85**, 116012 (2012).
7. S.R. Parker *et al.*, Phys. Rev. Lett. **106**, 180401 (2011).
8. J.R. Maze *et al.*, Nature **455**, 644 (2008).
9. D. Le Sage *et al.*, Nature **496**, 486 (2013).
10. D. Le Sage *et al.*, Phys. Rev. B **85**, 121202(R) (2012).
11. T. Gaebel *et al.*, Nature Phys. **2**, 408 (2006); B. Naydenov *et al.*, Appl. Phys. Lett. **96**, 163108 (2010).

USING A ROTATING MAGNETIC GUIDING FIELD FOR THE ^{3}HE-^{129}XE COMAGNETOMETER

F. ALLMENDINGER* and U. SCHMIDT
Physikalisches Institut, Ruprecht-Karls-Universität Heidelberg
69120 Heidelberg, Germany
**E-mail: allmendinger@physi.uni-heidelberg.de*

W. HEIL, S. KARPUK, A. SCHARTH, Y. SOBOLEV, K. TULLNEY and S. ZIMMER
Institut für Physik, Johannes Gutenberg-Universität
55099 Mainz, Germany

Our search for nonmagnetic spin-dependent interactions is based on the measurement of free precession of nuclear spin polarized ^{3}He and ^{129}Xe atoms in a homogeneous magnetic guiding field of about 400 nT. We report on our approach to perform an adiabatic rotation of the guiding field that allows us to modulate possible nonmagnetic spin-dependent interactions and to find an optimization procedure for long transverse relaxation times T_2^* both for helium and xenon.

1. Introduction

The most precise tests of nonmagnetic spin-dependent interactions like (i) the Lorentz invariance violating coupling of spins to a background field[1] or (ii) the search for a P- and T-violating short-range interaction mediated by light, pseudo-scalar bosons (axions)[2] are often performed in experiments that compare the transition frequencies of two co-located systems (clock comparison).[3–5] This type of experiment is only sensitive if the spin-dependent interaction in question is varying in time. In case of the search for a coupling of spins to a relic background field,[3,4] the Earth's daily rotation leads to a modulation of the orientation in space with the sidereal frequency. As the frequency of this modulation is unfavorably low, we developed a system to rotate the magnetic guiding field of our ^{3}He-^{129}Xe-comagnetometer ($\Omega_{rot} \approx 2\pi/(20\ \text{min})$) in order to have a much more preferable modulation frequency that may help to get rid of the correlated errors which at present limit our overall sensitivity. Inside the μ-metal shielded room BMSR-2 at PTB, Berlin, two square Helmholtz coil pairs (fixed perpendicular to each other) are driven by low noise, high resolution current sources in order to provide the homogeneous magnetic field of about 400 nT. The stability of the current is $\Delta I/I \approx 10^{-6}$, while the stability of the magnetic field is $\Delta B/B \approx 10^{-4}$ due to drifts of the μ-metal shielding, measured by analyzing Larmor frequency drifts. With this setup we can generate a guiding field, which points in any direction α in the horizontal plane.

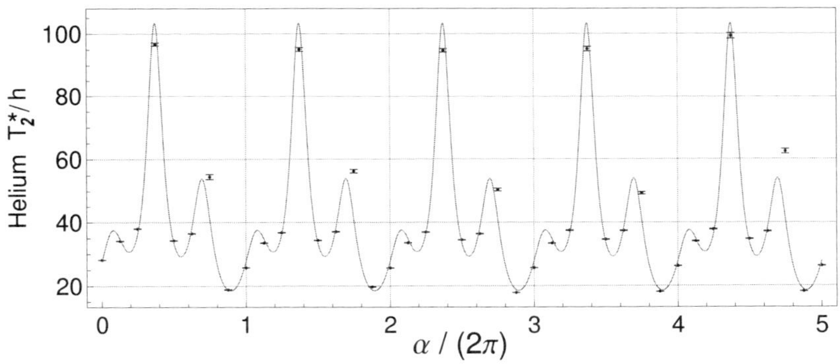

Fig. 1. The transverse relaxation time of helium as a function of the direction α of the magnetic guiding field in the horizontal plane, measured in steps of 45° for 5 turns. In total, the measurement took about 20 hours with ~30 min for each field setting to extract the T_2^* from the decay of the signal amplitude. Solid line: fit of a Fourier series to the relaxation rates. Corresponding xenon data in Ref. 6.

2. Results

Besides the proof of principle that a coherent measurement of spin precession with a slowly rotating guiding field is possible (i.e., we do not lose the phase information), our results have an unexpected impact on the measurements with static guiding fields: the transverse relaxation time strongly depends on the direction of the magnetic guiding field and varies between 20 h and 100 h for helium (Fig. 1), and between 6 h and 8.5 h for xenon. The characteristic pattern in Fig. 1 for $T^*_{2,He}$ repeats itself after every revolution and is reproduced in all successive runs over a period of at least two weeks. For $T^*_{2,Xe}$, the characteristic pattern is similar.[6] This effect of 'incidental shimming'[7] has the following explanation: the presence of magnetic field gradients across a sample cell causes an increased transverse relaxation rate. The origin of this relaxation mechanism is the loss of phase coherence. For a spherical sample cell of radius R the relaxation rate $1/T_2^*$ is[8]

$$\frac{1}{T_2^*} = \frac{1}{T_1} + \frac{4R^4\gamma^2}{175D}\left(|\vec{\nabla}B_y|^2 + |\vec{\nabla}B_z|^2 + 2|\vec{\nabla}B_x|^2\right), \qquad (1)$$

with the guiding field pointing into the x-direction. γ is the gyromagnetic ratio and D is the diffusion coefficient of the gas. It is useful to measure at low fields in order to minimize the absolute field gradients, which are of order pT/cm inside BMSR-2. There are two main sources of gradients: residual field gradients from the μ-metal shielding and gradients produced by the Helmholtz coils. The latter ones will change, as the magnetic guiding field is rotated. At some angle α the gradients from the chamber and coils almost cancel each other and T_2^* is maximized. At other angles the cancellation is less distinct with a minimum in T_2^* at a field orientation where the gradients add up constructively. This is consistent with the observation that by rotating the magnetic guiding field by 180° the transverse relaxation time goes from the global maximum to the global minimum.

3. Conclusion

These results gave us the possibility to optimize T_2^* and thus the observation time for the search for a Lorentz violating coupling of spins to a hypothetical background field. In our present experiments coherent spin precession can be monitored for more than 24 hours ($\approx 3 \cdot T_{2,Xe}^*$).[3] Longer observation times T cause a higher sensitivity in frequency measurement ($\sigma \propto T^{-3/2}$) according to the Cramer-Rao lower bound[4] and, furthermore, greatly reduce the correlated errors, especially if runs of coherent spin precession are substantially longer than the period of a sidereal day.[5]

References

1. *Data Tables for Lorentz and CPT Violation,* V.A. Kostelecký and N. Russell, 2013 edition, arXiv:0801.0287v6.
2. K. Tullney *et al.*, arXiv:1303.6612v2.
3. U. Schmidt *et al.*, these proceedings.
4. C. Gemmel *et al.*, Eur. Phys. J. D **57**, 303 (2010).
5. C. Gemmel *et al.*, Phys. Rev. D **82**, 111901(R) (2010).
6. W. Heil *et al.*, Ann. Phys. (Berlin), 1-11 (2013), in press.
7. H. Harcken *et al.*, J. Magn. Reson. **206**, 168-170 (2010).
8. G.D. Cates, S.R. Schaefer and W. Happer, Phys. Rev. A **37**, 2877 (1988).

CLASSIFICATION OF RELATIVITY THEORIES BY DISPERSION RELATION

M. JACKSON GOODMAN

Department of Physics and Astronomy, Swarthmore College
Swarthmore, Pennsylvania 19081, USA
E-mail: mjacksongoodman@gmail.com

We seek to distinguish and classify relativity theories based on their dispersion relations. Propositions for modified relativity theories can be made by deforming the dispersion relation of Special Relativity. We investigate the critical points of the dispersion relations in order to identify which of these modified dispersion relations represents a change of variables and which represents a significantly different theory.

1. Introduction

One of the obstacles towards uniting relativity and quantum mechanics is that the conjunction of quantum mechanics and General Relativity predicts a minimal length scale for the Universe. Since Special Relativity tells us that different observers measure lengths differently, it seems impossible to incorporate compatibly a fundamental length scale. One might consider modifying the Special Relativity dispersion relation to create a new relativity theory which avoids such incompatibility. We hope to classify such modified relativity theories, to see which are structurally identical to those we already know and which contain potential new physics.

2. Dispersion relations

A dispersion relation in a relativity theory is a function of the transformed variables that remains constant as the variables are transformed. In this study we limit ourselves to two transformation variables (which in this section we will call x and y, although in other sections we will also use E and p).

Given two relativity theories with dispersion relations $f_1(x, y)$ and $f_2(x, y)$, suppose there exists a diffeomorphism $\vec{h} : \mathbb{R}^2 \to \mathbb{R}^2$ such that

$f_1(\vec{h}(x,y)) = f_2(x,y)$. We say that f_1 and f_2 are related by the diffeomorphism $\vec{h}$. In this case the theories are structurally identical, and their differences arise from a renaming of the transformation variables.

A dispersion relation function $f(x,y)$ defines a 2D gradient vector field $(\partial_x f, \partial_y f)$. A critical point of f is a point at which both partial derivatives vanish. If two dispersion relations $f_1(x,y)$ and $f_2(x,y)$ are related by a diffeomorphism $\vec{h}$, the chain rule tells us that the gradients of f_1 and f_2 are mapped to each other by an invertible linear function at each point. Thus $\vec{h}$ and its inverse carry the critical points of each function onto the critical points of the other. If the two functions have a finite number of critical points, they will have the same number of critical points.

3. Winding number and index

If S is a closed curve in the plane, the winding number of a 2-dimensional vector field $\vec{V}$ about S is the number of counterclockwise revolutions that the vectors of $\vec{V}$ make as they move counterclockwise around S. If a curve is deformed continuously, without passing through a critical point, the winding number changes continuously. But the winding number is an integer and so must remain constant throughout the deformation. Thus a vector field will have the same winding number about any curve enclosing a single critical point. This number is the index of the critical point. It can be shown that the index of a critical point of a vector field is preserved by a diffeomorphism of the vector field, and so two relatable functions will have critical points with equal indices.

We can calculate the index of a critical point of a vector field from the eigenvalues of the Jacobian matrix. For a gradient vector field of a function f, the Jacobian is the Hessian matrix

$$\begin{bmatrix} \partial_x^2 f & \partial_y \partial_x f \\ \partial_x \partial_y f & \partial_y^2 f \end{bmatrix}. \tag{1}$$

We assume that f has continuous, and thus interchangeable derivatives. Then the Hessian matrix is symmetric. A symmetric matrix has real eigenvalues.

If both eigenvalues are positive or both eigenvalues are negative, the critical point is a local minimum or maximum of f. It follows that the index of the critical point is 1. If one eigenvalue is negative and the other positive, the critical point is a saddle point of f, and the index is -1. If one of the eigenvalues is zero, the Hessian is singular, and higher order derivatives must be used to determine if the index is 1, -1 , or 0.

Table 1 shows examples of this process. The Galilean Relativity dispersion relation is not relatable to the others, because it has no critical points while the others each have one. A 'deformed' Special Relativity is also shown. Critical point analysis does not reveal any difference between the deformed and conventional Special Relativities. Indeed, if we call the Special Relativity dispersion relation $f_1(E,p)$ and the deformed Special Relativity dispersion relation $f_2(E,p)$, then the diffeomorphism $\vec{h}(E,p) = (E+p, p-E)$ relates the two functions such that $f_2(E,p) = f_1(\vec{h}(E,p))$. The symmetry group of rotations provides a final example, with the invariant length acting as the dispersion relation. This dispersion relation is not relatable to that of the other examples since the critical points of each have different indices.

Table 1. Critical point and index analysis for examples of relativity theories.

Relativity theory	Dispersion relation	Critical points	Hessian eigenvalues	Index
Galilean Relativity	$E - p^2/2m$	none	–	–
Special Relativity	$E^2 - c^2p^2$	$(0,0)$	$2, -2c^2$	-1
deformed Special Relativity	$(1-c^2)(E^2+p^2)$ $+(2+2c^2)Ep$	$(0,0)$	$4, -4c^2$	-1
rotational symmetry	$x^2 + y^2$	$(0,0)$	$2, 2$	1

4. Conclusion

Analysis of critical points enables us to differentiate between relativity theories. We see that Galilean Relativity is structurally distinct from Special Relativity, while both are distinct from the symmetry group of rotations. This method does not distinguish between Special Relativity and a relatable deformed Special Relativity.

Further research will extend the process to concrete analysis of cases in which the Hessian is singular, and explore a possible link between the index of critical points and invariant scales like the speed of light or a minimal length scale.

Acknowledgments

I would like to thank the Swarthmore Physics Department for supporting my research, the Indiana University Center for Spacetime Symmetries for facilitating this presentation, and Matt Mewes for his help and guidance.

BIPARTITE FINSLER SPACES AND THE BUMBLEBEE MODEL

J. EUCLIDES G. SILVA* and C.A.S. ALMEIDA

Physics Department, Ceará Federal University
Fortaleza, Ceará ZIP 6030, 60455-760, Brazil
**E-mail: euclides@fisica.ufc.br*

We present a proposal to include Lorentz-violating effects in a gravitational field by means of Finsler geometry. In the Finsler setup, the length of an event depends both on the point and the direction in the spacetime. We briefly review the bumblebee model, where the Lorentz violation is induced by a spontaneous symmetry breaking due to the bumblebee vector field. The main geometrical concepts of Finsler geometry are outlined. Using a finslerian Einstein-Hilbert action we derive the bumblebee action from the bipartite Finsler function with a correction to the gravitational constant.

1. Bumblebee model

A model to include gravity in the Standard-Model Extension (SME)[1] is provided by a vector field B_μ, the so-called bumblebee field, which couples with the usual geometrical tensors through the action[2]

$$S_{LV} = K \int_M (uR + s^{\mu\nu} R_{\mu\nu})\sqrt{-g} d^4x, \tag{1}$$

where $s^{\mu\nu} = \xi \left(B^\mu B^\nu - \frac{1}{4} B^2 g^{\mu\nu} \right)$ and $u = \xi B^\mu B_\mu$.

Although this model introduces Lorentz violation into the geometry, the geometrical tensors remain Lorentz invariant. This drawback can be overcame by means of Finsler geometry.[2]

2. Finsler geometry

A Finsler geometry is an extension of the Riemann geometry where given a curve $\gamma : [0,1] \to M$, its arc length is given by[3]

$$s = \int_0^1 F(x,y) dt, \tag{2}$$

where $x \in M, y \in T_xM$. The function $F(x,y)$ is called the Finsler principal function. Note that the interval depends both on the position x and on the direction y.

As in the Riemann case, it is possible to define a Finsler metric by[3]

$$g^F_{\mu\nu}(x,y) = \frac{1}{2}\frac{\partial^2 F^2}{\partial y^\mu y^\nu}. \tag{3}$$

The Finsler metric (3) is a symmetric and anisotropic quadratic form on TTM. Differentiating the metric yields the so-called Cartan tensor $A_{\alpha\beta\gamma}(x,y) = \frac{1}{4}F\partial g_{\alpha\beta}/\partial y^\gamma$ from which it is possible to define a nonlinear connection by $N^\alpha_\delta = \gamma^\delta_{\alpha\beta}y^\beta - A^\delta_{\alpha\beta}\gamma^\beta_{\epsilon\xi}y^\epsilon y^\xi/F$ which decouples TTM into $TTM = hTTM \bigoplus vTTM$, where $\delta/\delta x^\alpha = \partial/\partial x^\alpha - N^\beta_\alpha\partial/\partial y^\beta$ is the basis for the horizontal section $hTTM$ and $F\partial/\partial y^\alpha$ is a base for the vertical section $vTTM$.

The compatible Cartan connection is given by $\omega^\delta_\alpha = \Gamma^\delta_{\alpha\beta}dx^\beta + A^\delta_{\alpha\beta}\delta y^\beta/F$, where $\Gamma^\delta_{\alpha\beta} = \frac{1}{2}g^{F\delta\epsilon}_{\alpha\beta}(\delta_\alpha g_{\epsilon\beta} + \delta_\beta g_{\epsilon\alpha} - \delta_\epsilon g_{\alpha\beta})$.[3]

Following the approach proposed by Pfeifer and Wohlfarth,[4] here we are concerned with only the Lorentz violation effects on tensor fields defined on $hTTM$. The horizontal Ricci curvature is defined as[3]

$$R_{\alpha\beta} = \delta_\gamma\Gamma^\gamma_{\alpha\beta} - \delta_\beta\Gamma^\gamma_{\alpha\gamma} + \Gamma^\gamma_{\epsilon\beta}\Gamma^\epsilon_{\alpha\gamma} - \Gamma^\gamma_{\epsilon\gamma}\Gamma^\epsilon_{\alpha\beta}, \tag{4}$$

and the scalar curvature is $R^F = g^{F\alpha\beta}R^F_{\alpha\beta}$.

3. The bipartite space

Based on a previous work on classical point particle lagrangians,[5] Kostelecký proposed a new Finsler function of the form[6]

$$F(x,y) = \sqrt{g_{\mu\nu}(x)y^\mu y^\nu} + l_P(a_\mu(x)y^\mu \pm \sqrt{s_{\mu\nu}(x)y^\mu y^\nu}), \tag{5}$$

where $s_{\mu\nu}(x) = b^2(x)g_{\mu\nu}(x) - b_\mu(x)b_\nu(x)$. The Planck length scale l_P provides a scale of length where the anisotropic effects have to be taken into account.

Kostelecký, Russell and Tso supposed $a_\mu = 0$ and enhanced the $s_{\mu\nu}$ tensor to be any symmetrical one.[7] This geometry is so-called bipartite.

The bipartite Finsler function yields the Finsler metric[7]

$$g^F_{\mu\nu} = \frac{F}{\alpha}g_{\mu\nu} + l_P^2\left(\frac{F}{\sigma}s_{\mu\nu} - \alpha\sigma k_\mu k_\nu\right), \tag{6}$$

where $k_\mu = \alpha^{-1}\partial\alpha/\partial y^\mu - \sigma^{-1}\partial\sigma/\partial y^\mu$. Furthermore, for dim $M = 4$, the relation between the finslerian and lorentzian volume elements is given by

$$\sqrt{-g^F} = \left(\frac{F}{\alpha}\right)^5 \left(\frac{S}{\sigma}\right)^2 \sqrt{-g}. \tag{7}$$

4. Finslerian Einstein-Hilbert action

Assuming the dynamics of the spacetime is governed by an Einstein-Hilbert action, then

$$\begin{aligned} S_F &= \kappa \int R^F \sqrt{-g^F} d^4x d^4y \\ &= \kappa \int R\sqrt{-g} d^4x + \kappa_F \int 8b^2 R\sqrt{-g} d^4x \\ &\quad + \kappa \int (4 + b^2 l_P^2) s^{\mu\nu} R_{\mu\nu} d^4x + \dots, \end{aligned} \tag{8}$$

and it is possible to regain some interaction terms of the bumblebee model action (1). As a perspective we expect to obtain the dynamical terms of the bumblebee model and some other interaction terms as well.

Acknowledgments

We are grateful to Alan Kostelecký and Yuri Bonder for useful discussions and to the Graduate Physics Program of Ceará Federal University and CNPq (the National Council for Scientific and Technological Development) for financial support.

References

1. D. Colladay and V.A. Kostelecký, Phys. Rev. D **55**, 6760 (1997).
2. V.A. Kostelecký, Phys. Rev. D **69**, 105009 (2004).
3. D. Bao, S.-S. Chern, Z. Shen, *An Introduction to Riemann-Finsler Geometry*, Springer, New York, 1991.
4. C. Pfeifer and M.N.R. Wohlfarth, Phys. Rev. D **85**, 064009 (2012).
5. V.A. Kostelecký and N. Russell, Phys. Lett. B **693**, 443 (2010).
6. V.A. Kostelecký, Phys. Lett. B **701**, 137 (2011).
7. V.A. Kostelecký, N. Russell, and R. Tso, Phys. Lett. B **716**, 470 (2012).

MICROLOCAL METHODS IN QUANTUM FIELD THEORY AND THEIR USE IN PROVING RENORMALIZABILITY IN SME AND TACHYONIC MODELS

M.J. RADZIKOWSKI

Dept. of Science and Mathematics, American University of Afghanistan
Kabul, Afghanistan
E-mail: mradzikowski@auaf.edu.af

The use of microlocal analysis is considered in proving renormalizability of a particular minimal SME model, as well as of a model of scalar tachyons.

1. Introduction

Microlocal analysis,[1,2] which has been invented as a tool to deal with questions concerning the solutions of partial differential equations, as well as to study properties of pseudo-differential operators and Fourier integral operators, may also be applied profitably to quantum field theory on curved spacetime (QFT on CST).[3,4] Since global Lorentz symmetry is typically lost for a QFT on CST, one may consider applying the ideas in this area to other unconventional quantum field models in which Lorentz violation (LV) may occur.

In the minimal Standard-Model Extension (mSME) initiated and developed by Colladay, Kostelecký and other collaborators,[5,6] the usual Standard-Model Lagrange density on Minkowski spacetime is supplemented by terms which break Lorentz symmetry at the classical level, while naïvely preserving the renormalizability of the model. Some partial proofs of renormalizability (e.g., to the one loop level for certain models) have been published, but as yet no general proof applying to all extension terms to all orders has been given. In this poster I allude to some of the definitions and results of microlocal analysis which have been useful in quantum field theory on curved spacetime, and indicate how mSME models may be treated by them in order to demonstrate existence of the distributions necessary to calculate Feynman diagrams for interacting models. The distributions are given in terms of the configuration space variable, instead of the momentum space variables. It is noted that unitarity is not shown to all orders; indeed, the distributions so constructed contain a smooth $\mathcal{C}^\infty$ ambiguity. Hence the results presented here, while promising, are partial. These methods are also applied to the separate case of a scalar tachyonic model.

2. Microlocal analysis in an mSME quantum field theory

The wave equation for an mSME model[5] with a^μ and b^μ terms is:

$$(i\gamma^\mu\partial_\mu - a_\mu\gamma^\mu - b_\mu\gamma_5\gamma^\mu - m)\psi = 0 \ . \tag{1}$$

Although in what follows, a^μ, b^μ are considered constant with respect to the spacetime variables x^μ, one may also consider using the following method to extend the results to the case in which a^μ, b^μ are functions of x^μ. Putting the mass term on the right and squaring both sides gives the analog of the Klein-Gordon equation [Eq. (7) of Ref. 5]:

$$[(i\partial - a)^2 - b^2 - m^2 + 2i\gamma_5\sigma^{\mu\nu}b_\mu(i\partial_\nu - a_\nu)]\psi = 0 \ . \tag{2}$$

The prescription to obtain the building blocks for the quantum field theory is as follows. (A) Find the scalar advanced, retarded and Feynman distinguished parametrices (Green's functions) of the above differential operator in brackets, and the Wightman two point function, obtained as $1/i$ times the difference of the Feynman and advanced parametrices. (B) Act on this operator by

$$\mathcal{O} \equiv i\gamma^\mu\partial_\mu - a_\mu\gamma^\mu - b_\mu\gamma_5\gamma^\mu + m \ , \tag{3}$$

to obtain the corresponding advanced, retarded, and Feynman distinguished parametrices, as well as the Wightman two point function for the analog of the Dirac equation, Eq. (1) above. See Refs. 7 for details for the Dirac case on curved spacetime. (C) Due to theorems ensuring the existence and uniqueness of the distinguished parametrices for the scalar case up to $\mathcal{C}^\infty$ with respect to the wave front sets for these parametrices,[2] the above construction should be unique up to $\mathcal{C}^\infty$ terms, and furthermore, the wave front set of the Wightman two point function should be similar in form to the case of the corresponding objects on flat or curved spacetime. More specifically, the wave front set vectors for the two point function should have a similar positivity property to the *wave front set spectrum condition* or *microlocal spectral condition* in Refs. 3 above. This is conjectured to be necessary and sufficient input to carry out a renormalization scheme similar to that devised by Brunetti and Fredenhagen.[4] More generally, in the mSME model with nonzero and small enough values of the $c, d, e, \ldots$ coefficients, the analog of the Klein Gordon equation should likewise have a similar leading order to Eq. (2), and therefore, the above procedure should work as for the usual Klein Gordon and Dirac cases.

3. Scalar quantum field theory of tachyons

I have argued that,[8] in order to maintain any chance of renormalizability of the free theory, Lorentz symmetry should be broken in a reasonable quantum field model satisfying the tachyonic Klein-Gordon equation,

$$(\Box - m^2)u = 0 \ , \tag{4}$$

where $\Box = \partial_t^2 - \nabla^2$. A construction of the QFT, including only energies $E' \geq 0$, is performed first in the preferred (primed) frame, and is extended to any other frame using Lorentz transformations. The two-point function has the same wave front set as for the regular massive case, and hence fits into the Brunetti-Fredenhagen renormalization scheme.[4] The unitarity equation has been verified to leading order (g^2 or λ^2) for both $g\varphi^3$ and $\lambda\varphi^4$ theory in 2-d, and for $g\varphi^3$ in 4-d.

Acknowledgments

The American University of Afghanistan is gratefully acknowledged for travel grants to this meeting and last year's summer school. Thanks are also due to the organizers and lecturers of the 2012 IUCSS Summer School on the Lorentz- and CPT-violating Standard-Model Extension.

References

1. L. Hörmander, *The Analysis of Linear Partial Differential Operators I–IV*, Springer, Berlin, Heidelberg, New York, Second Edition, 1990, 1985; Acta Math. **127**, 79 (1971).
2. J.J. Duistermaat and L. Hörmander, Acta Math. **128**, 183 (1972).
3. M. Radzikowski, Ph.D. Thesis, Princeton University, 1992; Commun. Math. Phys. **179**, 529 (1996); Commun. Math. Phys. **180**, 1 (1996); R. Brunetti, K. Fredenhagen, and M. Köhler, Commun. Math. Phys. **180**, 633 (1996).
4. R. Brunetti and K. Fredenhagen, Commun. Math. Phys. **208**, 623 (2000).
5. D. Colladay and V.A. Kostelecký, Phys. Rev. D **55**, 6760 (1997).
6. D. Colladay and V.A. Kostelecký, Phys. Rev. D **58**, 116002 (1998); V.A. Kostelecký, Phys. Rev. **69**, 105009 (2004).
7. K. Kratzert, Annalen Phys. **9**, 475 (2000); S. Hollands, Commun. Math. Phys. **216**, 635 (2001).
8. M. Radzikowski, arXiv:0801.1957; arXiv:0804.4534; arXiv:1007.5418.

RENORMALIZATION AND ASYMPTOTIC STATES IN LORENTZ-VIOLATING QUANTUM FIELD THEORY

M. CAMBIASO

Departamento de Ciencias Físicas, Universidad Andres Bello
Santiago, Chile
E-mail: mcambiaso@unab.cl

R. LEHNERT

Indiana University Center for Spacetime Symmetries
Bloomington, IN 47405, USA

R. POTTING

CENTRA, Departamento de Física, Universidade do Algarve
8005-139 Faro, Portugal

Radiative corrections in quantum field theories with small departures from Lorentz symmetry alter structural aspects of the theory, in particular the definition of asymptotic single-particle states. Specifically, the Dirac equation is radiatively modified by Lorentz-violating momentum-dependent operators not present in the Lagrange density, and the standard renormalization procedure needs to be adapted too.

In the presence of Lorentz breakdown, previous analyses involving the properties of freely propagating particles have been performed under the tacit assumption that the physics of free particles is determined by the quadratic pieces of the corresponding Lagrange density. However, by analyzing radiative corrections for a sample Lorentz-violating (LV) Lagrange density contained in the SME, we claim that the previous line of reasoning fails.

We consider the bare gauge-invariant Lagrange density for single-flavor QED within the minimal SME in the presence of the $c^{\mu\nu}$ and (the nonbirefringent sector of) $k_F^{\mu\nu\alpha\beta}$ coefficients only, the latter encoded in $\tilde{k}^{\mu\nu} = (k_F)^{\mu\alpha\nu}{}_{\alpha}$. One-loop multiplicative renormalizability of the model is assumed and dimensional regularization with minimal subtraction will be used.[1] To understand the external fermion states we need to consider radiative corrections to the fermion two-point function, from which the wavefunction renormalization 'constant' is read off as the residue of the one-particle pole. As a zeroth-order system on which perturbation theory is set up, we choose the full renormalized quadratic Lagrange density together with the LV part, the remaining nonquadratic contributions to the Lagrange density being taken as perturbations. Thus, the proper two-point function adopts the form $\Gamma^{(2)}(p) = \Gamma^\mu p_\mu - m - \Sigma(p^\mu)$, where $\Gamma^\mu = \gamma^\mu + c^{\mu\nu}\gamma_\nu$ and

$$\Sigma(p^\mu) = \Sigma_{\rm LI}(\not{p}) + \Sigma_{\rm LV}(p^2, c^p_\gamma, \tilde{k}^p_\gamma) + \delta\Sigma(p^\mu, c^{\mu\nu}, \tilde{k}^{\mu\nu}), \tag{1}$$

with $c^p_\gamma \equiv c^{\mu\nu} p_\mu \gamma_\nu$ and similarly for $\tilde{k}^p_\gamma$. The first term on the right-hand side of Eq. (1) denotes the usual Lorentz-symmetric contributions. The rest come from LV terms. The last, however, contains terms that are not present in the original Lagrange density. To linear order in LV and given that electromagnetic interactions preserve C, P, and T, we can write:

$$\Sigma_{\rm LV}(p^2, c^p_\gamma, \tilde{k}^p_\gamma) = f^c_2(p^2)\, c^p_\gamma + f^{\tilde{k}}_2(p^2)\, \tilde{k}^p_\gamma, \tag{2}$$

$$\delta\Sigma(p^\mu, c^p_p, \tilde{k}^p_p) = f^c_3(p^2)\, \frac{c^p_p}{m} + f^c_4(p^2)\, \frac{\not{p} c^p_p}{m^2} + f^{\tilde{k}}_3(p^2)\, \frac{\tilde{k}^p_p}{m} + f^{\tilde{k}}_4(p^2)\, \frac{\not{p}\tilde{k}^p_p}{m^2}. \tag{3}$$

The functions $f^c_i(p^2)$ and $f^{\tilde{k}}_i(p^2)$ are calculable to any order in the fine-structure constant α. Extracting the the one-particle propagator pole is nontrivial as $\Gamma^{(2)}(p)$ no longer depends on $\not{p}$ only, and the LV functions c^p_γ and $\tilde{k}^p_\gamma$ do not commute with the latter. Therefore including these terms is somewhat ambiguous, however, a generalization of the Lorentz-symmetric case for the one-particle propagator pole can be written as:

$$\Gamma^{(2)}(p) = \mathcal{Z}_R^{-1}\left((c,k)^p_p\right)\, \bar{P}(p) + \bar{P}(p)\, \Sigma_2\big(\not{p}, c^p_\gamma, \tilde{k}^p_\gamma, (c,k)^p_p\big)\, \bar{P}(p). \tag{4}$$

That this generalization is correct is endorsed by the fact that it yields the desired property $\Gamma^{(2)}(p)^{-1} = \mathcal{Z}_R\big((c,k)^p_p\big)\, \bar{P}(p)^{-1} + \text{finite}$, and most importantly, because it is in accordance with the general form for extracting the pole from the full fermion propagator as in Ref. 2.

Doing standard perturbative calculations to first order in LV coefficients and in the fine-structure constant α (see Ref. 3 for details), we obtain:

$$\mathcal{Z}_R^{-1}\big((c,k)^p_p\big) = (\mathcal{Z}_R)^{-1}_{\rm LI} - \frac{\alpha}{\pi m^2}\left[\frac{4}{3}\, c^p_p + \frac{1}{12}\tilde{k}^p_p\right], \tag{5}$$

$$\bar{P}(p) = \not{p} + (c_{\rm phys})^p_\gamma - m_{\rm phys} - \frac{\alpha}{3\pi m}\left(2(c_{\rm phys})^p_p - (\tilde{k}_{\rm phys})^p_p\right), \tag{6}$$

where $m_{\rm phys}$ is the usual loop-corrected mass in the MS scheme, and

$$(c_{\rm phys})^{\mu\nu} = c^{\mu\nu} + \frac{\alpha}{\pi}\left[-\frac{29}{36} + \frac{\gamma}{3} + \frac{1}{3}\ln\left(\frac{m^2}{4\pi\mu^2}\right)\right]\left(2c^{\mu\nu} - \tilde{k}^{\mu\nu}\right). \tag{7}$$

Equation (7) expresses the radiatively corrected, physical value of the Lorentz-violating parameter $c^{\mu\nu}$ in terms of its tree-level value. Note that both $c^{\mu\nu}$ and the mass scale μ are unphysical, renormalization-scheme-dependent quantities, unlike $(c_{\text{phys}})^{\mu\nu}$, which is in principle measurable. Similarly, the dispersion relation is obtained from $\bar{P}(p) = 0$ for which the $\mathcal{O}(\alpha)$ LV radiative corrections are proportional to $(2c^p_p - \tilde{k}^p_p)$ too, namely:

$$0 = p^2 - m^2_{\text{phys}} + 2c^p_p + \frac{2\alpha}{\pi}\left[-\frac{11}{36} + \frac{\gamma_E}{3} + \frac{1}{3}\ln\left(\frac{m^2}{4\pi\mu^2}\right)\right](2c^p_p - \tilde{k}^p_p). \quad (8)$$

Our results comply with the fact that $c^{\mu\nu}$ and $\tilde{k}^{\mu\nu}$ are observationally indistinguishable at leading order in any electron-photon system; only their difference $2c^{\mu\nu} - \tilde{k}^{\mu\nu}$ being observable.[4]

From our results we conclude that asymptotic single-particle states of fermions in LV QFT receive concrete modifications due to radiative corrections. Specifically, among the most important conclusions we find that the corresponding Dirac equation turns out to be modified by LV momentum-dependent operators not present in the Lagrange density, a novel feature of LV QFT as it does not occur in the Lorentz-symmetric case. Also, the fermion wavefunction renormalization is no longer a constant but rather depends on the LV coefficients under consideration and on the external momentum as well, in agreement with the expectation from Ref. 2. As shown in Ref. 3, we have also checked a nontrivial cancellation of infrared divergences in all physically observable quantities.

Acknowledgments

This work has been supported in part by UNAB grant number DI-27-11/R, FONDECYT grant number 11121633, the Portuguese Fundação para a Ciência e a Tecnologia, the Mexican RedFAE, and by the IUCSS.

References

1. V.A. Kostelecký, C.D. Lane, and A.G.M. Pickering, Phys. Rev. D **65**, 056006 (2002).
2. R. Potting, Phys. Rev. D **85**, 045033 (2012); R. Potting, these proceedings.
3. M. Cambiaso, R. Lehnert, and R. Potting, in preparation.
4. See, e.g., V.A. Kostelecký and R. Potting, Phys. Rev. D **51**, 3923 (1995); M.A. Hohensee *et al.*, Phys. Rev. Lett. **102**, 170402 (2009); Phys. Rev. D **80**, 036010 (2009); J.-P. Bocquet *et al.*, Phys. Rev. Lett. **104**, 241601 (2010).

CONSTRAINTS ON SPIN INDEPENDENT FORCES AT THE ~100 NM RANGE BY MEANS OF A MICROMECHANICAL OSCILLATOR

R.S. DECCA
Department of Physics, Indiana University-Purdue University Indianapolis
Indianapolis, IN46202, USA
E-mail:rdecca@iupui.edu

Preliminary data from improvements made in our experimental setup are presented. Forces measured with our setup are presented and possible origins for the systematics observed are discussed. The observed signal is most likely induced by an impulsive oscillation of the motor.

1. Introduction

Experiments searching for corrections to Newtonian gravity[1–3] by a direct measurement of the interaction between a source and a test mass are faced with systematic signals arising from electrostatic and Casimir interactions between the masses. While in some cases it is possible to screen the masses by using a shield, when the separation between them is small, this task becomes infeasible. Some experiments have developed a so called 'isoelectronic' approach, where the effect of the mean electrostatic contribution and the Casimir interaction are nullified by design,[2,3] by measuring differences between signals with the same mean electromagnetic distribution (for a discussion on the effect of the variance of electrostatic contributions see Ref. 4) but different gravitational signature.

In the experiment published in 2005,[3] the limits obtained in spin-independent contributions to hypothetical forces were limited by the presence of a spurious signal. This signal was likely associated with a Casimir residual due to the fact that the distance between the test and source mass was not kept constant as the test mass was moved over regions where the source mass had different density. This motivated a development of a new system, where the variations in height of the source mass are uncorrelated with the variations in density.

2. Results

The experimental setup is schematically shown in the inset of Fig. 1. The sample is made of sectors of Au and Si defined on a Si waffer. This is glued to a flat glass surface, the Si waffer is etched and the whole sample is covered with ~200 nm of Au. This sample is then rotated in close proximity to an Au-coated sapphire sphere, which in turn is glued to a microelectromechanical torsional oscillator. The actual sample has different regions with alternating Au-Si sectors, each subtending the same angle. Each region, which has a radial extent of 200μm, is separated by a solid Si region of 250μm from a different alternating Au-Si region with a different number of 'spokes.' The number of Au-Si regions varies from a minumum of 50 to a maximum of 250, in increments of 25.

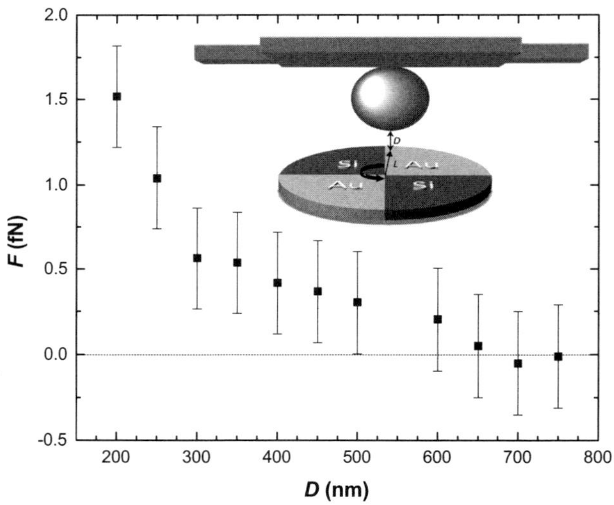

Fig. 1. Measured force as a funtion of separation. Inset: schematic of the experimental setup. The Au layer that covers the structured Au-Si test source mass is not shown.

With the sphere positioned over a structured Au-Si part of the sample, the system is rotated at $f = f_r/n$ where f_r is the resonant frequency of the oscillator plus sphere system and n is the number of Au-Si sectors. In this way, if there is a signal associated with the changes in mass in the source mass, the signal will be more easily detectable by the force sensitive mechanical oscillator.

The results obtained with the system are shown in Fig. 1. This signal, however, is not due to the mass density variation, since it appear also when the sphere is placed on top of the unstructured part of the rotating source mass. It can also be concluded that it is not associated with finite size effects,[5] or spatial fluctuations in the surface potential of the masses.[4] The fact that the observed signal is independent of f between 1 and 1000 Hz and increases with the distance L from the apex of the sphere to the axis of rotation indicates it is most likely associated with an impulsive motion of the axis of rotation of the system. If this is the explanation, the angular amplitude of the motion is smaller than 2 μrad.

3. Outlook

The results obtained will allow more stringent limits on spin independent contributions to hypothetical corrections to Newtonian gravity. These limits will improve by about three orders of magnitude existing limits at the $\sim$100 nm range. A similar setup can be used with a spin polarized mass replacing the Au. In this case the experiment would also provide the best limits for the monopole-dipole couplings at $\lambda \sim$100 nm.

Acknowledgments

Financial support from National Science Foundation grant PHY-0701636 is acknowledged. The author is also indebted to the IUPUI Nanoscale Imaging Center, Integrated Nanosystems Development Institute, Indiana University Collaborative Research Grants, and the Indiana University Center for Space Symmetries.

References

1. E. Fischbach and C. Talmadge, *The Search for Non-Newtonian Gravity,* Springer, New York, 1999; J.C. Long *et al.*, Nature **421**, 922 (2003); E.G. Adelberger *et al.*, Ann. Rev. Nucl. Part. Sci. **53**, 77 (2003); C.C. Speake *et al.*, Gen. Rel. Grav. **36**, 503 (2004); D.J. Kapner *et al.*, Phys. Rev. Lett. **98**, 021101 (2007); A.A. Geraci *et al.*, Phys. Rev. D **78**, 022002 (2008); E.G. Adelberger *et al.*, Prog. Part. Nucl. Phys. **62**, 102 (2009); M. Masuda and M. Sasaki, Phys. Rev. Lett. **102**, 171101 (2009); A.O. Sushkov *et al.*, Phys. Rev. Lett. **107**, 171101 (2011); I. Antoniadis *et al.*, C. R. Acad. Sci. (Physique) **12**, 755 (2011).
2. E.C. Chalkley *et al.*, in *Proceedings of the XLVIth Rencontres de Moriond and GPhyS Colloquium*, E. Augé *et al.*, eds., Thê Gioi Publishers, Vietnam, 2011.
3. R.S. Decca *et al.*, Phys. Rev. Lett. **94**, 240401 (2005).
4. R. Behunin *et al.*, submitted for publication.
5. C.R. Jamell and R.S. Decca, Int. J. Mod. Phys. A **26**, 3742 (2011).

CARDINAL GRAVITY

R. POTTING
CENTRA and Physics Department
Universidade do Algarve, FCT
8005-139 Faro, Portugal
E-mail: rpotting@ualg.pt

In this short review we describe a recently proposed model of gravity with spontaneously broken Lorentz invariance, with massless gravitons as Nambu-Goldstone modes. Interactions follow from imposing consistent coupling to the total energy-momentum tensor. At low energies the Einstein-Hilbert action is reproduced.

1. Linearized cardinal gravity

The masslessness of particles typically arises from the presence of some symmetry. In gauge theories each generator of (unbroken) gauge symmetry gives rise to a massless vector boson, while in general relativity the masslessness of gravitons has its origin in diffeomorphism invariance. However, another possibility exists as well. The Goldstone theorem states that the spontaneous breaking of a global continuous symmetry gives rise to a massless particle, the Nambu-Goldstone boson. Here we will review a model of gravity based on this mechanism, with gravitons arising through the spontaneous breaking of Lorentz invariance.

We start with the Lagrange density[1]

$$\mathcal{L} = \frac{1}{2} C^{\mu\nu} K_{\mu\nu\alpha\beta} C^{\alpha\beta} + V(C^{\mu\nu}, \eta_{\mu\nu}), \tag{1}$$

where the dynamical field $C^{\mu\nu}$ is a tensor density and $K_{\mu\nu\alpha\beta} = -\partial^2(\eta_{\mu\alpha}\eta_{\nu\beta} - \frac{1}{2}\eta_{\mu\nu}\eta_{\alpha\beta}) + \partial_\mu \eta_{\nu\alpha}\partial_\beta + \partial_\nu \eta_{\mu\alpha}\partial_\beta$ is the ghost-free quadratic kinetic operator for spin-2 particles. Of central importance is the scalar potential V, built out of the four independent scalars $X_1 = C^{\mu\nu}\eta_{\nu\mu}$, $X_2 = (C{\cdot}\eta{\cdot}C{\cdot}\eta)^\mu_\mu$, $X_3 = (C{\cdot}\eta{\cdot}C{\cdot}\eta{\cdot}C{\cdot}\eta)^\mu_\mu$, and $X_4 = (C{\cdot}\eta{\cdot}C{\cdot}\eta{\cdot}C{\cdot}\eta{\cdot}C{\cdot}\eta)^\mu_\mu$.

We will assume V has a local minimum for nonzero $C^{\mu\nu}$. Therefore, $C^{\mu\nu}$ can be expected to acquire a nonzero vacuum expectation value, $\langle C^{\mu\nu}\rangle =$

$c^{\mu\nu} \neq 0$, thereby spontaneously breaking Lorentz invariance. Generally this will break all six Lorentz generators, although special situations may arise with three or five broken generators.[2]

The fluctuations $\tilde{C}^{\mu\nu} = C^{\mu\nu} - c^{\mu\nu}$ consist of Nambu-Goldstone modes that keep the scalars X_i fixed, while the remaining degrees of freedom are massive modes. The former satisfy the linearized constraints

$$\tilde{C}^{\mu}_{\mu} = 0, \quad c^{\mu\nu}\tilde{C}_{\mu\nu} = 0, \quad (c\eta c)^{\mu\nu}\tilde{C}_{\mu\nu} = 0, \quad (c\eta c\eta c)^{\mu\nu}\tilde{C}_{\mu\nu} = 0. \tag{2}$$

It is reasonable to assume that at low energy we can neglect the massive modes. Thus the low-energy dynamics of $\tilde{C}_{\mu\nu}$-fluctuations around the vacuum expectation value is equivalent to linearized general relativity in the axial-type gauge (2). It can be shown that the equations of motion imply, apart from masslessness $\partial^2\tilde{C}_{\mu\nu} = 0$, the Lorenz conditions $\partial^{\mu}\tilde{C}_{\mu\nu} = 0$. Of the general six Nambu-Goldstone modes, four are nonpropagating auxiliary degrees of freedom, leaving two propagating massless degrees of freedom. There are generally four massive modes.

2. Bootstrap

The lagrangian (1) essentially describes free gravitons. A defining feature of gravity is that it acts on all forms of matter-energy. In General Relativity (GR) this property is expressed by requiring that the total energy-momentum tensor, including the graviton contribution, appear as a source in the equation of motion. For the free graviton lagrangian, this amounts to adding a quadratic term to the equation of motion, which in turn corresponds to a cubic term in the lagrangian. This in turn generates a quadratic term, and so forth. It can be shown that, starting with linearized GR, this recursive process yields the full nonlinear Einstein-Hilbert action.[3]

This process can be performed in one step by adopting a method devoped by Deser.[4] For the cardinal model this entails rewriting the original kinetic term in (1) using a Palatini formalism with auxiliary field $\Gamma^{\alpha}_{\mu\nu}$:

$$\mathcal{L}^L = \mathfrak{C}^{\mu\nu}(\Gamma^{\alpha}_{\mu\nu,\alpha} - \Gamma_{\mu,\nu}) + \eta^{\mu\nu}(\Gamma^{\alpha}_{\mu\nu}\Gamma_{\alpha} - \Gamma^{\alpha}_{\beta\mu}\Gamma^{\beta}_{\alpha\nu}), \tag{3}$$

where $\mathfrak{C}^{\mu\nu} = -C^{\mu\nu} + \frac{1}{2}\eta^{\mu\nu}C^{\alpha}_{\alpha}$ is the trace-reversed tensor density. Deser's procedure then yields the following final bootstrapped action:[5]

$$\begin{aligned} &\int d^4x \left(\mathfrak{C}^{\mu\nu}(\Gamma^{\alpha}_{\mu\nu,\alpha} - \Gamma_{\mu,\nu}) + (\eta + \mathfrak{C})^{\mu\nu}(\Gamma^{\alpha}_{\mu\nu}\Gamma_{\alpha} - \Gamma^{\alpha}_{\beta\mu}\Gamma^{\beta}_{\alpha\nu})\right) \\ &\quad \equiv \int d^4x (\eta + \mathfrak{C})^{\mu\nu} R_{\mu\nu}(\Gamma). \end{aligned} \tag{4}$$

We conclude that $(\eta+\mathfrak{C})^{\mu\nu}$ is naturally interpreted as a curved-space metric density. This bootstrap procedure can be applied also to scalar potential and matter energy-momentum (EM) tensor. The flat-space matter EM-tensor yields the conventional curved-space matter lagrangian with metric density $(\eta+\mathfrak{C})^{\mu\nu}$: $\mathcal{L}_{M,\mathfrak{C}} = \sqrt{|\eta+\mathfrak{C}|}\mathcal{L}^L_{M,\mathfrak{C}}|_{\eta\to\eta+\mathfrak{C}}$. The procedure is compatible with the scalar potential depending only on $C^{\mu\nu}$ and background Minkowski metric $\eta_{\mu\nu}$ if V satisfies integrability conditions. These conditions are satisfied only by particular, 'integrable' solutions.[5] For instance, a particular linear combination of these terms yields the cosmological constant $\sqrt{|\eta+\mathfrak{C}|}$. As it turns out, among the integrable potentials there are examples that have a local minimum at nonzero field. These are particularly interesting, as they may represent a stable vacuum.

3. Quantum effects

It is natural to integrate out the massive modes. Due to the presence of the scalar potential, the kinetic term not protected by gauge invariance (diffeomorphisms). Consequently, one can expect this to yield Lorentz-violating quantum corrections to the kinetic term for gravitons. Also, the four auxiliary modes may become propagating.[2] Possible effects include anisotropic graviton propagation, graviton birefringence and the possibility of cosmic rays emitting gravi-Cherenkov radiation.

Acknowledgments

It is a pleasure to thank Alan Kostelecký for collaboration. Financial support by the Portuguese Fundação para a Ciência e a Tecnologia is gratefully acknowledged.

References

1. V.A. Kostelecký and R. Potting, Gen. Rel. Grav. **37**, 1675 (2005).
2. S.M. Carroll, H. Tam and I.K. Wehus, Phys. Rev. D **80**, 025020 (2009).
3. R. Kraichnan, MIT thesis, 1947; Phys. Rev. **98**, 1118 (1955); A. Papapetrou, Proc. Roy. Irish Acad. **52A**, 11 (1948); S.N. Gupta, Proc. Phys. Soc. Lon. **A65**, 608 (1952); R.P. Feynman, *Lectures on Gravitation,* F.B. Morinigo and W.G. Wagner, eds., California Institute of Technology, Pasadena, 1963.
4. S. Deser, Gen. Rel. Grav. **1**, 9 (1970).
5. V.A. Kostelecký and R. Potting, Phys. Rev. D **79**, 065018 (2009).

BPS MAXWELL-CHERN-SIMONS-LIKE VORTICES IN A LORENTZ-VIOLATING FRAMEWORK

R. CASANA,* M.M. FERREIRA JR., E. DA HORA, and A.B.F. NEVES

Departamento de Física, Universidade Federal do Maranhão
São Luís, Maranhão 65080-805, Brazil
**E-mail: rodolfo.casana@gmail.com*

We have analyzed Maxwell-Chern-Simons-Higgs BPS vortices in a Lorentz-violating CPT-odd context. The Lorentz violation induces profiles with a conical behavior at the origin. For some combination of the coefficients for Lorentz violation there always exists a sufficiently large winding number for which the magnetic field flips its sign.

Lorentz-violating (LV) theories have caught much attention since the proposal of the Standard-Model Extension (SME),[1] whose properties have been intensively scrutinized in many areas of contemporary physics. The study of topological defects in LV scenarios was conducted initially for kinks,[2] monopoles.[3] Also, the formation of topological defects (monopoles) was analyzed in a broader framework of field theories endowed with tensor fields that spontaneously break Lorentz symmetry.[4] Recently, the existence of BPS vortex of type Abrikosov-Nielsen-Olesen has been shown in the presence of CPT-even LV terms of the SME.[5]

The aim of this contribution is to study Chern-Simons-like BPS vortices in an LV framework attained via the dimensional reduction of the Maxwell-Carroll-Field-Jackiw-Higgs model.[6] Such dimensional reduction provides a Maxwell-Chern-Simons-Higgs (MCSH) model modified by LV terms,

$$\begin{aligned}\mathcal{L} = &-\frac{1}{4}F_{\mu\nu}F^{\mu\nu} + \frac{1}{4}s\,\epsilon^{\nu\rho\sigma}A_\nu F_{\rho\sigma} + |\partial_\mu\phi - ieA_\mu\phi|^2 + \frac{1}{2}\partial_\mu\psi\partial^\mu\psi \\ &-e^2\psi^2|\phi|^2 - U(|\phi|,\psi) - \frac{1}{2}\epsilon^{\mu\rho\sigma}(k_{AF})_\mu\left(A_\rho\partial_\sigma\psi - \psi\partial_\rho A_\sigma\right), \qquad (1)\end{aligned}$$

where the LV parameter s plays the role of a Chern-Simons coupling, and $(k_{AF})_\mu$ is the (1+2)-dimensional Carroll-Field-Jackiw (CFJ) background. The potential $U(|\phi|,\psi) = (ev^2 - e|\phi|^2 - s\psi)^2/2$ provides the BPS vortices.

The corresponding Gauss law is

$$\partial_j \partial_j A_0 - sB - \epsilon_{ij} \left(k_{AF}\right)_i \partial_j \psi = 2e^2 A_0 \left|\phi\right|^2 . \tag{2}$$

To explicitly attain the vortex solutions, we use the ansatz $\phi = vg\left(r\right) e^{in\theta}$, $A_\theta = -(a(r) - n)/er$, $A_0 = \omega(r)$, $\psi = \mp\omega(r)$. The functions $a(r)$, $g(r)$ and $\omega(r)$ satisfy the following boundary conditions:

$$g(0) = 0,\ a(0) = n,\ \omega'(0) = \pm\omega_0 (k_{AF})_\theta/2, \tag{3}$$

$$g(\infty) = 1,\ a(\infty) = 0,\ \omega(\infty) = 0, \tag{4}$$

where n is the winding number of the vortex configuration, $\left(k_{AF}\right)_\theta$ is the angular component of $\left(k_{AF}\right)_i$, whereas the constant $\omega_0 = \omega\left(0\right)$ depends on the boundary conditions and it is numerically determined. In this ansatz the magnetic field is given by $B = -a'/er$. Then, the stationary energy density associated with the model (1) is written as

$$\mathcal{E} = \frac{1}{2}\left[B \mp \left(ev^2\left(1 - g^2\right) \pm s\omega\right)\right]^2 + v^2\left(g' \mp \frac{ag}{r}\right)^2 \pm ev^2 B + \partial_a \mathcal{J}_a, \tag{5}$$

where the condition $\left(k_{AF}\right)_0 = 0$ was adopted to insure positiveness of the energy density. The energy is minimized by requiring

$$g' = \pm\frac{ag}{r},\ B = \pm ev^2\left(1 - g^2\right) + s\omega, \tag{6}$$

which are the BPS equations, where the upper (lower) sign corresponds to $n > 0$ ($n < 0$). These and the Gauss law now written as

$$\omega'' + \frac{\omega'}{r} - sB \mp (k_{AF})_\theta \left(\omega' + \frac{\omega}{2r}\right) - 2e^2 v^2 g^2 \omega = 0 \tag{7}$$

describe topological vortices in this LV MCSH model. For $\left(k_{AF}\right)_\theta = 0$ one recovers the MCSH system. Under BPS equations and boundary conditions (3)-(4), the Eq. (5) provides the BPS energy $E_{BPS} = \pm 2\pi v^2 n$, which is proportional to the quantized magnetic flux, $\Phi_B = 2\pi n/e$.

Figure 1 depicts the profiles of the magnetic field obtained by numerical integration of Eqs. (6)-(7) for $s = 1$, and some values of $(k_{AF})_\theta$ and n. For large values of radial coordinate, the profiles behave in a similar way to the MCSH ones but with mass scales (β) satisfying $\beta_{(k_{AF})_\theta > 0} < \beta_{(k_{AF})_\theta = 0} < \beta_{(k_{AF})_\theta < 0}$. The role played by the LV parameter becomes more pronounced at the origin, where it generates a conical structure absent in the MCSH profiles. The numerical analysis shows that for $n > 0$ (analogous results are obtained for $n < 0$), a fixed s, and $(k_{AF})_\theta > 0$, there are always two well defined regions with positive magnetic flux, occurring no magnetic field reversion. On the other hand, for fixed s, and $(k_{AF})_\theta < 0$, there always

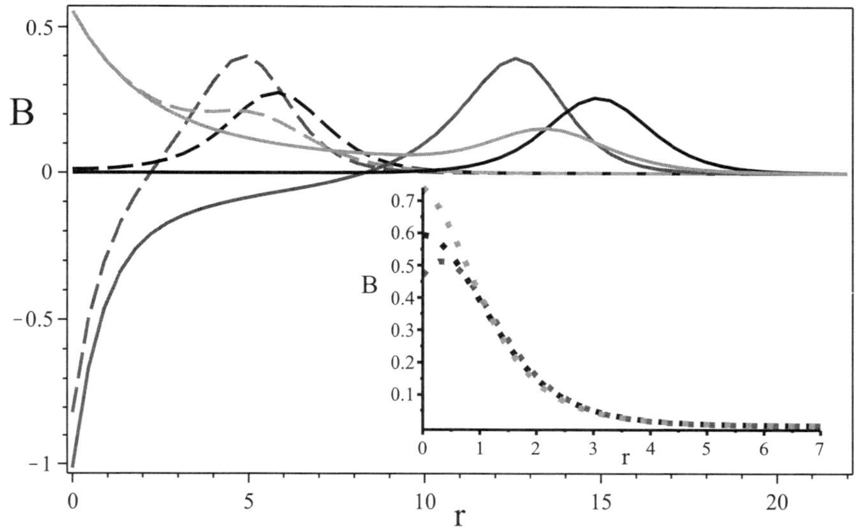

Fig. 1. Magnetic field $B(r)$. Here $s = 1$, $e = v = 1$. Dark gray lines, $(k_{AF})_\theta = -1$; black lines, $(k_{AF})_\theta = 0$, MCSH; slate gray lines, $(k_{AF})_\theta = +1$; dotted lines, $n = 1$; dashed lines, $n = 6$; and solid lines, $n = 15$.

exists a sufficiently large winding number n_0 such that for all $n > n_0$ the magnetic field reverses its sign. Consequently, there are always two well defined regions with opposite magnetic flux. The flipping of the magnetic flux represents another remarkable feature induced by Lorentz violation.

Acknowledgments

The authors thank FAPEMA, CNPq and CAPES for financial support.

References

1. D. Colladay and V.A. Kostelecký, Phys. Rev. D **55**, 6760 (1997); Phys. Rev. D **58**, 116002 (1998).
2. M.N. Barreto, D. Bazeia, and R. Menezes, Phys. Rev. D **73**, 065015 (2006).
3. N.M. Barraz Jr., J.M. Fonseca, W.A. Moura-Melo, and J.A. Helayël-Neto, Phys. Rev. D **76**, 027701 (2007).
4. M.D. Seifert, Phys. Rev. Lett. **105**, 201601 (2010); Phys. Rev. D **82**, 125015 (2010).
5. R. Casana, M.M. Ferreira Jr., E. da Hora, and C. Miller, Phys. Rev. D **86**, 065011 (2012); Phys. Lett. B **718**, 620 (2012).
6. H. Belich, M.M. Ferreira Jr, and J.A. Helayël-Neto, Eur. Phys. J. C **38**, 511 (2005).

OBSERVER AND PARTICLE TRANSFORMATIONS AND NEWTON'S LAWS

T.H. BERTSCHINGER,* NATASHA A. FLOWERS, and JAY D. TASSON

Department of Physics and Astronomy, Carleton College
Northfield, MN 55057, USA
**E-mail: bertschingert@carleton.edu*

A frequently confused point in studies of symmetry violation is the distinction between observer and particle transformations. In this work, we consider a model in which a coefficient in the Standard-Model Extension leads to violations of rotation invariance in Newton's second law. The model highlights the distinction between observer and particle transformations.

1. Introduction

The Standard-Model Extension (SME) provides a general field-theoretic framework for studying Lorentz violation,[1] including rotation-invariance violation. To highlight the basic ideas of Lorentz-symmetry breaking, we consider rotation-invariance violation in Newton's second law:[2]

$$F_j = m_{jk} a_k. \tag{1}$$

Here m_{jk} is a symmetric direction-dependent inertial mass (we consider conventional gravitational mass). This yields a valid and more general form that Newton himself could have chosen.

Our effective inertial mass can be generated as a low-energy limit of the SME:[3]

$$m_{jk} = m\left(\delta_{jk} + 2c_{jk}\right). \tag{2}$$

Here c_{jk} is a coefficient for Lorentz violation found in the fermion sector, taken as symmetric. It is depicted with background diamonds in Fig. 1. A similar construction can also be found associated with other SME coefficients for Lorentz violation.[4]

Using a block on an inclined plane, we show that observer rotation invariance holds, while a particle rotation changes the experiment's outcome, violating Lorentz symmetry.

2. Block on an inclined plane

Examine a block on an inclined plane devoid of friction. The x-axis points down the plane while the y-axis is perpendicular to the surface as shown in the left-hand diagram of Fig. 1. Let the block's effective inertial mass be diagonal:

$$m_{jk} = m \begin{pmatrix} 1+2c_{xx} & 0 & 0 \\ 0 & 1+2c_{yy} & 0 \\ 0 & 0 & 1+2c_{zz} \end{pmatrix}. \tag{3}$$

Solving for the particle's acceleration under the constraint $a_y = 0$ yields

$$a_x = a_R = (1 - 2c_{xx}) g \sin\theta + O(c^2), \tag{4}$$

where a_R is the acceleration down the ramp. The only difference here from the conventional problem is the presence of c_{xx}. Qualitatively, the motion is down the plane with constant acceleration, as in the absence of Lorentz violation.

3. Observer-rotation invariance remains

Perform an observer rotation on the original experiment; that is, consider the same problem in new coordinates as shown in the left-hand diagram of Fig. 1. Here,

$$m_{j'k'} = m \begin{pmatrix} 1+2c_{x'x'} & 2c_{x'y'} & 0 \\ 2c_{x'y'} & 1+2c_{y'y'} & 0 \\ 0 & 0 & 1+2c_{z'z'} \end{pmatrix} \tag{5}$$

is the mass in the new coordinates obtained via $m' = RmR^T$, R being a rotation matrix. Components are related by, for example,

$$c_{x'x'} = c_{xx}\cos^2\theta + c_{yy}\sin^2\theta. \tag{6}$$

Again solving, we find up to $O(c^2)$ the components

$$\begin{aligned} a_{x'} &= (1 - 2c_{x'x'}\cos^2\theta - 2c_{y'y'}\sin^2\theta + 4c_{x'y'}\sin\theta\cos\theta) g \sin\theta\cos\theta, \\ a_{y'} &= -(1 - 2c_{x'x'}\cos^2\theta - 2c_{y'y'}\sin^2\theta + 4c_{x'y'}\sin\theta\cos\theta) g \sin^2\theta. \end{aligned} \tag{7}$$

However, the acceleration is again along the ramp with the magnitude found in Eq. (4), which is obtained from $a_{x'}$ and $a_{y'}$ by $a_R = \sqrt{a_{x'}^2 + a_{y'}^2}$ and a substitution for $c_{j'k'}$ in terms of c_{jk}. Hence, both observers agree on the outcome of the experiment.

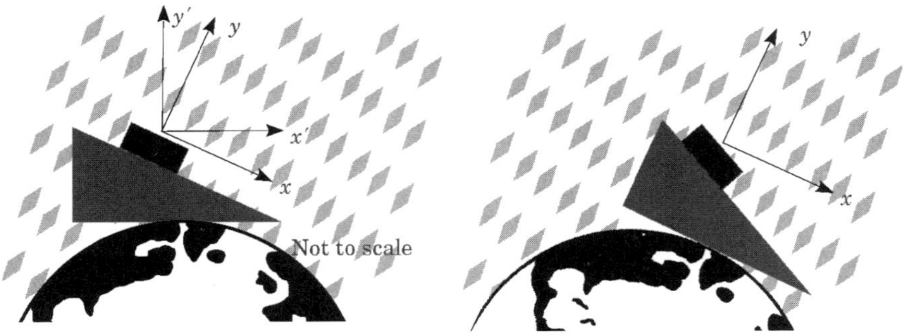

Fig. 1. Diagrams.

4. Particle-rotation invariance is violated

A particle rotation of the original system leaves the mass matrix as in Eq. (1), but alters the direction of the gravitational field relative to the background as shown in the right-hand diagram of Fig. 1. This produces an observably different acceleration. Solving for the motion of the particle subject to the constraint yields

$$\begin{aligned} a_y &= -(1 - 2c_{xx} - 2c_{yy}) g \sin^2\theta + O(c^2), \\ a_x &= (1 - 2c_{xx} - 2c_{yy}) g \sin\theta \cos\theta + O(c^2). \end{aligned} \tag{8}$$

Here the component along the ramp is

$$a_R = (1 - 2c_{xx} - 2c_{yy}) g \sin\theta + O(c^2), \tag{9}$$

which is different from the first cases, revealing observable Lorentz violation.

References

1. D. Colladay and V.A. Kostelecký, Phys. Rev. D **55**, 6760 (1997); Phys. Rev. D **58**, 116002 (1998).
2. T.H. Bertschinger *et al.*, in preparation.
3. V.A. Kostelecký and J.D. Tasson, Phys. Rev. D **83**, 016013 (2011).
4. D. Atkinson, M. Becker, and J.D. Tasson, these proceedings; in preparation.